International Conference on Applied Mathematics, Computational Science and Systems Engineering 2016

ITM Web of Conferences Volume 9 (2017)

Rome, Italy
5 - 7 November 2016

Editors:

Nikos Bardis
Joseph Quartieri
Claudio Guarnaccia
Nikos Doukas

ISBN: 978-1-5108-3512-2

Printed from e-media with permission by:

Curran Associates, Inc.
57 Morehouse Lane
Red Hook, NY 12571

Some format issues inherent in the e-media version may also appear in this print version.

Printed by Curran Associates, Inc. (2017)

For additional information, please contact EDP Sciences – Web of Conferences
at the address below.

EDP Sciences – Web of Conferences
17, Avenue du Hoggar
Parc d'Activité de Courtabœuf
BP 112
F-91944 Les Ulis Cedex A
France

Phone: +33 (0) 1 69 18 75 75
Fax: +33 (0) 1 69 28 84 91

contact-edps@webofconferences.org

Additional copies of this publication are available from:

Curran Associates, Inc.
57 Morehouse Lane
Red Hook, NY 12571 USA
Phone: 845-758-0400
Fax: 845-758-2633
Email: curran@proceedings.com
Web: www.proceedings.com

TABLE OF CONTENTS

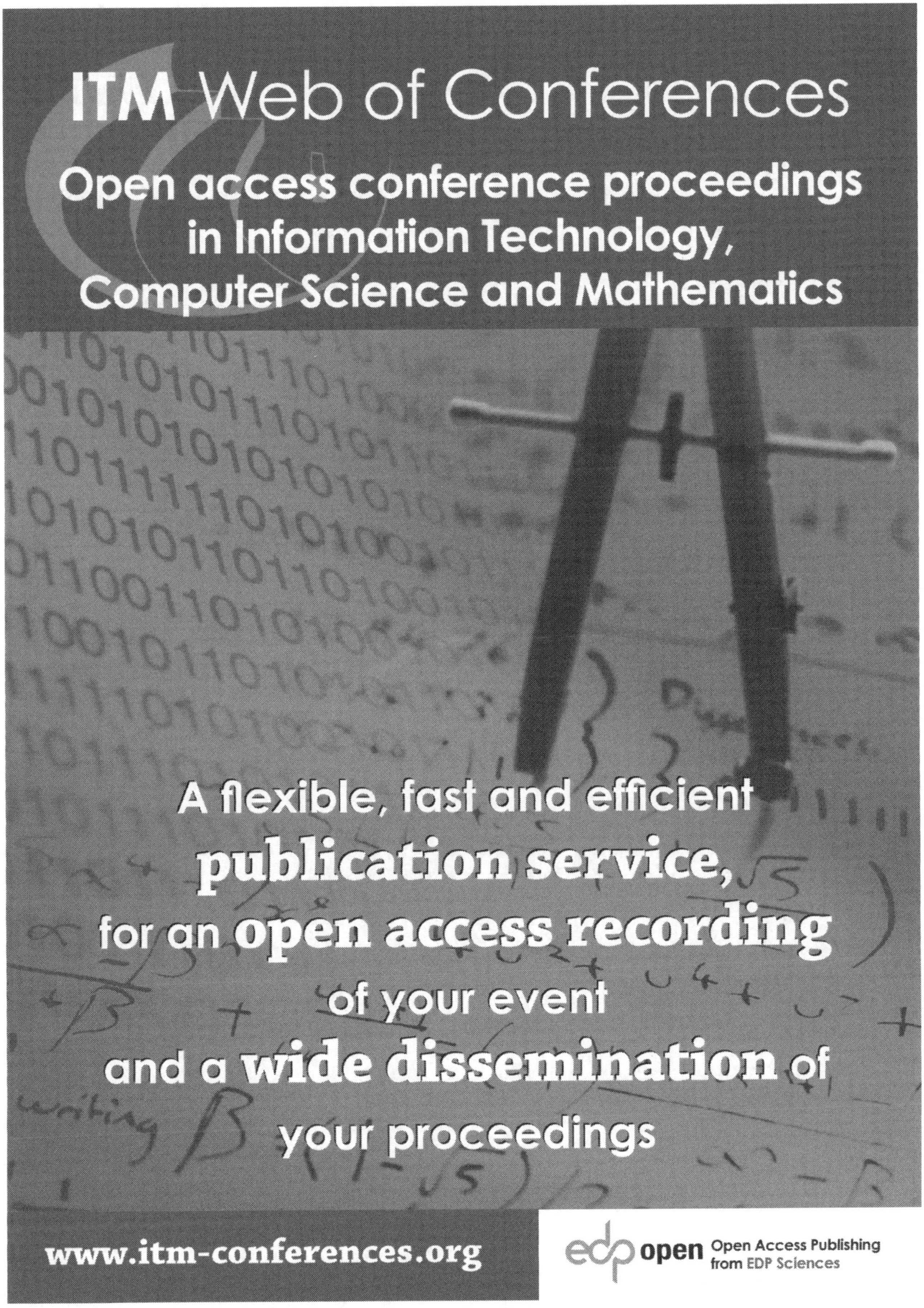
ITM Web of Conferences
Open access conference proceedings
in Information Technology,
Computer Science and Mathematics
A flexible, fast and efficient
publication service,
for an open access recording
of your event
and a wide dissemination of
your proceedings
www.itm-conferences.org
edp open
Open Access Publishing
from EDP Sciences

Aims & scope

ITM Web of Conferences is an open-access journal devoted to the publishing of conference proceedings dealing with all areas related to information technology, computer science and mathematics.

ITM Web of Conferences aims to make conference proceedings visible, citable and widely usable by the scientific community. It offers outstanding services for the publication and dissemination of contents in order to ensure long-term archiving and maximum exposure of scientific events.

Coverage includes the whole spectrum of pure and applied mathematics, as well as all domains related to computer science such as programming, operating systems, databases, software engineering, web technologies, human-computer interaction, etc.

Why publish with ITM Web of Conferences?

- **Open access:** conference papers available worldwide.
- **Fast publication:** less than 3 months.
- **Total flexibility:** no page restrictions, free unlimited use of colour in the online version, any kind of scientific document welcomed (traditional articles, posters, abstracts and slideshows).
- **High visibility:** all documents indexed in main databases and easily found online.
- **Fair author rights:** CC-BY Creative Commons license.
- **Range of additional publishing options:** printed proceedings, CD-ROMs, copy-editing, etc.

ISSN (electronic edition): pending

Contact: Isaline Augusto
itm@webofconferences.org

www.itm-conferences.org

edp open Open Access Publishing from EDP Sciences

Statement of Peer review

In submitting conference proceedings to ITM Web of Conferences, I certify to the Publisher that I adhere to the **Policy on Publishing Integrity** of the journal in order to safeguard good scientific practice in publishing.

1. All articles have been subjected to peer review administered by the proceedings editors.
2. Reviews have been conducted by expert referees, who have been requested to provide unbiased and constructive comments aimed, whenever possible, at improving the work.
3. Proceedings editors have taken all reasonable steps to ensure the quality of the materials they publish and their decision to accept or reject a paper for publication has been based only on the merits of the work and the relevance to the journal.

Title, date and place of the conference

AMCSE 2016, Rome, Italy, November 5-7, 2016

Proceedings editor(s):

Prof. Nikos Bardis, Hellenic Military Academy, Greece

Prof. Joseph Quartieri, University of Salerno, Italy

Prof. Claudio Guarnaccia, University of Salerno, Italy

Prof. Nikos Doukas, Hellenic Militaty Academy, Athens, Greece

Date and editor's signature 19 Dec 2016

A modified version of regularized meshless method for three dimensional potential problem

Cheng-Yang Lai[1], *Kue-Hong* Chen[1*], *Sheng-Wei* Lin[1], *Ren* Liu[1]

[1] Department of Civil Engineering, National Ilan University,Taiwan
[*] khc6177@niu.edu.tw (Corresponding author: K. H. Chen)

Abstract. In this study, three-dimensional potential problem is solved using a novel meshless method. Due to the singularity of the kernel functions, the diagonal terms of the influence matrices in the method of fundamental solutions (MFS) are unobtainable. A proposed approach in the literature of the past decade, namely regularized meshless method (RMM), is proposed to overcome such difficulties by using the proposed desingularization (subtracting and adding-back) technique. The main difficulty for the coincidence of the source and collocation points then disappears. However, the disadvantage of RMM is the order of precision of obtained solution is lower than other numerical methods. In this study, we present a novel technique to promote the order of precision of diagonal term; therefore, we can obtain a more precise solution. Finally, we introduce a typical 3-D numerical example to illustrate the technique. The numerical result is compared with those obtained by the RMM; and a more precise result is obtained.

Keywords: Method of fundamental solutions; Regularized meshless method; Desingularization; Diagonal terms.

1.Introduction

The method of fundamental solutions (MFS) [1] is a prospective meshless/meshfree boundary collocation method for the solution of real engineering problems. An increasing number of journal paper the growing attractiveness of the MFS. It was applied as a foolproof and accurate method to the solution of many engineering problems. In order to avoid the source singularity of fundamental solutions, the source points are distributed on the non-physical boundary (fictitious boundary) [2] outside the physical domain. Note that the kernel function is constituted of two-point function which is one kind of the radial basis functions (RBFs) [3] ; the independent variable of two-point function depends on point location only. A regular singularity-free formulation was obtained as a result, and realizes an attractive truly boundary type and mathematically simple meshfree method. The ease of implementation of the MFS makes it a designated numerical method in this study. However, because of the controversial artificial boundary (off-set boundary), which shift a distance from the real boundary, the MFS has not become a dominant numerical method. Notwithstanding its gain in singularity free, the influence matrix becomes ill-posed matrix [4] ; its result is very unstable since the condition number for the influence matrix becomes very large.

In 2005, Young, Chen, and Lee [5] developed a novel meshless method, namely the RMM, to overcome the drawback of MFS; it implement a subtracting and adding-back technique [6] to regularize the singularity of the kernel functions, and therefore the diagonal terms of influence matrices can be derived when the source points are located on the real boundary. Furthermore, unlike other meshless methods such as the boundary knot method (BKM) [7] which also places the source points on the physical boundary but results in a badly ill-conditioned interpolation matrix, the condition number of the coefficient matrix of the RMM remains gentle even with a large number of source nodes. A similar technique, namely singular boundary method (SBM), was proposed by Chen and his collaborators [8]. This SBM formulation keeps merits of RMM in which the source points can also be distributed on the physical boundary and makes it an advantage by using the single-layer fundamental solution. However the major drawback of the RMM or SBM is that the precision of solution is not enough for irregular domain problems due to the insufficient precision of the finite value of diagonal terms. In this study, we provide a novel technique to modify the diagonal coefficients of the coefficient matrix of the RMM; by introducing a defined auxiliary problem [9] in the technique, we can derive a more precise finite value of diagonal terms and can alleviate the ill condition of influence matrix.

The defined auxiliary problem is proposed a period of time [9]. The literature shows the meaning of the defined auxiliary problem and how to be applied it in a wide variety of problems. Due to the defined auxiliary problem and the original problem possess the same influence matrix under the distribution of the same mesh or source points. It's more important which the defined auxiliary problem has the analytical solution obtained by the complementary solutions, which satisfies the governing equation (G.E.) and it is similar to the real analytical solution of the original problem. Through the distinct identities, we can modify the diagonal coefficients of influence matrix by the known analytical solutions in the interesting domain of defined auxiliary problem; the obtained influence matrix can alleviate ill-posed, thus we can obtain a more excellent performance and higher convergence order than the RMM.

[*] Corresponding author: author@e-mail.org

2. Problem statement

Consider the boundary value problem (BVP) on a general domain $\Omega \in R^2$ with mixed-type boundary conditions (B.C.s) as:

$$\mathcal{L}\ [u(x)] = 0,\ \ x\ in\ \Omega \tag{1}$$

$$u(x) = \bar{u}(x),\ \ x\ on\ B_1 \tag{2}$$

$$t(x) = \frac{\partial u(x)}{\partial n_x} = \bar{t}(x),\ x\ on\ B_2 \tag{3}$$

Where $\mathcal{L}$ is the linearly differential of second order operator, $u(x)$ is the solution, Ω is the computational domain of the problem. The boundary conditions are described as following: B_1 is the essential boundary (Dirichlet boundary) in which the potential is prescribed by $\bar{u}(x)$, B_2 is the natural boundary (Neumann boundary) in which the normal derivative is prescribed as $\bar{t}(x)$; B_1 and B_2 construct the whole boundary of the domain Ω.

3. Formulation

3.1 Representation for numerical solution of problem

By employing radial basis function (RBF) concept, the representation of the solution for the interior problem can be approximated in terms of a set of interpolation functions as:

$$u(x_i) = \sum_{j=1}^{N} A(x_i, s_j)\alpha_j \tag{4}$$

$$t(x_i) = \sum_{j=1}^{N} B(x_i, s_j)\alpha_j \tag{5}$$

where x_i and s_j ,respectively, represent ith collocation point and jth source point, N is the number of source points and α_j is the generalized unknowns.

$$A(x_i, s_j) = \frac{y_i \bar{n}_i}{r^3} \tag{6}$$

$$B(x_i, s_j) = \frac{3 y_i y_j n_i \bar{n}_j}{r^5} - \frac{n_i \bar{n}_i}{r^3} \tag{7}$$

where $r = |x_i - s_j|$, $B(x_i, s_j) = \frac{\partial A(x_i, s_j)}{\partial n_x}$, in which n_x is the normal derivative of u at collocation point x_i , n_k is the kth component of the outward normal vector at s_j ; $\bar{n}_k$ is the kth component of the outward normal vector at x_i and $y_k = x_i - s_j$.

3.2 Adding-back technique of regularized meshless method

When the collocation point x_i approaches to the source point s_j, Equations (4) and (5) will become singular. Equations (4) and (5) for the interior problems need to be regularized by using special treatment of the desingularization of subtracting and adding-back technique [6] as follows

$$\begin{aligned} u(x_i) &= \sum_{j=1}^{N} A^{(i)}(s_j, x_i)\alpha_j - \sum_{j=1}^{N} A^{(e)}(s_j, x_i)\alpha^i \\ &= \sum_{j=1}^{i-1} A^{(i)}(s_j, x_i)\alpha_j + \sum_{j=i+1}^{N} A^{(i)}(s_j, x_i)\alpha_j \\ &+ [\sum_{m=1}^{N} A^{(i)}(s_m, x_i) - A^{(i)}(s_i, x_i)]\alpha_i,\ x_i \in B \end{aligned} \tag{8}$$

$$\begin{aligned} t(x_i) &= \sum_{j=1}^{N} B^{(i)}(s_j, x_i)\alpha_j - \sum_{j=1}^{N} B^{(e)}(s_j, x_i)\alpha_i \\ &= \sum_{j=1}^{i-1} B^{(i)}(s_j, x_i)\alpha_j + \sum_{j=i+1}^{N} B^{(i)}(s_j, x_i)\alpha_j \\ &- [\sum_{m=1}^{N} B^{(i)}(s_m, x_i) - B^{(i)}(s_i, x_i)]\alpha_i,\ x_i \in B \end{aligned} \tag{9}$$

in which

$$\sum_{j=1}^{N} A^{(e)}(s^j, x^i) = 0,\ x^i \in B \tag{10}$$

$$\sum_{j=1}^{N} B^{(e)}(s^j, x^i) = 0,\ x^i \in B \tag{11}$$

which the superscript (i) and (e) denotes the interior and exterior domain, respectively. The detail derivations of Equations (10) and (11) are given in the literature [6]. The original singular terms of $A^{(i)}(s^i, x^i)$ and $B^{(i)}(s^i, x^i)$ in Equations (4) and (5), have been transformed into regular terms $[\sum_{m=1}^{N} A^{(i)}(s^m, x^i) - A^{(i)}(s^i, x^i)]$ and $-[\sum_{m=1}^{N} B^{(i)}(s^m, x^i) - B^{(i)}(s^i, x^i)]$ in Equations (8) and (9), respectively. In which the terms of $\sum_{m=1}^{N} A^{(i)}(s^m, x^i)$ and $\sum_{m=1}^{N} B^{(i)}(s^m, x^i)$ are the adding-back terms and the terms of $A^{(i)}(s^i, x^i)$ and $B^{(i)}(s^i, x^i)$ are the subtracting terms in the two brackets for the special treatment technique. After using the desingularization of subtracting and adding-back technique [6] , we are able to remove the singularity and hypersingularity of the kernel functions. Therefore, the diagonal coefficients for the interior problems can be extracted out as:

 DOI: 10.1051/itmconf/20170901001

$$\{u_i\} = \begin{bmatrix} \sum_{m=1}^{N} a_{1,m} - a_{1,1} & a_{1,2} & \cdots & a_{1,N} \\ a_{2,1} & \sum_{m=1}^{N} a_{2,m} - a_{2,2} & \cdots & a_{2,N} \\ \vdots & \vdots & \ddots & \vdots \\ a_{N,1} & a_{N,2} & \cdots & \sum_{m=1}^{N} a_{N,m} - a_{N,N} \end{bmatrix} \{\alpha_i\} \quad (12)$$

$$\{t_i\} = \begin{bmatrix} -(\sum_{m=1}^{N} b_{1,m} - b_{1,1}) & b_{1,2} & \cdots & b_{1,N} \\ b_{2,1} & -(\sum_{m=1}^{N} b_{2,m} - b_{2,2}) & \cdots & b_{2,N} \\ \vdots & \vdots & \ddots & \vdots \\ b_{N,1} & b_{N,2} & \cdots & -(\sum_{m=1}^{N} b_{N,m} - b_{N,N}) \end{bmatrix} \{\alpha_i\}. \quad (13)$$

3.3 Modified technique for modifying diagonal term of influence matrics

The precision of the resingularized techniques in the above-section are not enough for irregular domain problems due to the insufficient accurancy in deriving valid value of diagonal terms of the influence matrices [5]. A modified technique is employed to modify the diagonal term of influence matrices in the Equations (12) and (13). By using the technique, we can obtain a more precise result in deriving a diagonal value of the influence matrix. The scheme of the modified technique is depicted as following section:

3.3.1 Defining auxiliary problem

(1) Specifying G.E., contour and B.C. type
The G.E., domain shape and B.C. type in the defined auxiliary problem is defined as:

$$\mathcal{L}'\left[u^q(x)\right] = 0 \quad x \ in \ \Omega' \quad (14)$$

$$u^q(x) = \bar{u}^q(x), on \ B_1' \quad (15)$$

$$\frac{\partial u^q(x)}{\partial n_x} = \bar{t}^q(x), on \ B_2' \quad (16)$$

where $\mathcal{L}'$ is the differential operator of $u^q(x)$, $u^q(x)$ is the potential of auxiliary problems, Ω' is the computational domain of the auxiliary problem. B_1' and B_2' construct the whole boundary of the domain Ω'. $\bar{u}^q(x)$ and $\bar{t}^q(x)$ are a specifying function $u^q(x)$.
In order to appropriate the original problem in the Equations (1)-(3), we choose $\mathcal{L}' = \mathcal{L}$, $\Omega' = \Omega, B_1' = B_1, B_2' = B_2$, namely, the G.E., domain shape and B.C. type in the defined auxiliary problem is a duplicate of the original problem.

(2) Giving quasi-analytical solution

The potential, $u^q(x)$, in the auxiliary problem at arbitrary point x in the domain is the linear combination of the complete set functions as follows:

$$u^q(x) = \sum_{j=1}^{M} \Phi_j(x) c_j, \ x \ in \ \Omega' \quad (17)$$

where $\Phi_j(x)$ choose complementary solution sets which satisfies the G.E. in Equation (14), M is the total number of the complementary solution set and c_j, denotes the jth undermined coefficient. Each of $\Phi_j(x)$ of complementary solution sets satisfies the governing equation in Equation (14) as:

$$\mathcal{L}\Phi_1(x) = 0, \mathcal{L}\Phi_2(x) = 0 \cdots, \\ \mathcal{L}\Phi_{M-1}(x) = 0, \mathcal{L}\Phi_M(x) = 0 \quad (18)$$

Because of the linear property of the differential equation operator in G.E., the potential, $u^q(x)$, satisfies the G.E. as:

$$\mathcal{L}u^q(x) = c_1\mathcal{L}\Phi_1(x) + c_2\mathcal{L}\Phi_2(x) + \\ \cdots + c_{M-1}\mathcal{L}\Phi_{M-1}(x) + c_M\mathcal{L}\Phi_M(x) = 0 \quad (19)$$

It is noted that the potential, $u^q(x)$, in the new defined problem is indeed an exact solution because it satisfies the G.E. as shown in Equation (19). In the procedure of numerical algorithm of M number of the undetermined coefficient, c_j in the Equation (17), of the quasi-analytic solution of the auxiliary problem, the value of B.C. in the Equations (15) and (16) is specified with the same value of B.C. in the Equations (2) and (3) in the original problem at certain preselected M_c number ($M_c \geq M$) of boundary points. By means of collocation scheme in the auxiliary problem, requiring that the value of quasi-analytic solution at preselected boundary points matches the selected boundary value obtain the numerical value of the undetermined coefficient. It is easy to see that the more boundary points are needed to obtain M number of the undetermined coefficient, c_j, for an accurate approximation of the real analytical solution. However, a certain M_c is implemented due the finite computer precision.

3.3.2 Modifying the diagonal term

Because of the same influence matrix between in the original problem and the defined auxiliary problem and the known analytical solution in the computational domain of the auxiliary problem we can modify it by solving the defined auxiliary problem. The process of treatment for modifying diagonal coefficients is presented as follows:
Step 1: By employing the expansion method of RBF in Equation (4), the representation of the solutions in the computational domain, Ω', of the defined auxiliary problem is

$$u^q(x) = \sum_{j=1}^{N} A(x, s_j)\alpha_j^q, x \ in \ \Omega' \quad (20)$$

Step 2: We distribute N' number of collocation points, $\{x_i\}_{i=1}^{N'}$ in the computational domain, Ω', of the auxiliary problem. The analytical solution, $u^q(x)$, of computational domain in the defined auxiliary problem is known in the Equation (14), therefore the known solution at N' number of $\{x_i\}_{i=1}^{N'}$ is $\{u_i^q\}_{i=1}^{N'}$. Therefore, the representation of the solutions in the

Equation (20) at N' number of collocation points can be represented as

$$u_i^q = \sum_{j=1}^{N} A(x_i, s_j)\alpha_j^q = \sum_{j=1}^{N} A_{ij}\alpha_j^q, i = 1, \ldots, N' \quad (21)$$

The above equations is belong to a linear algebraic equation system with $N' \times N$. We can derive the unknown coefficient α_j^q of the linear algebraic system in Equation (21) by a linear algebraic solver.

Step 3: We can obtain the following equation when the observation point is collocated at the N number of collocation points on the boundary; the location of the N number of collocation points is chosen the same as the original problem, as follows:

$$u_i^q = \sum_{j=1, j\neq i}^{N} A_{ij}\alpha_j^q + (A_{ii} + \delta_i)\alpha_i^q, i = 1, \ldots, N \quad (22)$$

where A_{ii} is the known diagonal term in the Equation (11), δ_i is the modified coefficient of the ith diagonal terms in the influence matrix, in which it is unknown. The above N number of linear equations is an uncouple system with N number of unknown, $\{\delta_i\}_{i=1}^{N}$, in the Equation (22).

Step 4: We can directly solve each the modified coefficient, δ_i, in each equation of the Equation (22) without any difficulty, as follows

$$\delta_i = \frac{\bar{u}_i^q(x_i) - \sum_{j=1, j\neq i}^{N} A_{ij}\alpha_j^q}{\alpha_i^q} - A_{ii}, i = 1, \ldots, N \quad (23)$$

Therefore, we can derive the new ith diagonal term, $A_{ii} + \delta_i$, in the Equation (23) to substitute for the origional ith diagonal term of the influence matrix in the Equation (11).

4. Numerical example

The case subjected with the Dirchlet B.C. as shown in Fig.1, by having an analytical solution as

$$u(x, y, z) = e^y \sin(z) + x \quad (24)$$

We solve this case by using the RMM and modified RMM. To see the discrepancy of the value of the diagonal terms for exact value, the diagonal values of the modified RMM and the RMM are shown in Fig.2. The curve of R.M.S error versus the number of source points N is shown in Fig.4. By observing the error curves in Fig.3, the results of the modified RMM are obviously better than the RMM. Finally, we obtain the field solution by using the modified RMM the RMM for N=2166 and compare the results with analytical solution as shown in Fig.3 (a)-(c).

5. Conclusion

Although the RMM can avoid the major issue in the MFS to choose the best location of a fictitious boundary and it makes the source nodes can be placed on the physical boundary instead at the fictitious boundary, the deficient precision of the solution of RMM is the major issue because of low precision of diagonal value. In this study, we propose a novel technique to modify the diagonal values of the RMM; the numerical results for a numerical example obtained by present scheme are highly acceptable. The proposed technique can thus successfully obtain a more precision solution than the RMM to model 3-D potential problem.

Acknowledgement

Financial support from the Ministry of Science and Technology of Taiwan, under Grant No: MOST 105-2221-E-197-006 to the National Ilan University is gratefully acknowledged.

References

[1] D. L. Young, S. J. Jane, C. M. Fan *et al.*, "The method of fundamental solutions for 2D and 3D Stokes problems," *Journal of Computational Physics*, **vol. 211, no. 1**, pp. 1-8,(2006).

[2] W. Chen, and F. Wang, "A method of fundamental solutions without fictitious boundary," *Engineering Analysis with Boundary Elements*, **vol. 34, no. 5**, pp. 530-532, (2010).

[3] J. Li, Y. C. Hon, and C. S. Chen, "Numerical comparisons of two meshless methods using radial basis functions," *Engineering Analysis with Boundary Elements*, **vol. 26, no. 3**, pp. 205-225,(2002).

[4] K. H. Chen, C. T. Chen, and J. F. Lee, "Adaptive error estimation technique of the Trefftz method for solving the over-specified boundary value problem," *Engineering Analysis with Boundary Elements*, **vol. 33, no. 7**, pp. 966-982, (2009).

[5] D. Young, K. Chen, and C. Lee, "Novel meshless method for solving the potential problems with arbitrary domain," *Journal of Computational Physics*, **vol. 209, no. 1**, pp. 290-321, (2005).

[6] W. Chen, J.-Y. Zhang, and Z.-J. Fu, "Singular boundary method for modified Helmholtz equations," *Engineering Analysis with Boundary Elements*, **vol. 44, no. 0**, pp. 112-119, (2014).

[7] F. Z. Wang, L. Ling, and W. Chen,

 DOI: 10.1051/itmconf/20170901001

"Effective condition number for boundary knot method," *Computers, Materials & Continua (CMC),* **vol. 12, no. 1**, pp. 57,(2009).

[8] W. Chen, "Singular boundary method: a novel, simple, meshfree, boundary collocation numerical method," *Chinese Journal of Solid Mechanics,* **vol. 30, no. 6**, pp. 592-599, (2009).

[9] K. H. Chen, and J. T. Chen, "Estimating the optimum number of boundary elements by error estimation in a defined auxiliary problem," *Engineering Analysis with Boundary Elements,* **vol. 39, no. 0**, pp. 15-22, (2014).

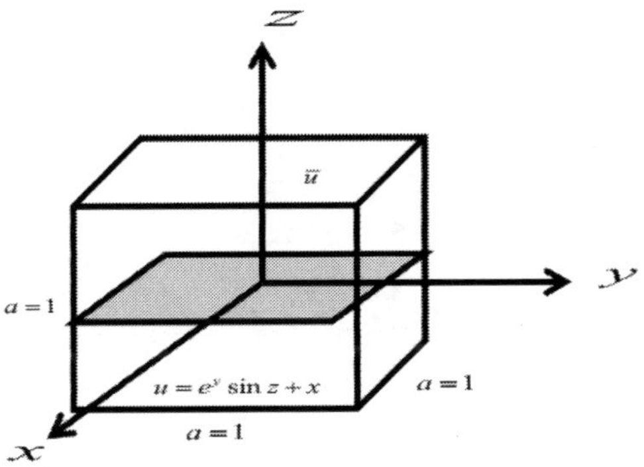

Fig.1 Problem sketch

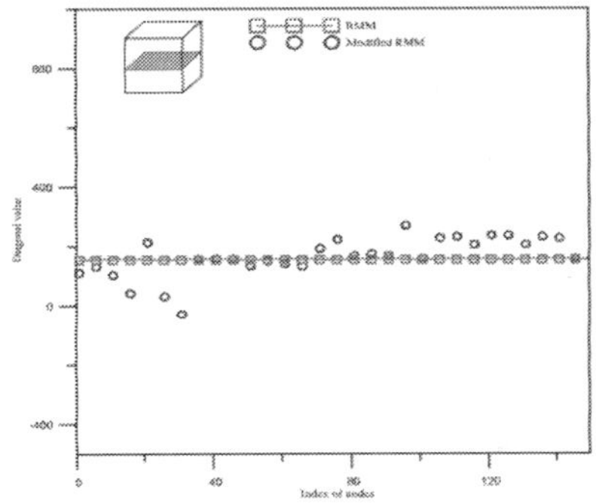

Fig.2 The diagonal coefficients of influence matrix versus the index of nodes

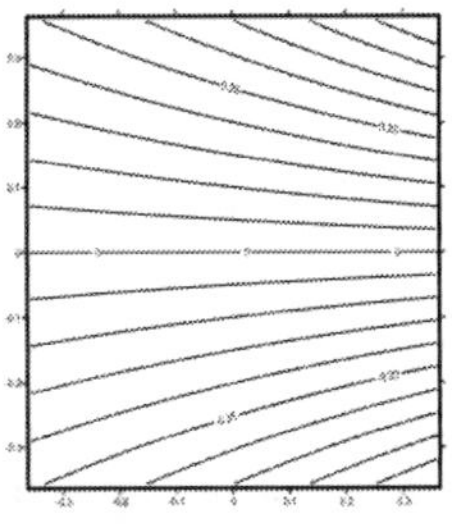

Fig.3 (a) Analytical solution

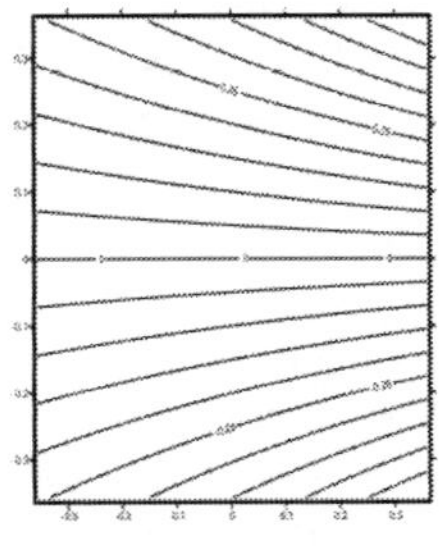

Fig.3 (b) RMM

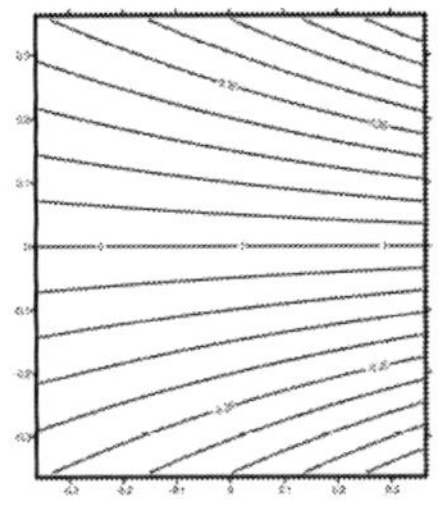

Fig.3 (c) Modified RMM

Fig.3 The field solution in the y-z plane (x=0), (a) Analytical solution, (b) RMM, (c) Modified

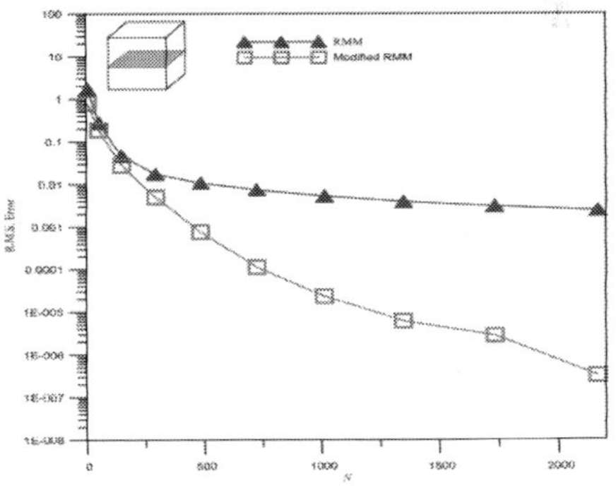

Fig.4 The error analysis versus the number of collocation points

ITM Web of Conferences **9**, 01002 (2017) DOI: 10.1051/itmconf/20170901002
AMCSE 2016

Minimax Normal Two-Armed Bandit with Indefinite Control Horizon

Alexander Kolnogorov[1],*

[1] *Yaroslav-the-Wise Novgorod State University*
B.St-Petersburgskaya Str, 41, Velikiy Novgorod, Russia, 173003

Abstract. We consider the two-armed bandit problem as applied to data processing if there are two alternative processing methods available with different a priori unknown efficiencies. On should determine the most effective method and provide its dominating application. The total number of data, which is interpreted as a control horizon, is assumed to have a priori known probability distribution.
The problem is considered in minimax (robust) setting. According to the main theorem of the theory of games minimax risk and minimax strategy are sought for as Bayesian ones corresponding to the worst-case prior distribution. We describe the properties of the worst-case prior and present a recursive Bellman-type equation for determination of both minimax strategy and minimax risk. Numerical results illustrating the proposed algorithm are given. The algorithm can be applied to optimization of parallel data processing if the number of processed data is not definitely known in advance.

1 Introduction

We consider the two-armed bandit problem (see, e.g. [1], [2]) which is also well-known as the problem of expedient behavior in a random environment (see, e.g. [3], [4]) and the problem of adaptive control (see, e.g. [5], [6]) in the following setting. Let ξ_n, $n = 1, \ldots, N$ be a controlled random process which values are interpreted as incomes, depend only on currently chosen actions y_n ($y_n \in \{1, 2\}$) and have Normal probability distribution densities

$$f(x|m_\ell) = (2\pi)^{-1/2} \exp\left\{-(x - m_\ell)^2/2\right\},$$

if $y_n = \ell$ ($\ell = 1, 2$). Normal two-armed bandit can be described by a vector parameter $\theta = (m_1, m_2)$. The set of parameters is assumed to be the following

$$\Theta = \{\theta : |m_1 - m_2| \leq 2C\},$$

where $0 < C < \infty$.

Control strategy σ at the point of time n assigns a random choice of the action y_n depending on the current history of the process, i.e. replies $x^{n-1} = x_1, \ldots, x_{n-1}$ to applied actions $y^{n-1} = y_1, \ldots, y_{n-1}$:

$$\sigma_\ell(y^{n-1}, x^{n-1}) = \Pr(y_n = \ell | y^{n-1}, x^{n-1}),$$

$\ell = 1, 2$. The set of strategies is denoted by Σ.

Generally, the goal is to maximize (in some sense) the total expected income. In this article, we consider the minimax approach. If parameter θ is known then the optimal strategy should always apply the action corresponding to the larger value of m_1, m_2. The total expected income would thus be equal to $N(m_1 \vee m_2)$. If parameter is unknown then the loss function

$$L_N(\sigma, \theta) = N(m_1 \vee m_2) - E_{\sigma,\theta}\left(\sum_{n=1}^{N} \xi_n\right) \quad (1)$$

describes expected losses of total income with respect to its maximal possible value due to incomplete information. Here $E_{\sigma,\theta}$ denotes the mathematical expectation calculated with respect to measure generated by a strategy σ and a parameter θ. According to the minimax approach the maximal value of the loss function on the set of parameters Θ should be minimized on the set of strategies Σ. The value

$$R_N^M(\Theta) = \inf_{\Sigma} \sup_{\Theta} L_N(\sigma, \theta) \quad (2)$$

is called the minimax risk and corresponding optimal strategy σ^M is called the minimax strategy. Note that application of the minimax strategy ensures that inequality holds

$$L_N(\sigma^M, \theta) \leq R_N^M(\Theta)$$

for all $\theta \in \Theta$ and this means the robustness of the control.

The minimax approach to the problem was proposed by H. Robbins in [7] for the so-called Bernoulli two-armed bandit. It is described by binary incomes $\{0, 1\}$ and probability distribution

$$\Pr(\xi_n = 1 | y_n = \ell) = p_\ell, \quad \Pr(\xi_n = 0 | y_n = \ell) = q_\ell,$$

$p_\ell + q_\ell = 1$, $\ell = 1, 2$. Bernoulli two-armed bandit is described by a parameter $\theta = (p_1, p_2)$ with the set of values $\Theta = \{\theta : 0 \leq p_\ell \leq 1; \ell = 1, 2\}$. The article [7] caused a significant interest to considered problem. It was shown in [8] that explicit determination of the minimax strategy and

*e-mail: Alexander.Kolnogorov@novsu.ru

DOI: 10.1051/itmconf/20170901002

minimax risk is practically impossible already for $N > 4$. However, the asymptotic minimax theorem was proved by W. Vogel in [9] which states that minimax risk has the order $N^{1/2}$ and provides lower and upper estimates for a factor. This theorem holds true for the Normal two-armed bandit as well.

A very popular approach to the problem is a Bayesian one. Let $\lambda(\theta)$ be a prior probability density. The value

$$R_N^B(\lambda) = \inf_{\Sigma} \int_{\Theta} L_N(\sigma, \theta)\lambda(\theta) d\theta \tag{3}$$

is called Bayesian risk and corresponding optimal strategy is called Bayesian strategy. Bayesian approach allows to find Bayesian strategy and risk by solving a recursive Bellman-type equation.

According to the main theorem of the theory of games the minimax risk (2) is equal to Bayesian risk (3) calculated with respect to the worst-case prior distribution, i.e.

$$R_N^M(\Theta) = R_N^B(\lambda_0) = \sup_{\lambda} R_N^B(\lambda) \tag{4}$$

This approach allows to obtain the following asymptotic estimate for the Normal two-armed bandit (see [10]–[13]):

$$\lim_{N\to\infty} N^{-1/2} R_N^M(\Theta) = R_0 \tag{5}$$

with $R_0 \approx 0.637$.

Let's explain the choice of Normal distribution of incomes. We consider the problem as applied to group control of processing a large amount of data. Let $T = NM$ data be given that can be processed using either of the two alternative methods. The result of processing of the t-th item of data is $\zeta_t = 1$ if processing is successful and $\zeta_t = 0$ if it is unsuccessful. Probabilities $\Pr(\zeta_t = 1 | y_t = \ell) = p_\ell$ ($\ell = 1, 2$) depend only on selected methods (actions). Let's assume that p_1, p_2 are close to p ($0 < p < 1$). We partition the data into N packets of M data in each packet and use the same method for data processing in the same packet. For the control, we use the values of the process $\xi_n = (D_p M)^{-1/2} \sum_{t=(n-1)M+1}^{nM} \zeta_t$, $n = 1, \dots, N$ with $D_p = p(1-p)$. According to the central limit theorem, distributions of ξ_n, $n = 1, \dots, N$ are close to Normal and their variances are close to unity as in considered setting.

Note that data in the same packet may be processed in parallel. However, there is a question of losses in the control performance as the result of such aggregation. It was shown in [10]–[13] that if N is large enough (e.g. $N \geq 30$) then parallel control is close to optimal. Therefore, say 30000 items of data can be processed in 30 steps by packets of 1000 data with almost the same maximal losses as if the data were processed optimally one-by-one.

Remark 1. There are some different approaches to robust control in the two-armed bandit problem, see, e.g. [6, 14–16]. In these articles stochastic approximation method and mirror descent algorithm are used for the control. Instead of minimax risk, the authors often consider the equivalent attitude called the guaranteed rate of convergency. The order of the minimax risk for these algorithms is $N^{1/2}$ or close to $N^{1/2}$. However, more precise estimates were not presented for these algorithms. The versions for parallel processing were not proposed as well.

Remark 2. Parallel control for the two-armed bandit problem was first proposed for the problem of treating a large group of patients by either of the two drugs with different unknown efficiencies. The discussion and bibliography of the problem can be found in [17].

The goal of this paper is to investigate the robust control of the Normal two-armed bandit with indefinite horizon. The structure of the paper is the following. In Section 2 we set the control problem, i.e. define the loss function and corresponding minimax and Bayesian risks and strategies if there is a priori known probability distribution on the set of horizons. We state the main theorem of the theory of games in this case. In Section 3 we analyze the properties of the worst-case prior distribution and present the recursive Bellman-type equation for calculation of the Bayesian risk with respect to this worst-case prior. We also present recursive Bellman-type equation for calculation of the expected losses. In Section 4 the results of numerical experiments are presented. Section 5 contains a conclusion.

2 Setup of the Control Problem with Indefinite Horizon

The disadvantage of approach considered in Section 1 is that the control horizon N is fixed. However, it is often more likely to consider the problem with indefinite control horizon. Let's consider $\{N_i\} = \{N_i; i = 1, \dots, I\}$ a finite set of control horizons s.t. $2 \leq N_1 < \cdots < N_I = N$. Let's define appropriate loss function as follows

$$L(\sigma, \theta, \{N_i\}) = \sum_{i=1}^{I} \beta_i L_{N_i}(\sigma, \theta), \tag{6}$$

where $L_{N_i}(\sigma, \theta)$, $i = 1, \dots, I$ are defined in (1). Factors β_i, $i = 1, \dots, I$ may be arbitrary chosen. However, it is natural to choose them so that all $\{\beta_i L_{N_i}(\sigma, \theta)\}$ have approximately equal maximal values. Taking into account (5) one can choose $\beta_i \asymp N_i^{-1/2}$, $i = 1, \dots, I$. If in addition the following condition holds

$$\sum_{i=1}^{I} \beta_i = 1$$

then one can say that the prior distribution $\Pr(N = N_i) = \beta_i$, $i = 1, \dots, I$ is assigned on the set of control horizons.

Given n satisfying condition $N_{i-1} < n \leq N_i$, let's put

$$\gamma_n = \sum_{j=i}^{I} \beta_j.$$

Using (1) and (6) one obtains

$$L(\sigma, \theta, \{N_i\}) = E_{\sigma,\theta}\left(\sum_{n=1}^{N} \gamma_n \left((m_1 \vee m_2) - \xi_n\right)\right) \tag{7}$$

In the sequel we'll denote $\hat{L}_N(\sigma, \theta) = L(\sigma, \theta, \{N_i\})$. Corresponding minimax and Bayesian risks are defined as

 DOI: 10.1051/itmconf/20170901002

follows

$$\hat{R}_N^M(\Theta) = \inf_{\Sigma} \sup_{\Theta} \hat{L}_N(\sigma, \theta), \tag{8}$$

$$\hat{R}_N^B(\lambda) = \inf_{\Sigma} \int_{\Theta} \hat{L}_N(\sigma, \theta)\lambda(\theta)d\theta. \tag{9}$$

Taking into account (7) an using reasonings similar to those in [10], one can prove that the main theorem of the theory of games holds in considered setting as well as in the case of definite horizon. It means that minimax risk (8) can be determined as Bayesian risk (9) calculated with respect to the worst-case prior distribution, i.e.

$$\hat{R}_N^M(\Theta) = \hat{R}_N^B(\lambda_0) = \sup_{\lambda} \hat{R}_N^B(\lambda) \tag{10}$$

and minimax strategy is equal to corresponding Bayesian strategy as well.

Bayesian risk can be calculated recursively. Denote by

$$f_D(x|M) = (2\pi D)^{-1/2} \exp\left\{-(x-M)^2/(2D)\right\}$$

the Normal distribution density with mathematical expectation M and variance D. Denote by $\lambda(m_1, m_2)$ the prior distribution density on the set of parameters Θ. Let history of control up to instant of time n be described by (X_1, n_1, X_2, n_2). Here n_1, n_2 are total numbers of applications of both actions ($n_1 + n_2 = n$) and X_1, X_2 are corresponding total incomes. Let $X_\ell = 0$ if $n_\ell = 0$. The posterior distribution density is thus equal to

$$\lambda(m_1, m_2|X_1, n_1, X_2, n_2) = \frac{f_{n_1}(X_1|n_1m_1)f_{n_2}(X_2|n_2m_2)\lambda(m_1, m_2)}{\iint_{\Theta} f_{n_1}(X_1|n_1m_1)f_{n_2}(X_2|n_2m_2)\lambda(m_1, m_2)dm_1, dm_2}.$$

If additionally it is assumed that $f_n(X|nm) = 1$ at $n = 0$ then this expression holds true if $n_1 = 0$ and/or $n_2 = 0$ as well.

Denote by $\hat{R}_{N-n}^B(\lambda; X_1, n_1, X_2, n_2)$ Bayesian risk at the latter $(N-n)$ steps calculated with respect to the posterior distribution density $\lambda(m_1, m_2|X_1, n_1, X_2, n_2)$. Let $x^+ = \max(x, 0)$. Then

$$\hat{R}_{N-n}^B(\cdot) = \min(\hat{R}_{N-n}^{(1)}(\cdot), \hat{R}_{N-n}^{(2)}(\cdot)), \tag{11}$$

where $\hat{R}_0^{(1)}(\cdot) = \hat{R}_0^{(2)}(\cdot) = 0$,

$$\begin{aligned}\hat{R}_{N-n}^{(1)}(\lambda; X_1, n_1, X_2, n_2) = \iint_{\Theta} (\ & \gamma_{n+1}(m_2 - m_1)^+ \\ & + E_x^{(1)} \hat{R}_{N-(n+1)}^B(\lambda; X_1 + x, n_1 + 1, X_2, n_2)\ ) \\ & \times \lambda(m_1, m_2|X_1, n_1, X_2, n_2)dm_1dm_2,\end{aligned} \tag{12}$$

$$\begin{aligned}\hat{R}_{N-n}^{(2)}(\lambda; X_1, n_1, X_2, n_2) = \iint_{\Theta} (\ & \gamma_{n+1}(m_1 - m_2)^+ \\ & + E_x^{(2)} \hat{R}_{N-(n+1)}^B(\lambda; X_1, n_1, X_2 + x, n_2 + 1)\ ) \\ & \times \lambda(m_1, m_2|X_1, n_1, X_2, n_2)dm_1dm_2\end{aligned} \tag{13}$$

and

$$E_x^{(\ell)} \hat{R}(x) = \int_{-\infty}^{+\infty} \hat{R}(x) f(x|m_\ell)dx, \quad \ell = 1, 2.$$

Bayesian strategy prescribes currently to choose action corresponding to the smaller value of $\hat{R}_{N-n}^{(1)}(\cdot)$, $\hat{R}_{N-n}^{(2)}(\cdot)$, the choice may be arbitrary if these values are equal.

3 Recursive Bellman-type Equation for Calculation of the Bayesian Risk and Expected Losses

Let's recall that Bayesian risk is a concave function of the prior distribution density, i.e.

$$\hat{R}_N^B(\alpha\lambda + \tilde{\alpha}\tilde{\lambda}) \ge \alpha\hat{R}_N^B(\lambda) + \tilde{\alpha}\hat{R}_N^B(\tilde{\lambda}),$$

if $\alpha + \tilde{\alpha} = 1$; $\alpha, \tilde{\alpha} > 0$. This property allows to specify the worst-case prior distribution. Like in [10], one can prove that the following transformations $\tilde{\lambda}$ of the prior distribution density λ do not change the Bayesian risk, i.e. $\hat{R}_N^B(\tilde{\lambda}) = \hat{R}_N^B(\lambda)$:

1. $\tilde{\lambda}^{(1)}(m_1, m_2) = \lambda(m_2, m_1)$ (for all m_1, m_2). This property means that expected losses do not change if one swaps the arms of the bandit.

2. $\tilde{\lambda}^{(2)}(m_1, m_2) = \lambda(m_1 + m, m_2 + m)$ (for all m_1, m_2 and any fixed m). This property means that expected losses do not change if one equally shifts both mathematical expectations.

So, if λ is the worst-case prior distribution then $\alpha\lambda + \tilde{\alpha}\tilde{\lambda}$ is the worst-case prior as well. It means that the worst-case prior distribution does not change if the above transformations are implemented.

In the sequel it is convenient to modify parameterization. Let $m_1 = u + v$, $m_2 = u - v$, then $\theta = (u + v, u - v)$ and $\Theta = \{\theta : |v| \le C\}$. Taking into account the Jacobian $|\partial(m_1, m_2)/\partial(u, v)| = 2$, a prior distribution density is equal to $\nu(u, v) = 2\lambda(u + v, u - v)$. Then the following transformations of the prior distribution densities $\tilde{\nu}^{(1)}(u, v) = \nu(u, -v)$, $\tilde{\nu}^{(2)}(u, v) = \nu(u + m, v)$ (for any fixed m) do not change the value of Bayesian risk. These properties allow to describe the worst-case prior. Namely, asymptotically the worst-case prior distribution density can be chosen the following one:

$$\nu_a(u, v) = \kappa_a(u)\rho(v), \tag{14}$$

where $\kappa_a(u)$ is the uniform density on the interval $|u| \le a$, $\rho(v)$ is a symmetric density (i.e. $\rho(-v) = \rho(v)$) on the interval $|v| \le C$ and $a \to \infty$. This prior does not change under the first transformation and asymptotically (as $a \to \infty$) does not change under the second transformation.

Now let's write the dynamic programming equation for calculation the Bayesian risk with respect to (14). This equation follows from (11)–(13) if the prior distribution density is formally assumed to be constant with respect to u and this gives true expressions for the posterior densities if $n_1 \ge 1$, $n_2 \ge 1$. At the former two steps actions should be chosen turn-by-turn. Note that equation is more simple for risks

$$\hat{R}_{n_1,n_2}(X_1, X_2) = \hat{R}_{N-n}^B(X_1, n_1, X_2, n_2)p_{n_1,n_2}(X_1, X_2)$$

with

$$p_{n_1,n_2}(X_1, X_2) = \iint_{\Theta} f_{n_1}(X_1|n_1m_1)f_{n_2}(X_2|n_2m_2)\lambda(m_1, m_2)dm_1dm_2.$$

Denote by $\hat{R}_{n_1,n_2}(Z) = \hat{R}_{n_1,n_2}(X_1, X_2)$ with $Z = X_1 n_2 - X_2 n_1$. Then

$$\hat{R}_{n_1,n_2}(\cdot) = \min(\hat{R}^{(1)}_{n_1,n_2}(\cdot), \hat{R}^{(2)}_{n_1,n_2}(\cdot)), \qquad (15)$$

with $\hat{R}^{(1)}_{n_1,n_2}(Z) = \hat{R}^{(2)}_{n_1,n_2}(Z) = 0$ at $n_1 + n_2 = N$,

$$\begin{aligned}\hat{R}^{(1)}_{n_1,n_2}(Z) &= \int_0^C 2\gamma_{n+1} v g_{n_1,n_2}(Z, v)\rho(v)dv \\ &+ n_2^{-1} \int_{-\infty}^{+\infty} \hat{R}_{n_1+1,n_2}(Z + z) h_{n_1}\left(\tfrac{Z - n_1 z}{n_2}\right) dz,\end{aligned} \qquad (16)$$

$$\begin{aligned}\hat{R}^{(2)}_{n_1,n_2}(Z) &= \int_0^C 2\gamma_{n+1} v g_{n_1,n_2}(Z, -v)\rho(v)dv \\ &+ n_1^{-1} \int_{-\infty}^{+\infty} \hat{R}_{n_1,n_2+1}(Z + z) h_{n_2}\left(\tfrac{Z - n_2 z}{n_1}\right) dz\end{aligned} \qquad (17)$$

at $n = n_1 + n_2 < N$, $n_1 \geq 1$, $n_2 \geq 1$. Here

$$\begin{aligned}g_{n_1,n_2}(Z, v) &= \frac{1}{(2\pi n_1 n_2 (n_1 + n_2))^{1/2}} \\ &\times \exp\left(-\frac{(Z + 2v n_1 n_2)^2}{2 n_1 n_2 (n_1 + n_2)}\right),\end{aligned} \qquad (18)$$

$$h_n(z) = \left(\frac{n+1}{2\pi n}\right)^{1/2} \times \exp\left(-\frac{z^2}{2n(n+1)}\right). \qquad (19)$$

Bayesian risk (9) is calculated according to the formula

$$\lim_{a\to\infty} \hat{R}^B_N(\nu_a(u, v)) = \hat{L}(\rho(v)) + \int_{-\infty}^{\infty} \hat{R}_{1,1}(z)dz \qquad (20)$$

where

$$\hat{L}(\rho(v)) = \int_0^C (\gamma_1 + \gamma_2) 2v\rho(v)dv.$$

Now let's present a recursive equation for expected losses corresponding to strategy $\sigma_\ell(Z, n_1, n_2) = \Pr(y_n = \ell | Z, n_1, n_2)$:

$$\begin{aligned}\hat{L}_{n_1,n_2}(Z) &= \sigma_1(Z, n_1, n_2)\hat{L}^{(1)}_{n_1,n_2}(Z) \\ &+ \sigma_2(Z, n_1, n_2)\hat{L}^{(2)}_{n_1,n_2}(Z),\end{aligned} \qquad (21)$$

with $\hat{L}^{(1)}_{n_1,n_2}(Z) = \hat{L}^{(2)}_{n_1,n_2}(Z) = 0$ at $n_1 + n_2 = N$,

$$\begin{aligned}\hat{L}^{(1)}_{n_1,n_2}(Z) &= \int_0^C 2\gamma_{n+1} v g_{n_1,n_2}(Z, v)\rho(v)dv \\ &+ n_2^{-1} \int_{-\infty}^{+\infty} \hat{L}_{n_1+1,n_2}(Z + z) h_{n_1}\left(\tfrac{Z - n_1 z}{n_2}\right) dz,\end{aligned} \qquad (22)$$

$$\begin{aligned}\hat{L}^{(2)}_{n_1,n_2}(Z) &= \int_0^C 2\gamma_{n+1} v g_{n_1,n_2}(Z, -v)\rho(v)dv \\ &+ n_1^{-1} \int_{-\infty}^{+\infty} \hat{L}_{n_1,n_2+1}(Z + z) h_{n_2}\left(\tfrac{Z - n_2 z}{n_1}\right) dz\end{aligned} \qquad (23)$$

if $n = n_1 + n_2 < N$, $n_1 \geq 1$, $n_2 \geq 1$. Expected losses are calculated according to the formula

$$\lim_{a\to\infty} L^B_N(\nu_a(u, v)) = \hat{L}(\rho(v)) + \int_{-\infty}^{\infty} \hat{L}_{1,1}(z)dz. \qquad (24)$$

4 Numerical Results

Minimax risk and minimax strategy were sought for the case $N_1 = 20$, $N_2 = N = 30$. Factors were chosen as follows: $\beta_1 = \beta^{-1} N_1^{-1/2}$, $\beta_2 = \beta^{-1} N_2^{-1/2}$ with $\beta = N_1^{-1/2} + N_2^{-1/2}$, so that $\beta_1 + \beta_2 = 1$. Bayesian risk was calculated by (15)–(20). The worst-case prior distribution corresponding to the maximum of the Bayesian risk is concentrated at $d \approx \pm 1.9$ and maximal value of the Bayesian risk is approximately equal to 0.594. Then expected losses were calculated by (21)–(24) using determined strategy. Results are presented on Figure 1. One can see that maximal value of expected losses does not exceed the value 0.594. Hence, the determined strategy is minimax strategy for considered problem.

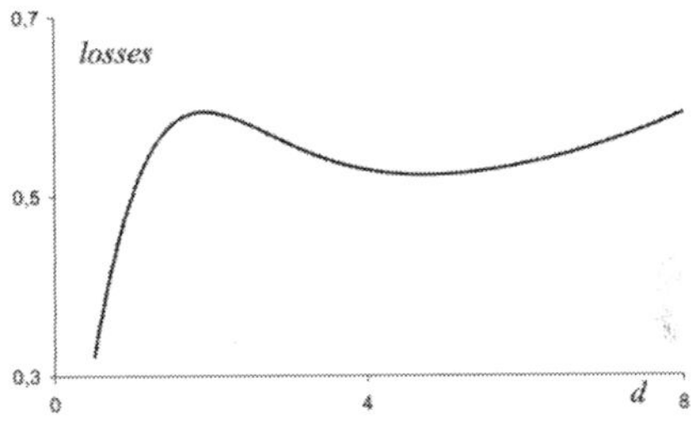

Figure 1. Expected losses corresponding to minimax strategy, $N_1 = 20$, $N_2 = 30$.

5 Conclusion

Minimax approach to the two-armed bandit problem is considered with a priori assigned distribution on the control horizon. The problem can be reduced to the classical two-armed bandit problem with discounted incomes. The algorithm of determination of the minimax strategy and minimax risk as Bayesian ones corresponding to the worst-case prior distribution is obtained and numerical results are presented.

This work was supported in part by the Project Part of the State Assignment in the Field of Scientific Activity by the Ministry of Education and Science of the Russian Federation, project no. 1.949.2014/K.

References

[1] D.A. Berry, B. Fristedt, *Bandit Problems: Sequential Allocation of Experiments* (Chapman and Hall, London, New York, 1985)

[2] E.L. Presman, I.M. Sonin, *Sequential Control with Incomplete Information* (Academic Press, New York, 1990)

[3] M.L. Tsetlin, *Automation Theory and Modeling of Biological Systems* (Academic Press, New York, 1973)

[4] V.I. Varshavsky, *Collective Behavior of Automata* (Nauka, Moscow, 1973) (In Russian)

DOI: 10.1051/itmconf/20170901002

[5] V.G. Sragovich, *Mathematical Theory of Adaptive Control* (World Scientific. Interdisciplinary Mathematical Sciences, New Jersey, London, **4**, 2006)

[6] A.V. Nazin, A.S. Poznyak, *Adaptive Choice of Alternatives* (Nauka, Moscow, 1986) (In Russian)

[7] H. Robbins, Bulletin AMS. **58**, 527–535 (1952)

[8] J. Fabius, W.R. van Zwet, Ann. Math. Statist. **41**, 1906–1916 (1970)

[9] W. Vogel, Ann. Math. Stat. **31**, 444–451 (1960)

[10] A.V. Kolnogorov, Automation and Remote Control **72**, 1017–1027 (2011)

[11] A.V. Kolnogorov, Automation and Remote Control **73**, 689–701 (2012)

[12] A.V. Kolnogorov, Automation and Remote Control **76**, 1229–1241 (2015)

[13] A.V. Kolnogorov, International Journal of Mathematics and Computers in Simulation **10**, 129–132 (2016)

[14] G. Lugosi, N. Cesa-Bianchi, *Prediction, Learning and Games* (Cambridge University Press, New York, 2006)

[15] A. Juditsky, A.V. Nazin, A.B. Tsybakov, N. Vayatis, Proc. 17th World Congress IFAC, 11560–11563 (2008)

[16] A.V. Gasnikov, Yu.E. Nesterov, V.G. Spokoiny, Computational Mathematics and Mathematical Physics **55**, 580–596 (2015)

[17] T.L. Lai, B. Levin, H. Robbins, D. Siegmund, Proc. Nati. Acad. Sci. USA **77**, 3135–3138 (1980)

DOI: 10.1051/itmconf/20170901003

Intensive wave power and steel quenching 3-D model for cylindrical sample. Time direct and reverse formulations and solutions

*Andris Buikis** and *Margarita Buike*
Institute of Mathematics and Computer Science, University of Latvia, Raina Blvd. 29, LV1459, Riga, LATVIA

Abstract. In this paper we develop mathematical models for three dimensional hyperbolic heat equations (wave equation or telegraph equation) with inner source power and construct their analytical solutions for the determination of the initial heat flux for cylindrical sample. As additional conditions the temperature and heat flux at the end time are given. In some cases we give expression of wave energy. In some cases we give expression of wave energy. Some solutions of time inverse problems are obtained in the form of first kind Fredholm integral equation, but others has been obtained in closed analytical form as series. We viewed both direct and inverse problems at the time. For the time inverse problem we use inversion in the time argument.

1 INTRODUCTION

Contrary to traditional method the intensive quenching process uses environmentally friendly highly agitated water or low concentration of water/mineral salt solutions [1]-[5]. This equation is important for wave energy generator [6, 7]. Traditionally for the mathematical description of the intensive quenching process, classical heat conduction equation is used. We have proposed to use hyperbolic heat equation [8-21] for more realistic description of the intensive quenching (IQ) process (especially for the initial stage of the process).

The idea of the usage of hyperbolic heat equation can be easily transferred to completely different sector of application - to the generation of electricity in sea or ocean by usage of wave energy [22, 23]. The first known patent to use energy from ocean waves dates back to 1799 and was filed in Paris by Girard and his son [22]. It is important to note, that Ekergard and his co-authors [23] examine the development of the system in time, describing the equipment with ordinary differential equation. Here we describe the equipment in development of both - in time as well as in spatial arrangement of equipment using the three-dimensional hyperbolic heat equation. Wave power plant has to work for long time period in moving environment – waves. Therefore it is important to examine not only the development of equipment in time, but also the movement of its different components [19, 21].

Wave energy generator models can be viewed both Cartesian coordinate and cylindrical co-ordinates. In papers [11], [12], [14], [19-21] we investigate the rectangular models. Generators of cylindrical or spherical forms with fin we investigate in papers [15] and [17].

In our previous papers we have constructed various one and two dimensional analytical exact and approximate [9-16], [19-21], [24, 25] solutions for IQ processes. We consider three-dimensional statements for non-homogeneous equation with non-homogeneous boundary conditions. Such statements allow constructing mathematical models for wave power plants in connection with other equipment, for example, with wind power. In recent years, we have been able to generalize the Green's function method to areas, which consist of several canonical connected sub-areas, and thus we have obtained the exact solutions for the L-, T- and П-type areas [9 - 11], [21].

2 MATHEMATICAL FORMULATION OF 3-D PROBLEM FOR IQP OR WAVE POWER

Already in the introduction we noted that Professor M. Leijon, see [23] examined the development of system in time. Here we offer to consider the description of system in time and space. For this purpose instead of the time depended ordinary differential equation, we consider the following partial differential equation:

$$\frac{\partial^2 U}{\partial t^2} = a_\tau^2\left[\frac{1}{r}\frac{\partial}{\partial r}\left(r\frac{\partial U}{\partial r}\right)+\frac{1}{r^2}\frac{\partial^2 U}{\partial \varphi^2}+\frac{\partial^2 U}{\partial z^2}\right]-$$
$$-CU+F(r,\varphi,z,t), r\in[0,R], \varphi\in[0,2\pi], \quad (1)$$
$$z\in[0,l], t\in[0,T], C\geq 0, a_\tau^2=\frac{a^2}{\tau_r}, a^2=\frac{k}{c\rho}.$$

* Corresponding author: buikis@latnet.lv

Here c is specific heat capacity, k - heat conductivity coefficient, ρ - density, τ_r - relaxation time. The source term $\Phi(r,\varphi,z,t)$ can be from different parts of the same device or outer source, for example, wind source.
In the case of wave energy we can assume different non-homogeneous boundary conditions. Important is to formulate boundary conditions (3), (4) and (5) in the heat energy transfer form [27]:

$$r\frac{\partial U}{\partial r}\bigg|_{r=0}=0, \tag{2}$$

$$\left(R\frac{\partial U}{\partial r}+k_1U\right)\bigg|_{r=R}=Rg_1(\varphi,z,t), k_1=\frac{Rh_3}{k}, \tag{3}$$

$$\left(\frac{\partial U}{\partial z}-k_2U\right)\bigg|_{z=0}=g_2(r,\varphi,t), k_i=\frac{h_i}{k}, i=2,3 \tag{4}$$

$$\left(\frac{\partial U}{\partial z}+k_3U\right)\bigg|_{z=l}=g_3(r,\varphi,t). \tag{5}$$

Here h_i is heat exchange coefficient. On all the other sides of device we have heat exchange with environment. In fact it is possible to look at other types of boundary conditions: first (Dirichlet) and second (Neumann) type. The initial conditions for the function $U(r,\varphi,z,t)$ are assumed in following form:

$$U\big|_{t=0}=U_0(r,\varphi,z), \tag{6}$$

$$\frac{\partial U}{\partial t}\bigg|_{t=0}=U_1(r,\varphi,z). \tag{7}$$

From the practical point of view in the steel quenching model the condition (7) can be unrealistic. The initial heat flux must be determined theoretically. As additional condition we assume that either the temperature distribution or the heat fluxes distribution at the end of process is given (known):

$$U\big|_{t=T}=U_T(r,\varphi,z), \tag{8}$$

$$\frac{\partial U}{\partial t}\bigg|_{t=T}=U_T^1(r,\varphi,z). \tag{9}$$

The formulation of the 3 dimensional mathematical model is important for wave energy generator [6], [7]. It is good see from the above point on the fig. 1 from patent [6].
For 3-D mathematical model is important that solution in $\varphi-$direction is continuous and smooth. These 2 conditions are important for

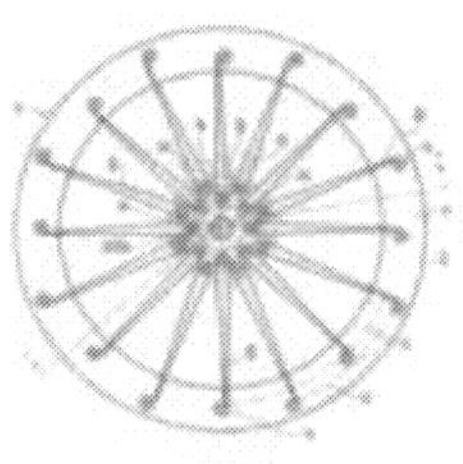

FIG. 2

Fig. 1. The look from the above point of cylindrical piezoelectric generator from patent [6].

the reduction of 3-D model to 2-D model by conservative averaging method [9], [24] and [25]:

$$U\big|_{\varphi=0}=U\big|_{\varphi=2\pi}, \tag{10}$$

$$\frac{\partial U}{\partial\varphi}\bigg|_{\varphi=0}=\frac{\partial U}{\partial\varphi}\bigg|_{\varphi=2\pi}. \tag{11}$$

3 SOLUTION OF 3-D PROBLEM

We assume that we have non-homogeneous Klein-Gordon equation-with source term: $C\geq 0.$ The solution in three-dimensional problem is in following form:

$$\begin{aligned}U(r,\varphi,z,t)&=H(r,\varphi,z,t)+\int_0^R\xi d\xi\int_0^{2\pi}d\varsigma\times\\&\int_0^l U_1(\xi,\eta,\varsigma)G(r,\varphi,z,\xi,\eta,\varsigma,t)d\eta+\int_0^{2\pi}d\varsigma\times\\&\int_0^R\xi d\xi\int_0^l U_0(\xi,\eta,\varsigma)\frac{\partial}{\partial t}G(r,\varphi,z,\xi,\eta,\varsigma,t)d\eta.\end{aligned} \tag{12}$$

Here

$$\begin{aligned}H(r,\varphi,z,t)&=a_\tau^2R^2\int_0^t d\tau\int_0^{2\pi}d\varsigma\times\\&\int_0^l g_1(\eta,\varsigma,\tau)G(r,\varphi,z,R,\eta,\varsigma,t-\tau)d\eta-a_\tau^2\int_0^t d\tau\times\\&\int_0^{2\pi}d\varsigma\int_0^R\xi g_2(\xi,\varsigma,\tau)G(r,\varphi,z,\xi,\varsigma,0,t-\tau)d\xi+a_\tau^2\times\\&\int_0^t d\tau\int_0^{2\pi}d\varsigma\int_0^R\xi g_3(\xi,\varsigma,\tau)G(r,\varphi,z,\xi,\varsigma,l,t-\tau)d\xi+\\&\int_0^t d\tau\int_0^{2\pi}d\varsigma\int_0^R\xi d\xi\times\end{aligned} \tag{13}$$

DOI: 10.1051/itmconf/20170901003

$$\int_0^l F(\xi,\eta,\varsigma,\tau)G(r,\varphi,z,\xi,\eta,\varsigma,t-\tau)d\eta.$$

The Green function [26], [27] for initial-boundary problem for Klein-Gordon equation is known; see [28]:

$$G(r,\varphi,z,\xi,\eta,\zeta,t)=\frac{1}{\pi}\sum_{n=0}^{\infty}\sum_{m=1}^{\infty}A_n\times$$

$$\sum_{s=1}^{\infty}\frac{\mu_{nm}^2 J_n(\mu_{nm}r)J_n(\mu_{nm}\xi)}{(\mu_{nm}^2R^2+k_1^2R^2-n^2)[J_n(\mu_{nm}R)]^2}\times \tag{14}$$

$$\frac{\cos[n(\varphi-\eta)]h_s(z)h_s(\zeta)\sin(\lambda_{nms}t)}{\|h_s\|^2\lambda_{nms}}.$$

Here $J_n(\xi)$ – is Bessel's function and

$$\lambda_{nms}=\sqrt{a_\tau^2(\mu_{nm}^2+\beta_s^2)+C},$$

$$A_n=\begin{cases}1, if\ n=0,\\ 2, if\ n>0;\end{cases}$$

$$h_s(z)=\cos(\beta_s z)+\frac{k_2}{\beta_s}\sin(\beta_s z),$$

$$\|h_s\|^2=\frac{k_3(\beta_s^2+k_2^2)}{2\beta_s^2(\beta_s^2+k_3^2)}+\frac{k_2}{2\beta_s^2}+\frac{l}{2}\left(1+\frac{k_2^2}{\beta_s^2}\right).$$

The eigenvalues μ_{nm},β_s are positive roots of the transcendental equations:

$$\mu J_n'(\mu R)+k_1J_n(\mu R)=0,\ \frac{tg(\beta l)}{\beta}=\frac{k_2+k_3}{\beta^2-k_2k_3}.$$

We assume that at final moment $t=T$ is known only one boundary condition (8). Then from solution (12) we easy obtain Fredholm first type integral equation with respect to function $U_1(r,\varphi,z)$:

$$\int_0^R\xi d\xi\int_0^{2\pi}d\varsigma\int_0^l U_1(\xi,\eta,\varsigma)G(r,\varphi,z,\xi,\eta,\varsigma,T)d\eta$$

$$=\Phi(r,\varphi,z).$$

(15)

The unknown right side function $\Phi(r,\varphi,z)$ is in the following form:

$$\Phi(r,\varphi,z)=U_T(r,\varphi,z)-H(r,\varphi,z,T)-$$

$$\int_0^{2\pi}d\varsigma\int_0^R\xi d\xi\int_0^l U_0(\xi,\eta,\varsigma)\frac{\partial}{\partial t}G(r,\varphi,z,\xi,\eta,\varsigma,t)\bigg|_{t=T}d\eta.$$

Similar situation is, if is done second boundary condition (9). We differentiate solution (12) regarding time:

$$\frac{\partial}{\partial t}U(r,\varphi,z,t)=\frac{\partial}{\partial t}H(r,\varphi,z,t)+$$

$$\int_0^R\xi d\xi\int_0^{2\pi}d\varsigma\int_0^l U_1(\xi,\eta,\varsigma)\frac{\partial}{\partial t}G(r,\varphi,z,\xi,\eta,\varsigma,t)d\eta+$$

$$\int_0^{2\pi}d\varsigma\int_0^R\xi d\xi\int_0^l U_0(\xi,\eta,\varsigma)\frac{\partial^2}{\partial t^2}G(r,\varphi,z,\xi,\eta,\varsigma,t)d\eta.$$

We obtain again 1st kind Fredholm integral equation for the determination of unknown initial heat flux:

$$\int_0^R\xi d\xi\int_0^{2\pi}d\varsigma\int_0^l U_1(\xi,\eta,\varsigma)\frac{\partial}{\partial t}G(r,\varphi,z,\xi,\eta,\varsigma,t)\bigg|_{t=T}d\eta$$

$$=\Phi_1(r,\varphi,z). \tag{16}$$

Here

$$\Phi_1(r,\varphi,z)=U_T^1(r,\varphi,z)-\frac{\partial}{\partial t}H(r,\varphi,z,t)\bigg|_{t=T}-$$

$$\int_0^{2\pi}d\varsigma\int_0^R\xi d\xi\int_0^l U_0(\xi,\eta,\varsigma)\frac{\partial^2}{\partial t^2}G(r,\varphi,z,\xi,\eta,\varsigma,t)\bigg|_{t=T}d\eta.$$

There is an interesting situation, if both additional conditions (8), (9) are known. In this case we introduce new time argument by formula

$$\tilde{t}=T-t. \tag{17}$$

The formulation for new function $V(r,\varphi,z,\tilde{t})$ with time variable $\tilde{t}$ is following:

$$\frac{\partial^2V}{\partial\tilde{t}^2}=a_\tau^2\left[\frac{1}{r}\frac{\partial}{\partial r}\left(r\frac{\partial V}{\partial r}\right)+\frac{1}{r^2}\frac{\partial^2V}{\partial\varphi^2}+\frac{\partial^2V}{\partial z^2}\right]-$$

$$-CV+F(r,\varphi,z,T-\tilde{t}),$$

$$\left(R\frac{\partial V}{\partial r}+k_1V\right)\bigg|_{r=R}=Rg_1(\varphi,z,T-\tilde{t}), \tag{18}$$

$$\left(\frac{\partial V}{\partial z}-k_2V\right)\bigg|_{z=l}=g_2(r,\varphi,T-\tilde{t}),$$

$$\left(\frac{\partial V}{\partial z}+k_3V\right)\bigg|_{z=0}=g_3(r,\varphi,T-\tilde{t}),$$

$$V|_{\tilde{t}=0}=U_T(r,\varphi,z),\ \frac{\partial V}{\partial\tilde{t}}\bigg|_{\tilde{t}=0}=-U_T^1(r,\varphi,z).$$

Similar to (10) the solution of inverse problem looks like:

$$V(x,y,z,\tilde{t})=H(x,y,z,\tilde{t})-\int_0^R\xi d\xi\int_0^{2\pi}d\varsigma\times$$

$$\int_0^l U_T^1(\xi,\eta,\varsigma)G(r,\varphi,z,\xi,\eta,\varsigma,\tilde{t})d\eta+\int_0^{2\pi}d\varsigma\times \tag{19}$$

$$\int_0^R\xi d\xi\int_0^l U_T(\xi,\eta,\varsigma)\frac{\partial}{\partial\tilde{t}}G(r,\varphi,z,\xi,\eta,\varsigma,\tilde{t})d\eta.$$

There is no problem to transform the expression for $H(x,y,z,\tilde{t})$ in following form:

$$H\left(x,y,z,\tilde{t}\right)=a_\tau^2 R\int_{T-\tilde{t}}^{T}d\tau\int_0^{2\pi}d\varsigma\times$$

$$\int_0^l g_1\left(\eta,\varsigma,\tau\right)G(r,\varphi,z,R,\eta,\varsigma,T-\tau)d\eta-a_\tau^2\int_{T-\tilde{t}}^{T}d\tau$$

$$\int_0^{2\pi}d\varsigma\int_0^R g_2\left(\xi,\varsigma,\tau\right)G(x,y,z,\xi,0,\varsigma,T-\tau)d\xi+a_\tau^2\times$$

$$\int_{T-\tilde{t}}^{T}d\tau\int_0^{2\pi}d\varsigma\int_0^R g_3\left(\xi,\varsigma,\tau\right)G(x,y,z,\xi,b,\varsigma,T-\tau)d\xi+$$

$$\int_{T-\tilde{t}}^{T}d\tau\int_0^{2\pi}d\varsigma\times$$

$$\int_0^R d\xi\int_0^l F\left(\xi,\eta,\varsigma,\tau\right)G(r,\varphi,z,\xi,\eta,\varsigma,T-\tau)d\eta.$$

For the heat flux in time from (19) we have the expression:

$$\frac{\partial}{\partial\tilde{t}}V(r,\varphi,z,\tilde{t})=\frac{\partial}{\partial\tilde{t}}H\left(r,\varphi,z,\tilde{t}\right)+$$

$$\int_0^R\xi d\xi\int_0^{2\pi}d\varsigma\int_0^l V_1(\xi,\eta,\varsigma)\frac{\partial}{\partial\tilde{t}}G(r,\varphi,z,\xi,\eta,\varsigma,\tilde{t})d\eta+$$

$$\int_0^{2\pi}d\varsigma\int_0^R\xi d\xi\int_0^l V_0(\xi,\eta,\varsigma)\frac{\partial^2}{\partial\tilde{t}^2}G(r,\varphi,z,\xi,\eta,\varsigma,\tilde{t})d\eta.$$

From last expression at $\tilde{t}=T$ and equality (17) we have solution for the time inverse problem:

$$U_T^1(r,\varphi,z)=-\frac{\partial}{\partial\tilde{t}}H\left(r,\varphi,z,\tilde{t}\right)\Big|_{\tilde{t}=T}-\qquad(20)$$

$$\int_0^R\xi d\xi\int_0^{2\pi}d\varsigma\int_0^l V_1(\xi,\eta,\varsigma)\frac{\partial}{\partial\tilde{t}}G(r,\varphi,z,\xi,\eta,\varsigma,\tilde{t})\Big|_{\tilde{t}=T}d\eta-$$

$$\int_0^{2\pi}d\varsigma\int_0^R\xi d\xi\int_0^l V_0(\xi,\eta,\varsigma)\frac{\partial^2}{\partial\tilde{t}^2}G(r,\varphi,z,\xi,\eta,\varsigma,\tilde{t})\Big|_{\tilde{t}=T}d\eta.$$

Very interesting is wave energy [29] as you can see in [21]:

$$I_0\left(t\right)=\sum_{n=0}^{\infty}\sum_{m=1}^{\infty}\sum_{s=1}^{\infty}\frac{\sin^2\left(\lambda_{nms}t\right)}{\lambda_{nms}^2}.$$

4 SOLUTION OF 3-D PROBLEM WITH CONSTANT INITIAL CONDITIONS

In the previous section we have constructed some three dimensional solutions for direct and time inverse problems for hyperbolic heat equation. Enough often initial conditions are constant functions [21], [25]. In this case it is to solve the solutions in the form of series. For simplicity we look the homogeneous boundary conditions:

$$U(r,\varphi,z,t)=U_1\int_0^R\xi d\xi\int_0^{2\pi}d\varsigma\times$$

$$\int_0^l G(r,\varphi,z,\xi,\eta,\varsigma,t)d\eta+U_0\int_0^{2\pi}d\varsigma\times\qquad(21)$$

$$\int_0^R\xi d\xi\int_0^l\frac{\partial}{\partial t}G(r,\varphi,z,\xi,\eta,\varsigma,t)d\eta=$$

$$=U_0G_0+U_1G_1.$$

We use the Green function form (14) in the little different form:

$$G(r,\varphi,z,\xi,\eta,\zeta,t)=\frac{1}{\pi}\times$$

$$\sum_{n=0}^{\infty}\sum_{m=1}^{\infty}\sum_{s=1}^{\infty}\frac{A_n\mu_{nm}^2J_n\left(\mu_{nm}r\right)J_n\left(\mu_{nm}\xi\right)}{\left(\mu_{nm}^2R^2+k_1^2R^2-n^2\right)\left[J_n\left(\mu_{nm}R\right)\right]^2}\times$$

$$\frac{\left[\cos(n\varphi)\cos(n\eta)+\sin(n\varphi)\sin(n\eta)\right]}{\left\|h_s\right\|^2}\times$$

$$\frac{h_s\left(z\right)h_s\left(\zeta\right)\sin\left(\lambda_{nms}t\right)}{\lambda_{nms}}.$$

The function G_0 after integration can be obtained in following form:

$$G_0=\frac{1}{\pi}\sum_{n=0}^{\infty}\sum_{m=1}^{\infty}\sum_{s=1}^{\infty}\frac{A_n\mu_{nm}^2J_n\left(\mu_{nm}r\right)\cos(\lambda_{nms}t)}{\left(\mu_{nm}^2R^2+k_1^2R^2-n^2\right)\left[J_n\left(\mu_{nm}R\right)\right]^2}\times$$

$$\frac{\left[\cos(n\varphi)\sin(nl)+\sin(n\varphi)(1-\cos(n\eta))\right]}{n\left\|h_s\right\|^2}h_s\left(z\right)\times$$

$$\frac{\sin(2\pi\beta_s)+\frac{k_2}{\beta_s}\left(1-\cos(2\pi\beta_s)\right)}{\beta_s}\int_0^R\xi J_n\left(\mu_{nm}\xi\right)d\xi.$$

Similarly we can transform the function G_1:

$$G_1=\frac{1}{\pi}\sum_{n=0}^{\infty}\sum_{m=1}^{\infty}\sum_{s=1}^{\infty}\frac{A_n\mu_{nm}^2J_n\left(\mu_{nm}r\right)\sin(\lambda_{nms}t)}{\left(\mu_{nm}^2R^2+k_1^2R^2-n^2\right)\left[J_n\left(\mu_{nm}R\right)\right]^2}\times$$

$$\frac{\left[\cos(n\varphi)\sin(nl)+\sin(n\varphi)(1-\cos(n\eta))\right]}{n\left\|h_s\right\|^2}h_s\left(z\right)\times$$

$$\frac{\sin(2\pi\beta_s)+\frac{k_2}{\beta_s}\left(1-\cos(2\pi\beta_s)\right)}{\beta_s}\int_0^R\xi J_n\left(\mu_{nm}\xi\right)d\xi.$$

In this paper we can shown that time reverse problem with two final time conditions is not ill-posed problem and can be solved similarly as time direct problem. It was shown in our paper [21] that for rectangular sample time reverse problem can be solved without some numerical problem. It is good known that for inverse problem is not easy to calculate the solution [30] - [34].

5 CONCLUSIONS

We have constructed some three dimensional solutions for direct and time inverse problems for hyperbolic heat equation. The solutions for determination of initial heat flux are obtained either in the form of Fredholm integral equation of 1st kind with continuous kernel or in the closed analytical form – in the form of series. In the future we will try to solve two dimensional and one dimensional problem for hyperbolic heat equation. These solutions can be obtained from of three dimensional solution by conservative averaging method. For the intensive steel quenching method with constant initial conditions we obtain the solution in the form of series. As second step we can use this method for different geometries.

This work has been supported by Latvian Council of Sciences (grant 623/2014).

References

1. N.I. Kobasko. *Steel Quenching in Liquid Media under Pressure.* – Kyiv, Naukova Dumka (1980).
2. N. I. Kobasko. Intensive Steel Quenching Methods, *Handbook "Theory and Technology of Quenching"*, Springer-Verlag (1992).
3. M.A. Aronov, N. Kobasko, J.A. Powell. Intensive Quenching of Carburized Steel Parts, IASME Transactions, Issue 9, **2**, p. 1841-1845 (2005).
4. N.I. Kobasko. Transient Nucleate Boiling Process to Be Widely for Super Strengthening of Materials and Obtaining Other Benefits in Heat Treating Industry. – UA Patent No. 109935, Published on Oct. 26, Bulletin No. 20, (2015).
5. N.I. Kobasko. Real and effective heat transfer coefficients (HTCs) used for computer simulation of transient nuclear boiling processes during quenching. Materials Performance and Characteristics, **1**, No. 1, p.1-20, (2012).
6. S.H. Salter. Apparatus for use in the extraction of energy from waves on water. US Patient 4,134,023. January 14, (1977).
7. C.B. Carroll. Piezoelectric rotary electrical energy generator. US Patient 6194815 B1. February 27, (2001).
8. M. Buike, A. Buikis. Approximate Solutions of Heat Conduction Problems in Multi- Dimensional Cylinder Type Domain by Conservative Averaging Method, Part 1. Proceedings of the 5th IASME/WSEAS Int. Conf. on Heat Transfer, Thermal Engineering and Environment, Vouliagmeni, Athens, August 25 -27, p. 15 – 20, (2007).
9. M. Buike, A. Buikis. Hyperbolic heat equation as mathematical model for steel quenching of L-shape samples, Part 1 (Direct Problem). Applied and Computational Mathematics. Proceedings of the 13th WSEAS International Conference on Applied Mathematics (MATH'08), Puerto De La Cruz, Tenerife, Canary Islands, Spain, December 15-17, WSEAS Press, p. 198-203, (2008).
10. T. Bobinska, M. Buike, A. Buikis. Hyperbolic Heat Equation as Mathematical Model for Steel Quenching of L-Shape Samples, Part 2 (Inverse Problem). Proceedings of 5th IASME/WSEAS International Conference on Continuum Mechanics (CM'10), University of Cambridge, UK, February 23-25, p. 21-26, (2010).
11. M. Buike, A. Buikis. Several Intensive Steel Quenching Models for Rectangular Samples. Proceedings of NAUN/WSEAS International Conference on Fluid Mechanics and Heat &Mass Transfer, Corfu Island, Greece, July 22-24. p. 88-93 (2010).
12. T. Bobinska, M. Buike, A. Buikis. Hyperbolic Heat Equation as Mathematical Model for Steel Quenching of L-and T-Shape Samples, Direct and Inverse Problems. Transactions of Heat and Mass Transfer. Vol.5, Issue 3, July 2010. p. 63-72.
13. S. Blomkalna, A. Buikis. Heat conduction problem for double-layered ball. *Progress in Industrial Mathematics at ECMI 2012.* Springer, p. 417-426, (2014).
14. T. Bobinska, M. Buike, A. Buikis. Comparing solutions of hyperbolic and parabolic heat conduction equations for L-shape samples. Recent Advances in Fluid Mechanics and Heat @Mass Transfer. Proceedings of the 9th IASME/WSEAS International Conference on THE'11. Florence, Italy, August 23-25, p. 384-389, (2011).
15. A. Piliksere, A. Buikis. Analytical solution for intensive quenching of cylindrical sample. Proceedings of 6th International Scientific Colloquium "Modelling for Material Processing", Riga, September 16-17, p. 181-186, (2010).
16. A. Piliksere, M. Buike, A. Buikis. Steel quenching process as hyperbolic heat equation for cylinder. Proceedings of 6th Baltic Heat Conference – BHTC2011, ISBN 978-952-15-2640-4 (CD-ROM), (2011).
17. S. Blomkalna, M. Buike, A. Buikis. Several intensive steel quenching models for rectangular and spherical samples. Recent Advances in Fluid Mechanics and Heat & Mass Transfer. Proceedings of the 9th

DOI: 10.1051/itmconf/20170901003

IASME/WSEAS International Conference on THE'11. Florence, Italy, August 23-25, p. 390-395, (2011).

18. A. Buikis, H. Kalis. Hyperbolic type approximation for the solutions of the hyperbolic heat conduction equation in 3-D domain. Mathematical and Computational Methods in Applied Sciences. Proceedings of the 3rd International Conference on Applied, Numerical and Computational Mathematics *(ICANCM'15)*. Sliema, Malta, August 17-19, pp. 42-51, (2015).
19. M. Buike, A. Buikis, H. Kalis. Wave energy and steel quenching models, which are solved exactly and approximately. Mathematical and Computational Methods in Applied Sciences. Proceedings of the 3rd International Conference on Applied, Numerical and Computational Mathematics (ICANCM'15). Sliema, Malta, August 17-19, pp. 72-81, (2015).
20. A. Buikis, H. Kalis. Hyperbolic Heat Equation in Bar and Finite Difference Schemes of Exact Spectrum. Latest Trends on Theoretical and Applied Mechanics, Fluid Mechanics and Heat & Mass Transfer. WSEAS Press, pp. 142-147, (2010).
21. M. Buike, A. Buikis, H. Kalis. Time Direct and Time Inverse Problems for Wave Energy and Steel Quenching Models, Solved Exactly and Approximately. WSEAS Transactions on Heat and Mass Transfer. Vol. 10, p. 30-43, (2015).
22. Wikipedia: http://en.wikipedia.org/wiki/Wave_power.
23. B. Ekergard, V. Castellucci, A. Savin, M. Leijon. Axial Force Damper in a Linear Wave Energy Convertor. Development and Applications of Oceanic Engineering. Vol. 2, Issue 2, May, (2013).
24. A. Buikis. Conservative averaging as an approximate method for solution of some direct and inverse heat transfer problems. Advanced Computational Methods in Heat Transfer, IX. WIT Press, p. 311-320, (2006).
25. R. Vilums, A. Buikis. Conservative averaging method for partial differential equations with discontinuous coefficients. WSEAS Transactions on Heat and Mass Transfer. 1, Issue 4, p. 383-390, (2006).
26. G.F. Roach. *Green's Functions.* Cambridge University Press, (1999).
27. H.S. Carslaw, C.J. Jaeger. *Conduction of Heat in Solids.* Oxford, Clarendon Press, (1959).
28. A.D. Polyanin. *Handbook of Linear Partial Differential Equations for Engineers and Scientists.* Chapman & Hall/CRC, (Russian edition, 2001). (2002).
29. L. Debnath. *Nonlinear Partial Differential Equations for Scientists and Engineers.* 2nd ed. Birkhäuser, (2005).
30. M. Buike, A. Buikis, R. Vilums. One-Dimensional Intensive Steel Quenching Models. Recent Advances in Mechanical Engineering. Proceedings of the 5th International Conference on Fluid Mechanics and Heat & Mass Transfer. Lisbon, Portugal, October 30-November 1, p. 54-62, (2014).
31. A. Buikis, S. Guseinov. Some one-dimensional coefficients inverse model problems of the heat transfer. – Proceedings of the Latvian Academy of Sciences, Sec. B, **57**, No 3/4 (626), pp.133-137, (2003).
32. A. Buikis, Sh. Guseinov. Solution of Reverse Hyperbolic Heat equation for intensive carburized steel quenching. Proceedings of ICCES'05 (Advances in Computational and Experimental Engineering and Sciences). December 1-6, IIT Madras. p. 741-746, (2005).
33. A. Buikis, Sh. Guseinov, M. Buike. Modelling of Intensive Steel Quenching Process by Time Inverse Hyperbolic Heat Conduction. Proceedings of the 4th International Scientific Colloquium "Modelling for Material Processing". Riga, June 8-9, p. 169-172, (2006).
34. A. Buikis, Sh. Guseinov. Conservative averaging method for solutions of inverse problems of mathematical physics. *Progress in Industrial Mathematics at ECMI 2002.* A. Buikis, R. Ciegis, A. D. Fitt (Eds.), Springer, p. 241-246, (2004).

ITM Web of Conferences **9**, 01004 (2017) DOI: 10.1051/itmconf/20170901004
AMCSE 2016

Hollow system with fin. Transient Green function method combination for two hollow cylinders

Andris Buikis*
Institute of Mathematics and Computer Science, University of Latvia, Raina Blvd. 29, LV1459, Riga, LATVIA

Abstract. In this paper we develop mathematical model for three dimensional heat equation for the system with hollow wall and fin and construct its analytical solution for two hollow cylindrical sample. The method of solution is based on Green function method for one hollow cylinder. On the conjugation conditions between both hollow cylinders we construct solution for system wall with fin. As result we come to integral equation on the surface between both hollow cylinders. Solution is obtained in the form of second kind Fredholm integral equation. The generalizing of Green function method allows us to use Green function method for regular non-canonical domains.

1 INTRODUCTION

Systems with extended surfaces (fins, spines) are related to very different branches of technique. Usually their mathematical modeling is realized by one dimensional steady-state assumptions [1]-[3]. In our previous papers we have constructed various two and three dimensional analytical approximate and exact solutions [4 -25].

In [2] the so-called Murray – Gardner assumptions are formulated. Some of them are:

1) The heat flow in the fin and the temperature at any point on the fin remains constant with time;
4) The temperature of the medium surrounding the fin is uniform;
5) The fin width is small compared to its height, so the temperature gradients across the fin width may be neglected;
6) The temperature at the base of the fin is uniform;
7) The heat transferred through the outmost edge of the fin (the fin tip) is negligible compared to that through the lateral surfaces (faces) of the fin.

In this paper we obtain exact transient solution by the Green function method [11 -13]. We give up all of these Murray – Gardner assumptions.

We consider three-dimensional statement for non-homogeneous equation with partly non-homogeneous boundary conditions.

In recent years, we have been able to generalize the Green's function method to areas, which consist of several canonical connected sub-areas, and thus we have obtained the exact solutions for the L-, T- and П-type areas [11 -13], [15,], 21, 23]. We have constructed two cylinders [17], and two-layer sphere [15].Cylinder was investigated in the paper [18].

2 FORMULATION OF 3-D PROBLEM

For this purpose instead of the time depended ordinary differential equation [1] – [3], we consider the following partial differential equation for the first cylinder:

$$\frac{1}{a^2}\frac{\partial U}{\partial t}=\left[\frac{1}{r}\frac{\partial}{\partial r}\left(r\frac{\partial U}{\partial r}\right)+\frac{1}{r^2}\frac{\partial^2 U}{\partial \varphi^2}+\frac{\partial^2 U}{\partial z^2}\right]+ F(r,\varphi,z,t), r\in[R_0,R], \varphi\in[0,2\pi], \quad (1)$$
$$z\in[0,l],\ a^2=\frac{k}{c\rho}.$$

Here c is specific heat capacity, k - heat conductivity coefficient, ρ - density. For the second cylinder which can be of different material:

$$\frac{1}{a_0^2}\frac{\partial U_0}{\partial t}=\left[\frac{1}{r}\frac{\partial}{\partial r}\left(r\frac{\partial U_0}{\partial r}\right)+\frac{1}{r^2}\frac{\partial^2 U_0}{\partial \varphi^2}+\frac{\partial^2 U_0}{\partial z^2}\right], \quad (2$$
$$r\in[R_0,R_1], \varphi\in[0,2\pi], z\in[l,l_0], a_0^2=\frac{k_0}{c_0\rho_0}.$$

We formulate the conjugations conditions as ideal thermal contact for $r\in[R_0,R_1]$:

$$U\big|_{z=l-0}=U_0\big|_{z=l+0}, \quad (3)$$

$$k\frac{\partial U}{\partial z}\bigg|_{z=l-0}=k_0\frac{\partial U_0}{\partial z}\bigg|_{z=l+0}. \quad (4)$$

We assume that the outer diameter of second cylinder is smaller as for primary cylinder: $R_1<R.$ The boundary conditions we assume partly homogeneous:

$$\left(\frac{\partial U}{\partial r}-k_0 U\right)\bigg|_{r=R_0}=0, k_0=\frac{h}{k}, \quad (5)$$

* Corresponding author: buikis@latnet.lv

DOI: 10.1051/itmconf/20170901004

$$\left(\frac{\partial U}{\partial r}+k_0U\right)\bigg|_{r=R}=g_0(\varphi,z,t), \quad (6)$$

$$\left(\frac{\partial U}{\partial z}-k_0U\right)\bigg|_{z=0}=g_1(r,\varphi,t), \quad (7)$$

$$\left(\frac{\partial U}{\partial z}+k_1U\right)\bigg|_{z=l,R_1<r<R}=g_2(r,\varphi,t), \quad (8)$$

$$\left(\frac{\partial U_0}{\partial r}-k_2U_0\right)\bigg|_{r=R_0}=0, k_2=\frac{h_0}{k_0}, \quad (9)$$

$$\left(\frac{\partial U_0}{\partial r}+k_2U_0\right)\bigg|_{r=R_1}=0, \quad (10)$$

$$\left(\frac{\partial U_0}{\partial z}+k_3U_0\right)\bigg|_{z=l_1}=g_3(r,\varphi,t). \quad (11)$$

Here h, h_0 is heat exchange coefficient. The initial conditions for the both cylinders are assumed in following form:

$$U|_{t=0}=U_0(r,\varphi,z), \quad (12)$$

$$U_0|_{t=0}=U_{00}(r,\varphi,z), \quad (13)$$

For 3-D mathematical model it is important that solution in $\varphi-$ direction is continuous and smooth. These 2 conditions are important for the reduction of 3-D model to 2-D model by conservative averaging method [9, 23] and [24]:

$$U|_{\varphi=0}=U|_{\varphi=2\pi}, \quad (14)$$

$$\frac{\partial U}{\partial \varphi}\bigg|_{\varphi=0}=\frac{\partial U}{\partial \varphi}\bigg|_{\varphi=2\pi}. \quad (15)$$

3 SOLUTION OF 3-D PROBLEM

The Green function for two cylinders is not known. The new idea is to combine the two conjugations conditions in the form of third type boundary condition with unknown non-homogeneity in the right hand side.

3.1. Solution for the wall

As for the fin we combine the conjugations conditions (3), (4) for $R_0<r<R_1$:

$$\left(\frac{\partial U}{\partial z}+k_1U\right)\bigg|_{z=l-0}=F_0(r,\varphi,t),$$

$$F_0(r,\varphi,t)=\left(\frac{k_0}{k}\frac{\partial U_0}{\partial z}+k_1U_0\right)\bigg|_{z=l+0}. \quad (16)$$

The Green function (1) is known; see [28]:

$$G(r,\varphi,z,\xi,\eta,\zeta,t)=$$
$$G_1(r,\varphi,\xi,\eta,t)G_2(z,\zeta,t),$$
$$G_1(r,\varphi,\xi,\eta,t)=\frac{1}{\pi}\sum_{n=0}^{\infty}\sum_{m=1}^{\infty}A_n\mu_{nm}^2\times$$
$$\frac{Z_n(\mu_{nm}r)Z_n(\mu_{nm}\xi)}{B_{nm}}\times$$
$$\frac{\cos[n(\varphi-\eta)]\sin(\lambda_{nms}t)}{\lambda_{nms}}\exp(-a_\tau^2\mu_{nm}^2t), \quad (17)$$
$$B_{nm}=(k_2^2R+\mu_{nm}^2R^2-n^2)Z_n^2(\mu_{nm}R)-$$
$$(k_2^2R_0+\mu_{nm}^2R_0^2-n^2)Z_n^2(\mu_{nm}R_0),$$
$$G_2(z,\zeta,t)=\sum_{s=1}^{\infty}\frac{h_s(z)h_s(\zeta)}{\|h_s\|^2}\exp(-a^2\lambda_s^2t).$$

The solution for three-dimensional problem for the wall (1) is in following form:

$$U(r,\varphi,z,t)=H(r,\varphi,z,t)+a^2\int_0^t d\tau\times$$
$$\int_0^{2\pi}d\varsigma\int_{R_0}^{R_1}F_0(\xi,\varsigma,\tau)G(r,\varphi,z,\xi,\varsigma,l,t-\tau)d\xi$$
$$+\int_{R_0}^{R_1}\xi d\xi\int_0^{2\pi}d\varsigma\times \quad (18)$$
$$\int_l^{l_1}U_0(\xi,\eta,\varsigma)G(r,\varphi,z,\xi,\eta,\varsigma,t)d\eta.$$

Here

$$H(r,\varphi,z,t)=a^2R\times$$
$$\int_0^t d\tau\int_0^{2\pi}d\eta\int_0^l g_0(\xi,\varsigma,\tau)G(r,\varphi,z,R,\eta,\varsigma,t-\tau)d\varsigma+$$
$$-a^2\times$$
$$\int_0^t d\tau\int_0^{2\pi}d\varsigma\int_{R_0}^{R_1}g_1(\xi,\varsigma,\tau)\xi G(r,\varphi,z,\xi,\varsigma,0,t-\tau)d\xi+$$
$$a^2\int_0^t d\tau\int_0^{2\pi}d\varsigma\int_{R_1}^{R}g_2(\xi,\varsigma,\tau)G(r,\varphi,z,\xi,\varsigma,l,t-\tau)d\xi.$$

Here $J_n(r), Y_n(r)-$ are Bessel's functions

$$Z_n(\mu_{nm}r)= \left[\mu_{nm}J_n'(\mu_{nm}R_0)-k_2J_n(\mu_{nm}R_0)\right]Y_n(\mu_{nm}r) -\left[\mu_{nm}Y_n'(\mu_{nm}R_0)-k_2Y_n(\mu_{nm}R_0)\right]J_n(\mu_{nm}r) \quad (19)$$

and μ_{nm} are the positive roots of the transcendental equation:

$$\frac{\left[\mu J_n'(\mu R_0)-k_0J_n(\mu R_0)\right]}{\left[\mu J_n'(\mu R_1)+k_0J_n(\mu R_1)\right]}= \frac{\left[\mu Y_n'(\mu R_0)-k_0Y_n(\mu_{nm}R_0)\right]}{\left[\mu Y_n'(\mu R_1)+k_0Y_n(\mu R_1)\right]}. \quad (20)$$

For the second part of the Green function $G_2(z,\zeta,t)$ we have:

$$h_s(z)=\cos(\lambda_s z)+\frac{k_0}{\lambda_s}\sin(\lambda_s z),$$

$$\|h_s\|^2=\frac{k_1}{2\lambda_s^2}\frac{\lambda_s^2+k_0^2}{\lambda_s^2+k_1^2}+\frac{k_0}{2\lambda_s^2}+\frac{l}{2}\left(1+\frac{k_0^2}{\lambda_s^2}\right), \quad (21)$$

$$A_n=\begin{cases}1, if\ n=0,\\ 2, if\ n>0.\end{cases}$$

The eigenvalues β_s are positive roots of the transcendental equations:

$$\frac{tg(\lambda l)}{\lambda}=\frac{k_0+k_1}{\lambda^2-k_0k_1}. \quad (22)$$

The wall solution (18) contains the fin temperature and its derivative $F_0(\xi,\varsigma,\tau)$ in the formula (16). It means that we must solve the solution for the fin.

3.2. Solution for the fin

The combination of conjugations conditions (3), (4) gives such third type boundary condition :

$$\left.\left(\frac{\partial U_0}{\partial z}-U_0\right)\right|_{z=l}=F(r,\varphi,t), \quad F(r,\varphi,t)=\left.\left(\frac{k}{k_0}\frac{\partial U}{\partial z}-U\right)\right|_{z=l-0}. \quad (23)$$

The solution in three-dimensional problem for the fin (2) is in following form:

$$U_0(r,\varphi,z,t)=H_0(r,\varphi,z,t)+\int_{R_0}^{R_1}\xi d\xi\int_0^{2\pi}d\eta\times \int_l^{l_1}U_{00}(\xi,\eta,\varsigma)G_0(r,\varphi,z,\xi,\eta,\varsigma,t)d\varsigma-a^2\int_0^t d\tau\times \int_0^{2\pi}d\eta\int_{R_0}^{R_1}F(\xi,\eta,\tau)\xi G_0(r,\varphi,z,\xi,\eta,l,t-\tau)d\xi. \quad (24)$$

Here in the function we have collected all known boundary right hand side functions:

$$H_0(r,\varphi,z,t)=a^2\times \int_0^t d\tau\int_0^{2\pi}d\eta\int_{R_0}^{R_1}g_3(\xi,\eta,\tau)\xi G_0(r,\varphi,z,\xi,\eta,l_1,t-\tau)d\xi. \quad (25)$$

The Green function is similar as (17) for initial-boundary problem for Klein-Gordon equation (2).It is known; see [27]:

$$G_0(r,\varphi,z,\xi,\eta,\zeta,t)= G_{01}(r,\varphi,\xi,\eta,t)G_{02}(z,\zeta,t),$$

$$G_{01}(r,\varphi,\xi,\eta,t)=\frac{1}{\pi}\sum_{n=0}^{\infty}\sum_{m=1}^{\infty}A_n\mu_{0nm}^2\times \frac{Z_n(\mu_{0nm}r)Z_n(\mu_{0nm}\xi)}{B_{0nm}}\times \frac{\cos[n(\varphi-\eta)]\sin(\lambda_s t)}{\lambda_s}\exp(-a_\tau^2\mu_{0nm}^2t), \quad (26)$$

$$B_{0nm}=\left(k_2^2R_1+\mu_{0nm}^2R_1^2-n^2\right)Z_n^2(\mu_{0nm}R_1)- \left(k_2^2R_0+\mu_{0nm}^2R_0^2-n^2\right)Z_n^2(\mu_{0nm}R_0),$$

$$G_2(z,\zeta,t)=\sum_{s=1}^{\infty}\frac{h_{0s}(z)h_{0s}(\zeta)}{\|h_{0s}\|^2}\exp(-a_0^2\lambda_s^2t).$$

Here μ_{0nm} are the positive roots of the transcendental equation:

$$\left[\mu J_n'(\mu R_0)-k_2J_n(\mu R_0)\right]\times \left[\mu Y_n'(\mu R_1)+k_2Y_n(\mu R_1)\right]= \left[\mu Y_n'(\mu R_0)-k_2Y_n(\mu_{nm}R_0)\right]\times \left[\mu J_n'(\mu R_1)+k_2J_n(\mu R_1)\right].$$

For the second part of the Green function $G_{02}(z,\zeta,t)$ we have:

 DOI: 10.1051/itmconf/20170901004

$$h_{0s}(z)=\cos(\lambda_{0s}z)+\frac{\sin(\lambda_{0s}z)}{\lambda_{0s}},$$

$$\|h_{0s}\|^2=\frac{k_3}{2\lambda_{0s}^2}\frac{\lambda_{0s}^2+1}{\lambda_{0s}^2+k_3^2}+\frac{1}{2\lambda_{0s}^2}+\frac{l_1-l}{2}\left(1+\frac{1}{\lambda_{0s}^2}\right),$$

$$A_n=\begin{cases}1, if\ n=0,\\ 2, if\ n>0.\end{cases}$$

The eigenvalues β_s are positive roots of the transcendental equations:

$$\frac{tg\left[\lambda(l_1-l)\right]}{\lambda}=\frac{1+k_3}{\lambda^2-k_3}.$$

4 SOLUTION AS COMBINATION OF WALL AND FIN SOLUTIONS

The solution in three-dimensional problem for the fin (24) is in following form:

$$U_0(r,\varphi,z,t)=$$
$$-\bar{H}_0(r,\varphi,z,t)-a^2\int_0^t d\tau\int_0^{2\pi}d\eta\times$$
$$\int_{R_0}^{R_1}F(r,\varphi,t)G_0(r,\varphi,z,\xi,\eta,l,t-\tau)d\xi, \quad (27)$$

$$\bar{H}_0(r,\varphi,z,t)=H_0(r,\varphi,z,t)+\int_{R_0}^{R_1}\xi d\xi\int_0^{2\pi}d\eta\times$$
$$\int_l^{l_1}U_{00}(\xi,\eta,\varsigma)G_0(r,\varphi,z,\xi,\eta,\varsigma,t)d\varsigma.$$

Unfortunately, the representation (27) is unusable as solution for the fin because of the unknown function $F(r,\varphi,t)$, i.e. temperature in the wall $U(r,\varphi,z=l,t)$ and derivative of temperature.
From equation (16) we can write the combination $F_0(r,\varphi,t)$:

$$F_0(r,\varphi,t)=-\hat{\bar{H}}_0(r,\varphi,t)-a^2\int_0^t d\tau\int_0^{2\pi}d\eta\times$$
$$\int_{R_0}^{R_1}F(\xi,\eta,\tau)\hat{G}_0(r,\varphi,\xi,\eta,\varsigma,t-\tau)d\xi. \quad (28)$$

Here

$$\hat{\bar{H}}_0(r,\varphi,t)=$$
$$\left.\left(\frac{k_0}{k}\frac{\partial}{\partial z}\bar{H}_0(r,\varphi,z,t)-\bar{H}_0(r,\varphi,z,t)\right)\right|_{z=l},$$

$$\hat{G}_0(r,\varphi,\xi,\eta,\varsigma,t)=$$
$$\left.\left[\frac{k_0}{k}\frac{\partial}{\partial z}G_0(r,\varphi,z,\xi,\eta,\varsigma,t)-G_0(r,\varphi,z,\xi,\eta,\varsigma,t)\right]\right|_{z=l}.$$

Similarly we do with equations (23) and (18):

$$U(r,\varphi,z,t)=\bar{H}(r,\varphi,z,t)+a^2\int_0^t d\tau\times$$
$$\int_0^{2\pi}d\varsigma\int_{R_0}^{R_1}F_0(\xi,\varsigma,\tau)G(r,\varphi,z,\xi,\varsigma,l,t-\tau)d\xi,$$

$$\bar{H}(r,\varphi,z,t)=H(r,\varphi,z,t)+\int_{R_0}^{R_1}\xi d\xi\int_0^{2\pi}d\varsigma\times$$
$$\int_l^{l_1}U_0(\xi,\eta,\varsigma)G(r,\varphi,z,\xi,\eta,\varsigma,t)d\eta.$$

From equation (23) we can write the combination $F(r,\varphi,t)$:

$$F(r,\varphi,t)=\hat{\bar{H}}(r,\varphi,t)+a^2\int_0^t d\tau\times$$
$$\int_0^{2\pi}d\eta\int_{R_0}^{R_1}F_0(\xi,\eta,\tau)\hat{G}(r,\varphi,l,\xi,\eta,l,t-\tau)d\xi. \quad (29)$$

Here

$$\hat{\bar{H}}(r,\varphi,t)=\left.\left[\frac{k}{k_0}\frac{\partial}{\partial z}\bar{H}(r,\varphi,t)-\bar{H}(r,\varphi,t)\right]\right|_{z=l},$$

$$\hat{G}(r,\varphi,l,\xi,\varsigma,l,t)=$$
$$\left.\left[\frac{k}{k_0}\frac{\partial}{\partial z}G(r,\varphi,z,\xi,\varsigma,l,t)-G(r,\varphi,z,\xi,\varsigma,l,t)\right]\right|_{z=l}.$$

As the last step we substitute the combination $F_0(\xi,\eta,\tau)$ from (28)

$$F_0(\xi,\eta,\tau)=-\hat{\bar{H}}_0(\xi,\eta,\tau)-a^2\int_0^\tau d\tau_1\int_0^{2\pi}d\eta_1\times$$
$$\int_{R_0}^{R_1}F(\xi_1,\eta,\tau_1)\hat{G}_0(\xi,\eta,\xi,\eta_1,\varsigma,\tau-\tau_1)d\xi_1,$$

DOI: 10.1051/itmconf/20170901004

in the equation (29):

$$F(r,\varphi,t)=\hat{\hat{H}}(r,\varphi,t)+a^2\int_0^t d\tau\times$$

$$\int_0^{2\pi} d\eta\int_{R_0}^{R_1} F_0(\xi,\eta,\tau)\hat{G}(r,\varphi,l,\xi,\eta,l,t-\tau)d\xi.$$

Finally we obtain following non-homogeneous Fredholm integral equation of 2nd kind:

$$F(r,\varphi,t)=\Phi(r,\varphi,t)+a^2\int_0^t d\tau\int_0^{\tau} d\tau_1\int_0^{2\pi} d\eta_1\times$$
$$\int_0^{2\pi} d\eta\int_{R_0}^{R_1} F(\xi_1,\eta,\tau_1)\Gamma(r,\varphi,l,\xi,\eta,l,t-\tau)d\xi. \tag{30}$$

Here

$$\Gamma(r,\varphi,l,\xi,\eta,l,t)=\int_{R_0}^{R_1}\hat{G}(r,\varphi,l,\xi,\eta,l,t)d\xi\times$$

$$\int_{R_0}^{R_1}\hat{G}_0(\xi,\eta,\xi_1,\eta_1,\varsigma,\tau-\tau_1)d\xi_1,$$

$$\Phi(r,\varphi,t)=\hat{\hat{H}}(r,\varphi,t)-a^2\int_0^t d\tau\times$$

$$\int_0^{2\pi} d\eta\int_{R_0}^{R_1}\hat{\hat{H}}_0(\xi,\eta,\tau)\hat{G}(r,\varphi,l,\xi,\eta,l,t-\tau)d\xi.$$

We finish our paper with the following remark. The problem with non-homogeneous environment temperatures (6), (7), (8) and (11) and its solution allow conjugating temperature field with hydrodynamic field.

5 CONCLUSIONS

We have constructed exact three-dimensional transient analytical solution for one element with hollow cylinder fin. The solution is obtained in the form of Fredholm integral equation of 2nd kind and has continuous kernel. As the result we come to integral equation on the surface between both hollow cylinders. The generalizing of Green function method allows us to use Green function method for regular non-canonical domains.

This work has been supported by Latvian Council of Sciences (grant 623/2014).

References

1. Kern, D.Q, Kraus, A.D. Extended Surface Heat Transfer. – McGraw-Hill Book Company (1972)
2. Kraus, A.D., Analysis and evaluation of extended surface thermal systems. Hemisphere publishing Corporation (1982)
3. Manzoor, M., Heat Flow Through Extended Surface Heat Exchangers. Springer-Verlag: Berlin and New York (1984)
4. Buikis, A., Two-dimensional solution for heat transfer in a regular fin assembly. Latvian Journal of Physics and Technical Sciences, No. 5, pp. 51-58 (1995)
5. Buikis, A., Buike, M. Approximate analytical two-dimensional solution for a longitudinal fin of rectangular profile. Acta Universitatis Latviensis, 612, pp.33-44 (1998)
6. Buikis A., Buike M. Closed two-dimensional solution for heat transfer in a periodical system with a fin. Proceedings of the Latvian Academy of Sciences. Section B, 52, No.5, pp. 218-222. (1998)
7. Buike M. Simulation of steady-state heat process for the rectangular fin-containing system, Mathematical Modelling and Analysis, 4, pp. 33-43 (1999)
8. Malik, M.Y., Wood, A.S., Buikis, A. An approximate analytical solution to a familiar conjugate heat transfer problem. International journal of Pure and Applied Mathematics, Vol.10, Nr.1, pp. 91-107. (2004)
9. Buike M., Buikis A. Approximate Solutions of Heat Conduction Problems in Multi- Dimensional Cylinder Type Domain by Conservative Averaging Method, Part 1. Proceedings of the 5th IASME/WSEAS Int. Conf. on Heat Transfer, Thermal Engineering and Environment, Vouliagmeni, Athens, August 25 -27, p. 15 – 20, (2007).
10. Buike M., Buikis A. Several Intensive Steel Quenching Models for Rectangular Samples. Proceedings of NAUN/WSEAS International Conference on Fluid Mechanics and Heat &Mass Transfer, Corfu Island, Greece, July 22-24. p. 88-93 (2010).
11. Buike M., Buikis A. Hyperbolic heat equation as mathematical model for steel quenching of L-shape samples, Part 1 (Direct Problem). Applied and Computational Mathematics. Proceedings of the 13th WSEAS International Conference on Applied Mathematics (MATH'08), Puerto De La Cruz, Tenerife, Canary Islands, Spain, December 15-17, WSEAS Press, p. 198-203, (2008).
12. Bobinska T., Buike M., Buikis A. Hyperbolic Heat Equation as Mathematical Model for Steel Quenching of L-Shape Samples, Part 2 (Inverse Problem). Proceedings of 5th IASME/WSEAS International Conference on Continuum Mechanics (CM'10), University of Cambridge, UK, February 23-25, p. 21-26, (2010).
13. Bobinska T., Buike M., Buikis A. Hyperbolic Heat Equation as Mathematical Model for Steel Quenching of L-and T-Shape Samples, Direct and Inverse Problems. Transactions of Heat and Mass Transfer. Vol.5, Issue 3, July 2010. p. 63-72.
14. Blomkalna S., Buikis A. Heat conduction problem for double-layered ball. Progress in Industrial Mathematics at ECMI 2012. Springer, p. 417-426, (2014).

15. Bobinska T., Buike M., Buikis A. Comparing solutions of hyperbolic and parabolic heat conduction equations for L-shape samples. Recent Advances in Fluid Mechanics and Heat @Mass Transfer. Proceedings of the 9th IASME/WSEAS International Conference on THE'11. Florence, Italy, August 23-25, p. 384-389 (2011).
16. M. Buike, A. Buikis, R. Vilums. One-Dimensional Intensive Steel Quenching Models. Recent Advances in Mechanical Engineering. Proceedings of the 5th International Conference on Fluid Mechanics and Heat & Mass Transfer. Lisbon, Portugal, October 30- November 1, p. 54-62, (2014).
17. Piliksere A., Buikis A. Analytical solution for intensive quenching of cylindrical sample. Proceedings of 6th International Scientific Colloquium "Modelling for Material Processing", Riga, September 16-17, p. 181-186, (2010).
18. Piliksere A., Buike M., Buikis A. Steel quenching process as hyperbolic heat equation for cylinder. Proceedings of 6th Baltic Heat Conference – BHTC2011, ISBN 978-952-15-2640-4 (CD-ROM), (2011).
19. Buike M., Buikis. Exact Transient Solution for System with Rectangular Fin, Proceedings of the 5th WSEAS Int. Conf. on Heat and Mass Transfer, Acapulco, Mexico, January 25-27, 2008, pp. 25-30.
20. Lencmane M. and Buikis A. Analytical solution of a two-dimensional double-fin assembly. Recent Advances in Fluid Mechanics and Heat & Mass Transfer, Proceedings of the 9th IASME/WSEAS International Conference on Heat Transfer, Thermal Engineering and Environment (HTE'11), Florence, Italy, August 23 – 25, pp. 396 – 401, (2011).
21. Lencmane M. and Buikis A. Analytical solution for steady stable and transient heat process in a double-fin assembly. International Journal of Mathematical Models and Methods in Applied Sciences 6, No. 1. p.81 – 89, (2012).
22. Lencmane M., Buikis A. Some new mathematical models for the Transient Hot Strip method with thin interlayer. Proc. of the 10th WSEAS Int. Conf. on Heat Transfer, Thermal engineering and environment (HTE '12), WSEAS Pres, p. 283-288, (2012).
23. Buikis A., Buike M., Ulanova N., Solutions for the Transient Heat Transfer in Periodical Systems with Fins. Numerical Mathematics and Mathematical Problems of Mechanics. Ukrainian Mathematical congress – 2001, Section 8, Kiev, 2002, pp.15-23 (in Russian).
24. Buikis A. Conservative averaging as an approximate method for solution of some direct and inverse heat transfer problems. Advanced Computational Methods in Heat Transfer, IX. WIT Press, p. 311-320, (2006).
25. Vilums R., Buikis A. Conservative averaging method for partial differential equations with discontinuous coefficients. WSEAS Transactions on Heat and Mass Transfer. 1, Issue 4, p. 383-390, (2006).
26. Roach G.F. Green's Functions. Cambridge University Press, (1999).
27. Carslaw H.S., Jaeger C.J. Conduction of Heat in Solids. Oxford, Clarendon Press (1959).
28. Polyanin A.D. Handbook of Linear Partial Differential Equations for Engineers and Scientists. Chapman & Hall/CRC, (Russian edition, 2001). (2002).

ITM Web of Conferences **9**, 01005 (2017) DOI: 10.1051/itmconf/20170901005
AMCSE 2016

Notes about the linear complexity of Ding-Helleseth generalized cyclotomic sequences of length pq over the finite field of order p or q

Vladimir Edemskiy[1,*] and *Nikita* Sokolovskiy[1]

[1] *Novgorod State University, Veliky Novgorod, Russia*

Abstract. We investigate three classes of Ding-Helleseth-generalized cyclotomic sequences of length pq. We derive the linear complexity and the minimal polynomial of above-mentioned sequences over the finite fields of orders p and q, where p and q are two odd distinct primes, and obtain series of sequences with high linear complexity.

1 Introduction

Pseudo-random sequences are widely used in many fields, in particular in stream ciphers [3]. The linear complexity is defined to be the length of the shortest linear feedback shift register that can generate the sequence [9]. The concept of the linear complexity of sequence is very useful in the study of the security of stream ciphers.

Cyclotomic sequences have good pseudo-random properties and have been widely used as keystreams in private-key cryptosystems. In [7], Ding and Helleseth introduced a generalized cyclotomy of order 2 with respect to odd modulo N, in particularly for $N = pq$, where p and q are two odd distinct primes. Further, this cyclotomy was generalized in series of papers. There are many works devoted to the study of the properties of Ding-Helleseth-generalized cyclotomic sequences and their use in cryptography and coding. So, the linear complexity of Ding-Helleseth-generalized cyclotomic sequences of length pq over the finite field of order two have been calculated in [4, 5, 10, 14, 15](see also references here). Further, Ding [6] investigated the linear complexity of the series of these sequences over the finite field $GF(l^m)$ where $\gcd(l, pq) = 1$, and constructed several classes of cyclic codes with optimal or almost optimal property. Wang et al. [11] derive the linear complexity of Ding-Helleseth sequences of order 2 over $GF(l)$ where $\gcd(l, pq) = 1$. Also, there are a long lot of articles devoted to the investigation the linear complexity of of Ding-Helleseth sequences of another orders over the different finite fields when a field's order and a sequence's period are relatively prime. At the same time, the linear complexity of cyclotomic sequences with period p over $GF(p)$ was investigated in [1, 2].

The purpose of this paper is to study the linear complexity of Ding-Helleseth sequences length pq over the finite fields of orders p and q. We find the linear complexity and the minimal polynomial of these generalized cyclotomic sequences.

2 Preliminaries

First, we briefly repeat the basic definitions and the general information. Let p and q be two odd distinct primes with $\gcd(p-1, q-1) = d$. Define $N = pq$, $e = (p-1)(q-1)/d$. We denote by Z_N the residue class ring modulo N and by Z_N^* the unit group of Z_N. The Chinese Remainder Theorem guarantees that there exists a common primitive root, g, of both p and q, and the order of g modulo N is e [8]. There also exists an integer x satisfying $x \equiv g(\bmod\ p)$, and $x \equiv 1(\bmod\ q)$. By [13] we have

$$Z_N^* = \{g^i x^j : i = 0, 1, ...e-1; j = 0, 1, ..., d-1\},$$

where the multiplication is that of Z_N^*.

Ding-Helleseth generalized cyclotomic classes of order d with respect to p and q are defined as

$$D_i = \{g^{i+dt} x^j : t = 0, 1, ...e/d - 1; j = 0, 1, ..., d-1\},$$

$i = 0, 1, ...d-1$. Then

$$Z_N^* = \bigcup_{i=0}^{d-1} D_i, \quad D_i \cap D_j = \varnothing \text{ for } i \neq j,$$

where $\varnothing$ denotes the empty set.

By definition, put $P = \{p, 2p, \ldots, (q-1)p\}$, $Q = \{q, 2q, \ldots, (p-1)q\}$, and $R = \{0\}$. Define $C_0 = \bigcup_{i=0}^{d/2-1} D_{2i}$, $C_1 = \bigcup_{i=0}^{d/2-1} D_{2i+1}$. Then C_0, C_1 is a partition of Z_N^* and C_0, C_1, P, Q, and R is a partition of Z_N. Further, we investigate the linear complexity of three types of sequences based on the partition $Z_N^* = C_0 \cup C_1$.

It is well known that if $s = \{s_i\}$ is a sequence of period N, then the linear complexity $LC_n(s)$ over the finite field $GF(n)$ of order n and the minimal polynomial $m_n(x)$ of this

*e-mail: Vladimir.Edemsky@novsu.ru This work was supported by the Ministry of Education and Science of the Russian Federation as a part of state-sponsored project no 1.949.2014/K.

 DOI: 10.1051/itmconf/20170901005

sequence are defined by

$$LC_n(s) = N - \deg\gcd(x^N - 1, S(x)),$$
$$m_n(x) = (x^N - 1)/\gcd(x^N - 1, S(x)) \quad (1)$$

where $S(x) = s_0 + s_1x + ... + s_{N-1}x^{N-1}$ [3]. It is worth pointing out that the minimal polynomials of $m_n(x)$ defined here may be the reciprocals of the minimal polynomials defined in other references. From (1) we can establish that

$$LC_p(s) = N - \deg\gcd((x^q - 1)^p, S(x)),$$
$$m_p(x) = (x^q - 1)^p/\gcd((x^q - 1)^p, S(x)) \quad (2)$$

and

$$LC_q(s) = N - \deg\gcd((x^p - 1)^q, S(x)),$$
$$m_q(x) = (x^p - 1)^q/\gcd((x^p - 1)^q, S(x)) \quad (3)$$

Let α be a primitive q-th root of unity and β be a primitive p-th root of unity in the extensions of the fields $GF(p)$ and $GF(q)$, respectively. Then, according to (2), (3), in order to find the minimal polynomial and the linear complexity of $\{s_i\}$ over $GF(q)$ or $GF(p)$ it is sufficient to find the roots of $S(x)$ in the set $\{1, \alpha, \ldots, \alpha^{q-1}\}$ or $\{1, \beta, \ldots, \beta^{p-1}\}$ and determine their multiplicity.

Let $H_i^{(p)} = C_i \bmod p$, and $H_i^{(q)} = C_i \bmod q, i = 0, 1$. Here and hereafter $x \bmod n$ denotes the least nonnegative integer that is congruent to x modulo n. Then $H_0^{(p)}, H_0^{(q)}$ are the quadratic residue classes modulo p and q, respectively.

Legendre sequences $\{l_i^{(q)}\}$ and $\{l_i^{(p)}\}$ of lengths q and p defined as

$$l_i^{(q)} = \begin{cases} 1, & \text{if } i \bmod q \in H_0^{(q)}, \\ 0, & \text{otherwise} \end{cases}$$

and

$$l_i^{(p)} = \begin{cases} 1, & \text{if } i \bmod q \in H_0^{(p)}, \\ 0, & \text{otherwise.} \end{cases}$$

The linear complexity and the minimal polynomial of Legendre sequence over the finite field of any order were studied in [12].

Let us introduce subsidiary polynomials $S^{(q)}(x) = \sum_{i=0}^{q-1} l_i^{(q)}x^i$ and $S^{(p)}(x) = \sum_{i=0}^{p-1} l_i^{(p)}x^i$. By [12] over $GF(q)$ we have

$$S^{(q)}(\alpha) = \begin{cases} S^{(q)}(\alpha^j), & \text{if } j \bmod q \in H_0^{(q)}, \\ S^{(q)}(\alpha^g), & \text{if } j \bmod q \in H_1^{(q)}, \end{cases}$$
$$\text{and } S^{(q)}(\alpha) + S^{(q)}(\alpha^g) = -1. \quad (4)$$

Similarly, over $GF(p)$ we see that

$$S^{(p)}(\beta) = \begin{cases} S^{(q)}(\beta^j), & \text{if } j \bmod p \in H_0^{(p)}, \\ S^{(p)}(\alpha^g), & \text{if } j \bmod p \in H_1^{(p)}, \end{cases} \text{ and}$$
$$S^{(p)}(\alpha) + S^{(p)}(\alpha^g) = -1. \quad (5)$$

The following statement was discussed in [12].

Lemma 1 *$S^{(q)}(\alpha)$ and $S^{(q)}(\alpha^g)$ are zeros of the polynomial*

$$\begin{cases} z^2 + z - (q-1)/4, & \text{if } q \equiv 1(\text{mod}4), \\ z^2 + z + (q+1)/4, & \text{if } q \equiv 3(\text{mod}4). \end{cases}$$

If there exist $j : j \neq 0$ and $S^{(q)}(\alpha^j) = 0$ then without loss of generality, we can choose α such that $S^{(q)}(\alpha) = 0$. Thus, by (4) and lemma 1 we have

$$\{j \mid S^{(q)}(\alpha^j) = 0, j = 1, \ldots, q-1\} =$$
$$\begin{cases} H_0^{(q)}, & \text{if } q \equiv 1(\text{mod}p) \text{ and } q \equiv 1(\text{mod}4), \\ & \text{or } q \equiv -1(\text{mod}p) \text{ and } q \equiv 3(\text{mod}4), \\ \varnothing, & \text{otherwise.} \end{cases} \quad (6)$$

3 Subsidiary lemmas

The following lemmas are needed for the sequel.

According to the Chinese Remainder Theorem

$$Z_N \cong Z_p \times Z_q$$

relatively to isomorphism $\varphi(x) = (x \bmod p, x \bmod q)$. Put, by definition $F_k = \varphi^{-1}(\{k\} \times H_0^{(q)}), k = 1, \ldots, p-1$. From our definitions it follows that

$$C_0 = \bigcup_{k=1}^{p-1} F_k. \quad (7)$$

Define $S_0(x) = \sum_{i \in C_0} x^i$. Then, by (7) we have $S_0(x) = \sum_{k=1}^{p-1} \sum_{i \in F_k} x^i$

Lemma 2 *Let $F_k = \varphi^{-1}(\{k\} \times H_0^{(q)})$. Then there exist polynomials $g_k(x) = \sum_{j=1}^{(q-1)/2} x^{b_{j,k}}, k = 0, 1 \ldots, p-1, \ 0 \leq b_{j,k} < q$ such that $\sum_{i \in F_k} x^i = x^k g_k(x^p)$.*

Proof. Suppose $i \in F_k$; then by definition of F_k it follows that $i \equiv k \pmod p$. So, there exist $b_{i,k}, \ 0 \leq b_{j,k} < q$ such that $i = k + b_{i,k}p$.

By Lemma 2 we obtain that

$$S_0(x) = \sum_{k=1}^{p-1} x^k g_k(x^p). \quad (8)$$

Lemma 3 *Let $S_0(x) = \sum_{i \in C_0} x^i$. Then:*

(i) $S_0(x) \equiv (q-1)/2(1 - (x-1)^{p-1}) \pmod{(x-1)^p}$ in the ring $GF(p)[x]$;

(ii) $S_0(x) \equiv (p-1)\left(-1/2 + T(x)(x-1)^{(q-1)/2}\right) \pmod{(x-1)^q}$ in the ring $GF(q)[x]$, where $T(1) \neq 0$.

Proof. At the beginning, we prove the first statement. It is clear that $g_k(x^p) \equiv (q-1)/2 \pmod{(x-1)^p}$; then by (4) we see that $S_0(x) \equiv -(q-1)(x + x^2 + \cdots + x^{p-1})/2 \pmod{(x-1)^p}$ in $GF(p)[x]$. The statement (i) of this lemma follows from the last congruence.

Further, $S_0(x) = \sum_{k=1}^{p-1} \sum_{i \in F_k} x^i$. From the definition of F_k it follows that $F_k \bmod q = H_0^{(q)}$, hence $\sum_{i \in F_k} x^i \equiv \sum_{i \in H_0^{(q)}} x^i \pmod{(x-1)^q}$, i.e. $\sum_{i \in F_k} x^i \equiv S^{(q)}(x) \pmod{(x-1)^q}$ over $GF(q)[x]$. To conclude the proof, it remains to note that by [1] we have $S^{(q)}(x) \equiv -1/2 + T(x)(x-1)^{(q-1)/2} \pmod{(x-1)^q}$, where $T(1) \neq 0$.

Lemma 4 *Let $S_0^{(n)}(x)$ be a formal derivative of order n of the polynomial $S_0(x)$ over $GF(p)$ and $0 \leq n < p$. Then*

$$S_0^{(n)}(x) = \sum_{k=n}^{p-1} k(k-1)\ldots(k-n+1)x^{k-n} g_k(x^p).$$

Proof. Since $\left(x^k g_k(x^p)\right)' = kx^{k-1}g_k(x^p)$ we have

$$\left(x^k g_k(x^p)\right)^{(n)} = \begin{cases} k(k-1)\dots(k-n+1)x^{k-n}g_k(x^p), & \text{if } k \geq n,\\ 0, & \text{if } k < n.\end{cases}$$

The conclusion of this lemma then follows from (8).

4 The linear complexity of the first class of sequences

First, we consider Ding-Helleseth-generalized cyclotomic sequence $t = \{t_i\}$ defined by

$$t_i = \begin{cases} 1, & \text{if } i \bmod N \in C_0 \cup P \cup R,\\ 0, & \text{otherwise.}\end{cases} \quad (9)$$

4.1 The linear complexity of sequences over $GF(p)$

In this subsection we regard t as a sequence over the finite field $GF(p)$. Let $S_t(x) = \sum_{i=0}^{N} t_i x^i$. From our definitions it follows that $S_t(x) = S_0(x) + 1 + x^p + \dots + x^{(q-1)p}$. Since $1 + x^p + \dots + x^{(q-1)p} = (x^q-1)^p/(x-1)^p$, by (8) we see that

$$S_t(x) = S_0(x) + 1 + x^p + \dots + x^{(q-1)p} = \sum_{k=1}^{p-1} x^k g_k(x^p) + \left(1 + x + \dots + x^{(q-1)}\right)^p. \quad (10)$$

Before we give the main result of this section, we establish the following lemma.

Lemma 5 *Let $F_k = \varphi^{-1}(\{k\} \times H_0^{(q)})$. Then $\sum_{i\in F_k} \alpha^{vi} = S^{(q)}(\alpha^v)$ for any $v:\ 0 < v < p$.*

Proof. From definition of F_k it follows that $F_k \bmod q = H_0^{(q)}$. So, $\sum_{i\in F_k}\alpha^{vi} = \sum_{i\in H_0^{(q)}}\alpha^{vi}$. To conclude the proof, it remains to note that $\sum_{i\in H_0^{(q)}}\alpha^{vi} = S^{(q)}(\alpha^v)$.

Further, using Lemmas 5 and 2, we can write the following statement.

Corollary 6 *If $S^{(q)}(\alpha^v) = 0$ then $g_k(\alpha^{pv}) = 0$.*

Denote $\omega_0(x) = \prod_{i\in H_0^{(q)}}(x-\alpha^i)$ and $\omega_1(x) = \prod_{i\in H_1^{(q)}}(x-\alpha^i)$.

Theorem 7 *Let $\{t_i\}$ be defined by (9). Then we have:*

1. *$LC_p(t) = p(q+1)/2$ and $m(x) = (x-1)^p\omega_1^p(x)$ if $q \equiv 1 \pmod p$ and $q \equiv 1 \pmod 4$,*
2. *$LC_p(t) = p(q-1)/2 + 1$ and $m(x) = (x-1)\omega_1^p(x)$ if $q \equiv -1 \pmod p$ and $q \equiv 3 \pmod 4$,*
3. *$LC_p(t) = pq - p + 1$ and $m(x) = (x^q-1)^p/(x-1)^{p-1}$ if $q \equiv -1 \pmod p$ and $q \equiv 1 \pmod 4$,*
4. *$LC_p(t) = pq$ and $m(x) = (x^q-1)^p$ otherwise.*

Proof. First of all, we note that $S_t(1) = (p-1)(q-1)/2 + q = (q+1)/2$. Thus, if $q \not\equiv -1 \pmod p$ then $S_t(1) \neq 0$. Otherwise, if $q \equiv -1 \pmod p$ then $S_t(1) = 0$ and by Lemma 3 and (10) we obtain that $S_t(x) \equiv 1 - (x-1)^{p-1} + (1+x+\dots+x^{(q-1)})^p \pmod{(x^p-1)}$ or $S_t(x) \equiv (x-1)^{p-1} + \left((1-1)+(x-1)+\dots+(x^{(q-1)}-1)\right)^p \pmod{(x^p-1)}$.

So, we have that in this case 1 is a root of the polynomial $S_t(x)$ of multiplicity $p-1$.

Suppose $v = 1,\dots,q-1$; then by Lemma 3 and (10) we obtain $S_t(\alpha^v) = (p-1)S^{(q)}(\alpha^v)$. So, by (6) $S_t(\alpha^v) = 0$ if and only if $v \in H_0^{(q)}$ and $q \equiv 1 \pmod p$ and $q \equiv 1 \pmod 4$ or $q \equiv -1 \pmod p$ and $q \equiv 3 \pmod 4$. If $S_t(\alpha^v) = 0$ then by Lemma 4 and Corollary 6 we obtain that α^v is a root of $S_t(x)$ of multiplicity p. The conclusion of this theorem then follows from (2).

Our examples $p = 7, q = 29$; $p = 3, q = 13$; $p = 3, q = 17$; $p = 5, q = 3$ show that all the cases of Theorem 7 are possible.

4.2 The linear complexity of sequences over $GF(q)$

Now, we consider the linear complexity of $\{t_i\}$ over $GF(q)$. In this case we see that

$$S_t(x) = S_0(x)+1+x^p+\dots+x^{(q-1)p} = \sum_{k=1}^{p-1} x^k g_k(x^p) + (x^p-1)^{q-1}. \quad (11)$$

Theorem 8 *Let $\{t_i\}$ be defined by (9). Then:*

1. *$LC_q(t) = pq - q + 1$ and $m_q(x) = (x-1)(1+x+\dots+x^{p-1})^q$ if $p \equiv 1 \pmod q$,*
2. *$LC_q(t) = pq$ and $m_q(x) = (x^q-1)^p$ otherwise.*

Proof. By definition, $S(1) = (p-1)(q-1)/2 + q$, thus $S(1) = -(p-1)/2$. If $p \not\equiv 1 \pmod q$ then $S_t(1) \neq 0$. Suppose $p \equiv 1 \pmod q$; then $S_t(1) = 0$ and by (11) and Lemma 3 we can establish that $S_t(x) \equiv (x^p-1)^{q-1} \pmod{(x-1)^q}$. So, 1 is a root of multiplicity $q-1$ of $S_t(x)$ for $p \equiv 1 \pmod q$.

Let $v:\ 0 < v < q$. Then by (11) $S_t(\beta^v) = (q-1)(\beta+\beta^2+\dots+\beta^{p-1})/2 + q = -(\beta^p-1)/(2(\beta-1)) + 1/2 = 1/2$, i.e. $S_t(\beta^v) \neq 0$. The conclusion of this theorem then follows from (3).

5 The linear complexity of the second class of sequences

Now, we can also consider the sequence u defined by

$$u_i = \begin{cases} 1, & \text{if } i \bmod N \in C_0 \cup Q \cup R,\\ 0, & \text{otherwise.}\end{cases} \quad (12)$$

Let $S_u(x) = \sum_{i=0}^{N} u_i x^i$. By definition we see that $S_u(x) = S_0(x) + 1 + x^q + \dots + x^{(p-1)q}$.

5.1 The linear complexity of sequences over $GF(p)$

Here $1+x^q+\cdots+x^{(p-1)q} = (x^{qp}-1)/(x^q-1) = (x^q-1)^{p-1}$, by (8) we see that

$$S_t(x) = S_0(x) + 1 + x^q + \cdots + x^{(p-1)q} = \sum_{k=1}^{p-1} x^k g_k(x^p) + (x^q-1)^{p-1}. \quad (13)$$

With similar arguments as above we obtain the following results for u.

Theorem 9 *Let $\{u_i\}$ be defined by (12). Then:*

1. *$LC_p(u) = pq-(p-1)(q+1)/2$ and $m_p(x) = (x-1)\omega_0(x)\omega_1^p(x)$ if $q \equiv 1 \pmod p$ and $q \equiv 1 \pmod 4$,*
2. *$LC_p(u) = pq-(q-1)(p-1)/2$ and $m_p(x) = (x-1)^p\omega_0(x)\omega_1^p(x)$ if $q \equiv -1 \pmod p$ and $q \equiv 3 \pmod 4$,*
3. *$LC_p(u) = pq-p+1$ and $m_p(x) = (x^{pq}-1)/(x-1)^{p-1}$ if $q \equiv 1 \pmod p$ and $q \equiv 3 \pmod 4$,*
4. *$LC_p(u) = pq$ and $m(x) = (x^q-1)^p$ otherwise.*

Proof. First of all, we note that $S_u(1) = (p-1)(q-1)/2 + p = -(q-1)/2$. Thus, if $q \not\equiv 1 \pmod p$ then $S_u(1) \neq 0$. Otherwise, if $q \equiv 1 \pmod p$ then $S_u(1) = 0$ and by Lemma 3 and (13) we obtain that $S_u(x) \equiv (x^q-1)^{p-1} \pmod{(x^p-1)}$.

So, we have that in this case 1 is a root of the polynomial $S_u(x)$ of multiplicity $p-1$.

Suppose $v = 1,\dots,q-1$; then by Lemma 4 and (13) we obtain $S_u(\alpha^v) = (p-1)S^{(q)}(\alpha^v)$. So, by (6) $S_u(\alpha^v) = 0$ if and only if $v \in H_0^{(q)}$ and $q \equiv 1 \pmod p$ and $q \equiv 1 \pmod 4$ or $q \equiv -1 \pmod p$ and $q \equiv 3 \pmod 4$. If $S_t(\alpha^v) = 0$ then by Lemma 4, Corollary 6, and (13) we obtain that α^v is a root of $S_u(x)$ of multiplicity $p-1$. By (2) this completes the proof of Theorem 9.

Our examples $p=5, q=41$; $p=3, q=11$; $p=5, q=11$; $p=3, q=17$ show that all the cases of Theorem 9 are possible.

5.2 The linear complexity of sequences over $GF(q)$

Here $1+x^q+\cdots+x^{(p-1)q} = (x^{qp}-1)/(x^q-1) = (1+x+\cdots+x^{p-1})^q$, by (8) we see that

$$S_t(x) = S_0(x) + 1 + x^q + \cdots + x^{(p-1)q} = \sum_{k=1}^{p-1} x^k g_k(x^p) + (1+x+\cdots+x^{p-1})^q. \quad (14)$$

Theorem 10 *Let $\{u_i\}$ be defined by (12). Then:*

1. *$LC_q(u) = pq-(q-1)/2$ and $m_p(x) = (x^{pq}-1)/(x-1)^{(q-1)/2}$ if $p \equiv -1 \pmod q$,*
2. *$LC_q(u) = pq$ and $m(x) = (x^q-1)^p$ otherwise.*

Proof. By definition, $S_u(1) = (p-1)(q-1)/2 + p = (p+1)/2$, such that $S_u(1) \neq 0$ for $p \not\equiv -1 \pmod q$. Suppose $p \equiv -1 \pmod q$; then $S_u(1) = 0$ and by (14) and Lemma 3 we can establish that $S_u(x) \equiv 1 - 2T(x)(x-1)^{(q-1)/2}) + (1+x+\cdots+x^{p-1})^q \pmod{(x-1)^{q-1}}$, i.e. $S_u(x) \equiv -2(T(x)x-1)^{(q-1)/2}) + (x-1+\cdots+x^{p-1}-1)^q \pmod{(x-1)^{q-1}}$. So, 1 is a root of multiplicity $(q-1)/2$ of $S_u(x)$ for $p \equiv -1 \pmod q$.

Let $v: 0 < v < q$. Then by (14) $S_t(\beta^v) = (q-1)(\beta+\beta^2+\cdots+\beta^{p-1})/2 + 0 = -(\beta^p-1)/(2(\beta-1)) + 1/2 = 1/2$, i.e. $S_t(\beta^v) \neq 0$. The conclusion of this theorem then follows from (3).

6 The linear complexity of the third class of sequences

In conclusion, we study the linear complexity of balanced sequence v.

Let $C_0^{(q)} = qH_0^{(p)}$, $C_0^{(p)} = pH_0^{(q)}$. We consider a sequence v defined by

$$u_i = \begin{cases} 1, & \text{if } i \bmod N \in C_0 \cup C_0^{(q)} \cup C_0^{(p)}, \\ 0, & \text{otherwise.} \end{cases} \quad (15)$$

Then v is balanced sequence. Denote $S_v(x) = \sum_{i=0}^{N-1} v_i x^i$. Then, by definition we have

$$S_v(x) = \sum_{k=0}^{p-1} \sum_{i \in F_k} x^i + \sum_{i \in C_0^{(q)}} x^i + \sum_{i \in C_0^{(p)}} x^i. \quad (16)$$

For any $a \in H_0^{(q)}$ there exist $b_a, 0 < b_a < p$ such that $a + b_a q \equiv 0 \pmod p$.

Lemma 11 *Let $E^{(q)} = \{c \mid pc = a + b_a q,\ a, c \in H_0^{(q)}\}$. Then*

$$C_0 = \bigcup_{a \in H_0^{(q)}} \{a, a+q, \dots, a+(p-1)q\} \setminus pE^{(q)}.$$

The statement of this lemma follows immediately from our definitions.

6.1 The linear complexity of sequences over $GF(p)$

In this subsection we regard t as a sequence over the finite field $GF(p)$.

Lemma 12 *Let $S_0(x) = \sum_{i \in C_0} x^i$. Then*

$$S_0(x) = (x^q-1)^{p-1} S^{(q)}(x) - \sum_{i \in pE^{(q)}} x^i$$

Proof. By Lemma 11 we obtain $S_0(x) = \sum_{a \in H_0^{(q)}} (x^a + x^{a+q} + \cdots + x^{a+(p-1)q}) - \sum_{i \in pE^{(q)}} x^i$. To conclude the proof, it remains to note that $\sum_{a \in H_0^{(q)}} x^a = S^{(q)}(x)$ and $1 + x^q + \cdots + x^{(p-1)q} = (x^q-1)^{p-1}$.

Lemma 13 *Let α be a primitive q-th root of unity in the extension $GF(p)$. Then*

$$\sum_{i \in C_0^{(p)}} \alpha^{ji} = \begin{cases} S^{(q)}(\alpha^j), & \text{if } \left(\frac{p}{q}\right) = 1, \\ S^{(q)}(\alpha^{jq}), & \text{if } \left(\frac{p}{q}\right) = -1. \end{cases}$$

Proof. Suppose $\left(\frac{p}{q}\right) = 1$; then $p \in H_0^{(q)}$ and $pH_0^{(q)} = H_0^{(q)} \bmod q$. Combining this with $C_0^{(p)} = pH_0^{(q)}$, we get

$$\sum_{i\in C_0^{(p)}} \alpha^{ji} = \sum_{m\in H_0^{(q)}} \alpha^{jm} = S^{(q)}(\alpha^j).$$

The second formula in this lemma may be proved similarly.

Lemma 14 *Let $\{v_i\}$ be defined by* (15). *Then $S\left(\alpha^j\right) \neq 0$ for $j = 1, 2 \ldots, p-1$.*

Proof. If exist $j : j \neq 0$ and $S(\alpha^j) = 0$ then without loss of generality, we can choose α such that $S(\alpha) = 0$. We consider two cases.

(i) Suppose $\left(\frac{p}{q}\right) = 1$; then by (16), Lemmas 5,13 we have

$$(p-1)S^{(q)}(\alpha) + S^{(q)}(\alpha) + (p-1)/2 = 0.$$

It is impossible.

(ii) Suppose $\left(\frac{p}{q}\right) = -1$; then in this case we obtain

$$(p-1)S^{(q)}(\alpha) + S^{(q)}(\alpha^g) + (p-1)/2 = 0.$$

Hence, by (4) $S^{(q)}(\alpha) = -3/4$.

Let $q \equiv 1 (\bmod 4)$. Therefore, since by Lemma 1 $S^{(q)}(\alpha)$ is a zero of $z^2 + z - (q-1)/4 = 0$ we obtain $q \equiv 1/4 (\bmod p)$. Then $\left(\frac{q}{p}\right) = 1$. Further, by the law of quadratic reciprocity we see

$$\left(\frac{p}{q}\right) = (-1)^{\frac{p-1}{2}\frac{q-1}{2}}\left(\frac{q}{p}\right) = 1.$$

We have a contradiction here.

Now, let $q \equiv 3 (\bmod 4)$. In this case by Lemma 1 $S^{(q)}(\alpha)$ is a zero of $z^2 + z + (q+1)/4 = 0$. Hence $9/16 - 3/4 + (q+1)/4 = 0$ or $q \equiv -1/4 (\bmod p)$. Then

$$\left(\frac{q}{p}\right) = \left(\frac{-1}{p}\right) = (-1)^{(p-1)/2}.$$

By the law of quadratic reciprocity and the condition we obtain

$$-1 = \left(\frac{p}{q}\right) = (-1)^{\frac{p-1}{2}\frac{q-1}{2}}\left(\frac{q}{p}\right) = (-1)^{p-1} = 1.$$

So, we have a contradiction. This completes the proof of Lemma 14.

Theorem 15 *Let $\{v_i\}$ be defined by* (15). *Then $LC_p(v) = pq$ and $m(x) = x^{pq} - 1$.*

Proof. By definition, $S_v(1) = (pq-1)/2$, so that $S_v(1) \neq 1$. Further, by Lemma 14 $S_v(\alpha) \neq 0$ and the statements of this theorem follows from (2).

6.2 The linear complexity of sequences over $GF(q)$

With similar arguments as above we obtain the following results for v.

Theorem 16 *Let $\{v_i\}$ be defined by* (15). *Then*

1. $LC_q(v) = q(p+1)/2$ if $p \equiv 1(\bmod 4)$ and $p \equiv 1(\bmod q)$ or $p \equiv 3(\bmod 4)$ and $p \equiv -1(\bmod q)$.

2. $LC_q(v) = pq$ otherwise.

7 Conclusion

For additive stream ciphering, the linear complexity (span) of the keystream sequence must be large enough. We study the linear complexity of three classes of Ding-Helleseth-generalized cyclotomic sequences of length pq. We derive the linear complexity and the minimal polynomial of above-mentioned sequences over the finite fields of orders p and q, where p and q are two odd distinct primes and obtain series of sequences with high linear complexity. Long periods can also be obtained easily. The sequences from the third class have almost ideal balance property.

References

[1] H. Aly, A.Winterhof, On the k-error linear complexity over Fp of Legendre and Sidelnikov sequences, Des. Codes Cryptogr. **40** (2006) 369–374.

[2] H. Aly, W. Meidl, A. Winterhof, On the k-Error Linear Complexity of Cyclotomic sequences, J. Math. Crypt. **1** (2007) 1–14.

[3] T.W. Cusick, C. Ding, A. Renvall. *Stream Ciphers and Number Theory*, North-Holland Publishing Co., Amsterdam 1998

[4] Bai E, Liu X, Xiao G. Linear complexity of new generalized cyclotomic sequences of order two of length pq. IEEE Transactions on Information Theory, **51**(5) (2005) 1849-1853.

[5] C. Ding, Linear Complexity of Generalized Cyclotomic Binary Sequences of Order 2. Finite Fields and Their Appl., **3**(1997) 159-174

[6] C. Ding, Cyclic Codes From the Two-Prime Sequences. IEEE Trans. Inf. Theory, **58** (6) (2012) 3881-3891

[7] C. Ding, T. Helleseth, New generalized cyclotomy and its applications, Finite Fields and Their Applications. **4**(2)(1998) 140–166

[8] K. Ireland, M. Rosen. *A Classical Introduction to Modern Number Theory*, Springer, Berlin 1982

[9] R. Lidl, H. Niederreiter. *Finite Fields*. Addison-Wesley 1983

[10] W. Meidle, Remarks on a cyclotomic sequence. Des. Codes Cryptogr., **51**(2009) 33-43

[11] Q. Wang, Y. Jiang, D. Lin, Linear complexity of Ding-Helleseth sequences of order 2 over *GF(l)*. Crypto. Commun., DOI 10.1007/s12095-015-0138-5

[12] Q. Wang, D. Lin, X. Guang, On the Linear Complexity of Legendre Sequences Over F_q, IEICE Trans. Fundamentals, **E97-A** (7)(2014) 1627- 1630.

[13] A. L. Whiteman, A family of difference sets,. Illinois J. Math., **6** (1962) 107-121

[14] Yan T, Chen Z, Xiao G. Linear complexity of Ding generalized cyclotomic sequences. Journal of Shanghai University, **11**(1) (2007) 22-26.

[15] Yan T, Hong L, Xiao G. The linear complexity of new generalized cyclotomic binary sequences of order four. Information Sciences, **178**(3)(2008) 807-815.

 DOI: 10.1051/itmconf/20170901006

Empirical Bayes Credibility Models for Economic Catastrophic Losses by Regions

Pavla Jindrová[1,*] and *Lucie* Kopecká[1]

[1]University of Pardubice, Faculty of Economics and Administration, Institute of Mathematics and Quantitative Methods, Studentská 95, 532 10 Pardubice, Czech Republic

Abstract. Catastrophic events affect various regions of the world with increasing frequency and intensity. The number of catastrophic events and the amount of economic losses is varying in different world regions. Part of these losses is covered by insurance. Catastrophe events in last years are associated with increases in premiums for some lines of business. The article focus on estimating the amount of net premiums that would be needed to cover the total or insured catastrophic losses in different world regions using Bühlmann and Bühlmann-Straub empirical credibility models based on data from Sigma Swiss Re 2010-2016. The empirical credibility models have been developed to estimate insurance premiums for short term insurance contracts using two ingredients: past data from the risk itself and collateral data from other sources considered to be relevant. In this article we deal with application of these models based on the real data about number of catastrophic events and about the total economic and insured catastrophe losses in seven regions of the world in time period 2009-2015. Estimated credible premiums by world regions provide information how much money in the monitored regions will be need to cover total and insured catastrophic losses in next year.

1 Introduction

Based on *Sigma Swiss Re* terminology [19] the term *natural catastrophe* refers to an event caused by natural forces.

Table 1. The Sigma event selection criteria 2015

Insured losses:	
Maritime disasters	19.7 USD million
Aviation	39.3 USD million
Other losses	48.8 USD million
or Total economic losses:	97.7 USD million
or Casualties:	
Dead or missing	20
Injured	50
Homeless	2000

Source: Sigma Swiss Re, No 1/2016, [19]

Such an event generally results in a large number of individual losses involving many insurance policies. An event is classified as a catastrophe and included in the *Sigma* database when insured claims, total losses or the number of casualties exceed certain thresholds, detailed in Table 1.

The occurrences of catastrophic events are becoming more frequent (Fig. 1) and also grow indemnity of insurance and reinsurance companies at these events (Fig. 2).

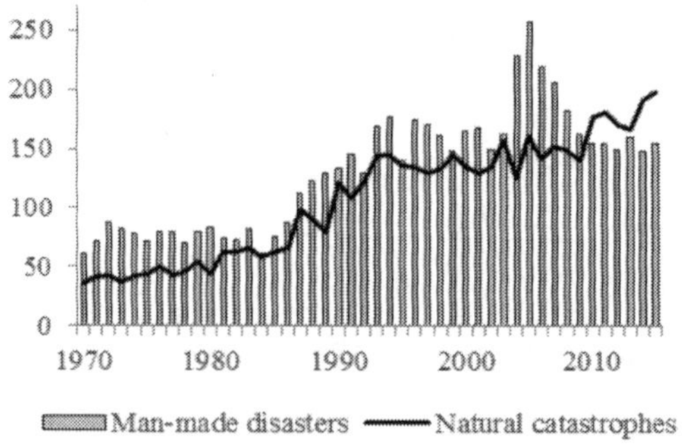

Source: Sigma Swiss Re, No 1/2016, [19]
Fig. 1. Number of catastrophes events 1970-2015

By *Sigma* [19] there were 198 natural catastrophes in 2015, and 155 man-made disaster events. The estimated total economic losses from natural catastrophes and

* Corresponding author: Pavla.Jindrova@upce.cz

man-made disasters across the world were USD 92 billion in 2015. This is less than the USD 113 billion total loss in 2014, and is well below the inflation-adjusted average of the previous 10 years (USD 192 billion).

The insurance industry covered almost USD 37 billion – less than half – of the total losses from natural and man-made disasters in 2015, and well below the inflationadjusted previous 10-year annual average of USD 62 billion. Natural catastrophes resulted in claims of USD 28 billion, the lowest since 2009 and again much lower than the previous 10-year inflation-adjusted annual average (USD 55 billion). Large man-made disasters led to claims of USD 9 billion, up from USD 7 billion in 2014.

Fig. 2 shows also the difference between insured and total losses over time period 1970-2015, termed the insurance protection or funding gap. It is the amount of financial loss generated by catastrophes not covered by insurance. In this figure it is evident growing trend both total and insured catastrophe losses in time period 1970-2015, determined by 10-years moving averages. The rate of growth of total losses has outpaced the growth of insured losses.

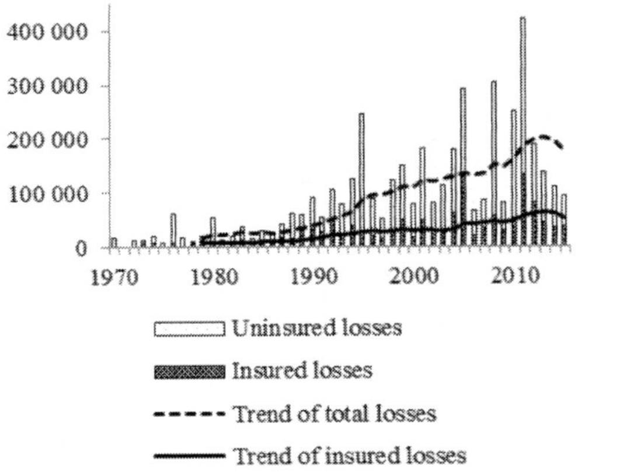

Source: Sigma Swiss Re, No 1/2016

Fig. 2. Insured and uninsured losses in 1970-2015 in USD million in 2015 prices

Catastrophic events affect various regions of the world with increasing frequency and intensity. Large catastrophic events can be caused by natural phenomena or are caused by man.

The number of catastrophic events is also varying in different regions of the world. The catastrophic events by Sigma Swiss Re are monitored for seven regions - North America, Latin America, Europe, Africa, Asia, Oceania/Australia and Seas/space. Number of catastrophe events by regions in 2009-2015 present Fig. 3 and Table 2.

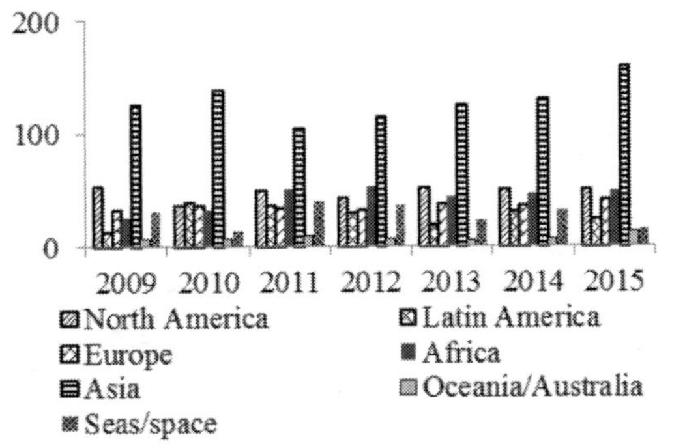

Source: Own processing by Sigma Swiss Re, 2010-2016

Fig. 3. Number of catastrophe events by regions for 2009-2015

The worldwide insurance industry has been rocked by the increasing catastrophes in recent years and increased demand for catastrophe cover (e.g., per occurrence excess of loss reinsurance), leading to a capacity shortage in catastrophe reinsurance. Catastrophe events in last years are associated with increases in premiums for some lines of business.

Table 2. Number of catastrophe events by regions (2009-2015)

	2009	**2010**	**2011**	**2012**	**2013**	**2014**	**2015**
North America	54	36	50	43	52	51	51
Latin America	13	39	36	30	20	31	25
Europe	32	37	34	33	38	37	41
Africa	26	32	51	53	44	47	49
Asia	125	139	104	115	125	130	159
Oceania/ Australia	7	7	10	7	6	7	13
Seas / space	31	14	40	37	23	33	15

Source: Own processing by Sigma Swiss Re, 2010-2016

Regional differences can be observed not only in the number of catastrophic events, but also in the amount of economic losses. Part of these losses is covered by insurance. The share of insured losses on total losses again varies by region (see Table 3 and Table 4).

 DOI: 10.1051/itmconf/20170901006

Table 3. Total economic losses by region in 2009-2015 in USD billion.

	2009	2010	2011	2012	2013	2014	2015
North America	20.1	20.6	63.5	118.5	32.0	28.6	28.6
Latin America	0.6	53.4	5.6	4.2	9.0	8.2	7.5
Europe	20.1	35.2	8.7	26.8	33.0	15.9	12.6
Africa	0.5	0.3	1.6	1.5	1.0	1.5	1.2
Asia	16.7	78.8	260.1	30.5	62.0	51.7	37.7
Oceania/ Australia	2.1	13.1	27.8	1.1	3.0	2.3	3.0
Seas / space	2.0	20.6	3.6	3.1	1.0	1.7	1.1

Source: Own processing by Sigma Swiss Re, 2010-2016

The collected premiums used to cover the insured losses not only in the region of insured, but if necessary, in other regions of the world. It is caused by the globalization of the insurance and reinsurance market.

Table 4. Insurance losses by region in 2009-2015 in USD billion.

	2009	2010	2011	2012	2013	2014	2015
North America	12.7	15.3	39.8	64.6	19.0	17.5	17.3
Latin America	0.1	9.0	0.6	0.9	2.0	2.3	3.2
Europe	7.7	6.3	4.3	5.5	15.0	6.6	6.2
Africa	0.2	0.1	0.3	0.2	1.0	0.8	0.0
Asia	2.4	2.2	49.2	3.4	6.0	5.2	7.0
Oceania/ Australia	1.3	8.9	19.1	0.3	1.0	1.0	2.1
Seas / space	2.0	1.6	2.4	2.4	1.0	1.3	0.9

Source: Own processing by Sigma Swiss Re, 2010-2016

The article will therefore focus on estimating the amount of net premiums that would be needed to cover the total catastrophic losses in each region. For this purpose we will use empirical credibility models.

2 Application of Bühlmann credibility model

Credibility theory is a technique, or set of techniques, for calculating premiums for short term insurance contracts. The technique calculates a premium for a risk (in our case for a region) using two ingredients: past data from the risk itself and collateral data, i.e. data from other sources considered to be relevant, in our cases from other regions. The essential features of a credibility premium are that it is a linear function of the past data from the risk itself and that it allows for the premium to be regularly updated as more data are collected in the future [1, 2, 12, 13, 15, 18].

A credibility premium represents a compromise between the two above mentioned sources of information. The credibility formula for estimation of pure premium P_c in next year is [6, 7]:

$$P_c = Z\,P_r + (1-Z)\,\mu \tag{1}$$

where P_r is estimation based on own past data in risk, or region, μ is estimation based on collateral data and Z is a number between zero and one, known as the *credibility factor*. Credibility factor Z is a measure of how much reliance the company is prepared to place on the own data [8, 9, 10].

Empirical Bayes credibility theory is the collective name for the vast literature which has developed since Bühlmann [2] and Bühlmann and Straub [3].

The form of empirical Bayes credibility premium as a result of Bühlmann credibility model has been derived in [2] as

$$E(m(\theta)/x) = Z\,\bar{x} + (1-Z)E(m(\theta)) \tag{2}$$

with factor credibility

$$Z = \frac{n}{n + \dfrac{E(s^2(\theta))}{D(m(\theta))}} \tag{3}$$

In empirical Bayes credibility model we use the available data (Tab. 3) to estimate the quantities $E(m(\theta))$, $D(m(\theta))$, $E(s^2(\theta))$, and hence obtain a Bühlman type credibility premium estimate [2] for a particular risk.

Derived relations necessary for estimates are:

$$est\,E(m(\theta)) = \overline{X} \tag{4}$$

$$estE(s^2(\theta)) = \frac{1}{N}\sum_{i=1}^{N}\frac{1}{n-1}\sum_{j=1}^{n}(X_{ij} - \overline{X}_i)^2 \tag{5}$$

$$estD(m(\theta)) = \frac{1}{N-1}\sum_{i=1}^{N}(\overline{X}_i - \overline{X})^2 - \frac{1}{Nn}\sum_{i=1}^{N}\frac{1}{n-1}\sum_{j=1}^{n}(X_{ij} - \overline{X}_i)^2 \tag{6}$$

The derivation of these relations can be found in Bühlmann [2] and Waters [18].

Table 3 and Fig. 4 give present the total economic losses (in USD billion) to each of seven regions of the world over seven years.

DOI: 10.1051/itmconf/20170901006

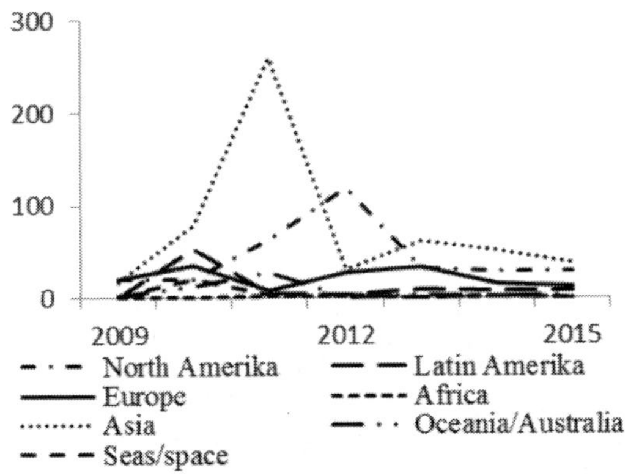

Source: Own processing by Sigma Swiss Re, 2010-2016
Fig. 4. Times series plot of total economic losses in USD bn

Using the Bühlmann credibility model we have calculated the credibility factor and credibility premium for next year for each of the seven regions. For each of the regions we calculate means $\overline{X_i} = \sum_{j=1}^{n} X_{ij} / n$ .These values contains second column in Table 5.

Table 5. Results of Bühlmann credibility model.

	$\overline{X_i}$	**Credibility premium**
North America	44.54	39.66
Latin America	12.63	15.38
Europe	21.76	22.33
Africa	1.08	6.60
Asia	76.80	64.21
Oceania / Australia	7.48	11.47
Seas/space	4.73	9.38

Source: Own calculations

Using expressions (4) - (6) we have estimated $E(m(\theta))$, $E(s^2(\theta))$, $D(m(\theta))$:

$$estE(m(\theta)) = 24.1481;$$
$$estE(s^2(\theta)) = 1259.777;$$
$$estD(m(\theta)) = 572.2954.$$

Therefore using (3) we have calculated the credibility factor Z which is the same for each region:

$$Z = 0{,}7608$$

The credibility premium for each of seven regions in USD billion by (2) is in last column of Table 5.

3 Application of Bühlmann - Straub credibility model

Contrary to the Bühlmann model in section 3 Bühlmann-Straub credibility model accounts for the varying annual volume of business, as is for example observed number of claims. We have applied this model on insured catastrophe losses (Table 4) and figure 5 with acceptation of the number of catastrophe events (Table 2).

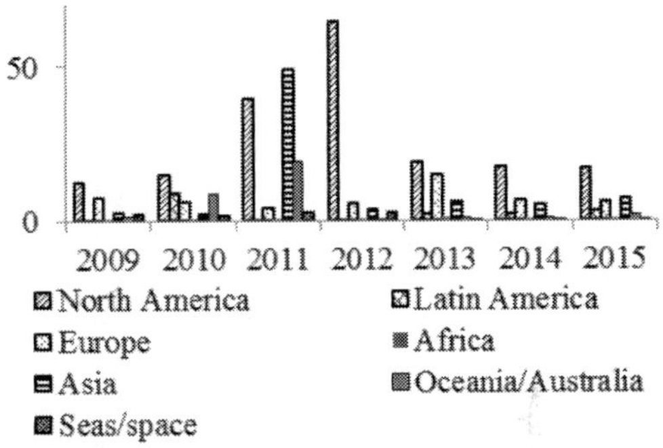

Source: Own processing by Sigma Swiss Re, 2010-2016
Fig. 5. Insured catastrophe losses by regions in 2009-2015 in USD billion

Let region, for which the estimates of insurance catastrophic loses are computed, is one of N regions. Insured losses in previous n years are known for all these regions.

Let P_{ij} is a variable, describing number of catastrophic events of the i-th observed region ($i = 1,\ldots, N$; $N = 7$) in the j-th year ($j = 1, \ldots, n$; $n = 7$).

Let Y_{ij} is the size of insured catastrophe losses for each region $i = 1, \ldots, N$ and each year $j = 1, \ldots, n$.

According to [4, 5, 11, 13, 16, 17, 18] for the estimation of credible net insured premium for the i-th risk, the following formula is used

$$E(m(\theta)/X) = Z_i\overline{X}_i + (1 - Z_i)E(m(\theta)) = Z_i\overline{X}_i + (1 - Z_i)\overline{X} \quad (7)$$

Credibility factor for the i-th risk is calculated in form [3, 18]:

$$Z_i = \frac{P_i}{P_i + \dfrac{E(s^2(\theta))}{D(m(\theta))}} \quad (8)$$

The rules for estimating the parameters $E(m(\theta))$, $E(s^2(\theta))$, $D(m(\theta))$ are by [3, 18]:

$$estE(m(\theta)) = \overline{X} \quad (9)$$

$$estE(s^2(\theta)) = \frac{1}{N(n-1)}\sum_{i=1}^{N}\sum_{j=1}^{n} P_{ij}\left(X_{ij} - \overline{X}_i\right)^2 \quad (10)$$

DOI: 10.1051/itmconf/20170901006

$$estD(m(\theta)) = \frac{1}{P^*}\left\{\begin{array}{l}\frac{1}{Nn-1}\sum_{i=1}^{N}\sum_{j=1}^{n}P_{ij}\left(X_{ij}-\overline{X}\right)^2 \\ -\frac{1}{N(n-1)}\sum_{i=1}^{N}\sum_{j=1}^{n}P_{ij}\left(X_{ij}-\overline{X}_i\right)^2\end{array}\right\} \quad (11)$$

when it is mean, that

$$\overline{X}_i = \frac{1}{P_i}\sum_{j=1}^{n}P_{ij}X_{ij} = \frac{1}{P_i}\sum_{i=1}^{n}Y_{ij} \quad (12)$$

$$\overline{X} = \frac{1}{P}\sum_{i=1}^{N}\sum_{j=1}^{n}P_{ij}X_{ij} = \frac{1}{P}\sum_{i=1}^{N}P_i\overline{X}_i \quad (13)$$

$$P^* = \frac{1}{Nn-1}\sum_{i=1}^{N}P_i\cdot\left(1-\frac{P_i}{P}\right) \quad (14)$$

The value of credibility factor shows the rate of reliability of own data for each region.

The estimates of parameters $E(m(\theta))$, $D(m(\theta))$, $E(s^2(\theta))$ are the same for all the regions. Credibility factor Z_i differs for each region. The higher is the value of risk rate P_i, the higher is the value of credibility factor Z_i.

Bühlmann-Straub model was applied to compute credible net premium for seven regions of the world based insured catastrophe losses in time period 2009-2015 published in Sigma Swiss Re [20] (Table 4).

Total amount of insured losses Y_i, where $Y_i = \sum_{j=1}^{n} Y_{ij}$, insured by each region i in the time 2009-2015, total number of catastrophic events P_i, where $P_i = \sum_{j=1}^{n} P_{ij}$ by each region in the whole period, and average insured catastrophe losses per insurance events $\overline{X}_i$ for each region were computed according to (12). These characteristics present Table 6.

Table 6. Computed characteristics of the regions

	P_i	Y_i	$\overline{X}_i$
North America	337	186.16	0.5524
Latin America	194	18.06	0.0931
Europe	252	51.64	0.2049
Africa	302	2.67	0.0088
Asia	897	75.53	0.0842
Oceania / Australia	57	33.66	0.5906
Seas/space	193	11.59	0.0600

Source: Own calculations

Computed values in Table 6 have used to estimate parameters of the Bühlmann-Straub credibility model according to (9)-(11):

$$estE(m(\theta)) = 0.1699;$$
$$estE(s^2(\theta)) = 2.4757;$$
$$estD(m(\theta)) = 0.0366.$$

According to (8), values of credibility factors Z_i are computed for each region. It is necessary to find these values to be able to estimate credible net premium per region according to (7). Values of credibility factors and values of credible net premiums by regions contain Table 7.

Table 7. Credibility factor Z_i and estimates of net insurance premium per region in 2016.

	Z_i	Net insurance premium
North America	0.8327	0.4884
Latin America	0.7413	0.1130
Europe	0.7882	0.1975
Africa	0.8168	0.0383
Asia	0.9298	0.0902
Oceania / Australia	0.4570	0.3622
Seas/space	0.7403	0.0886

Source: Own calculations

Estimates of credible net premiums by regions provide information how much money in the region will be needed to cover insured catastrophe losses in next year.

4 Conclusion

The enormous impact of catastrophic events on our society is deep and long. Not only we need to investigate the causes of such events and develop plans to protect against them, but also we will have to resolve the resulting huge financial losses.

Results of application Bühlmann credibility empirical model in Table 5 provide an estimate of the total amount of credible premium in the regions of the world in the sum of USD 169,03 billion in year 2016, the highest in Asia and North America.

Development of the financial consequences of natural and man-made catastrophes have a major impact on the global insurance market and forcing the insurance

DOI: 10.1051/itmconf/20170901006

and reinsurance companies to seek for new approaches and ways to cover these risks [14].

Raises the concern that the capacity of the world's insurance and reinsurance markets in the future will not be sufficient to cover these risks and aims to seek alternative options for their transfer.

Acknowledgement

This paper was supported by SGS University of Pardubice, FEA, No. SGS_2016_023, "Ekonomický a sociální rozvoj v soukromém a veřejném sektoru".

References

1. P. J. Boland, Statistical and Probabilistic Methods in Actuarial Science. (New York: Taylor & Francis Group, 2007)
2. H. Bühlmann, Experience rating and credibility. *ASTIN* Bull. **4**, pp. 199-207 (1967)
3. H. Bühlmann, E. Straub, Glaubwürdigkeit für Schadensätze. *Mitteilungen der Vereinigung Schweizerischer Versicherungs-mathematiker* 70, pp. 111-133 (1970)
4. J. Gogola, Method for permanent adjustments of premium in non-life insurance (Spôsob permanentnej úpravy výšky pojistného v neživotnom poistení). *E + M Economics and Management*, **16** (4), pp.134-141 (2013)
5. P. Jindrová, Credibility Risk Models in Accident Insurance. In *Proceedings of the 7th International Scientific Conference: Managing and Modelling of Financial risk*, pp. 307-316. (2014)
6. P. Jindrová, V. Pacáková, Bayesian Probability Models for Critical Illness Insurance. In *Recent Advances in Mathematical Methods in Applied Sciences. Athens*, pp. 218-221 (2014)
7. P. Jindrová, V. Pacáková, Actuarial Models for Valuation of Critical Illness Insurance. *International Journal of Mathematical Models and Methods in Applied Sciences*, **9**, pp. 218-222 (2015)
8. P. Jindrová, K. Seinerová, Bayesian Estimates of the Regional Costs in Public Health System of the Czech Republic. In *Proceedings of the 12th International Scientific Conference: European Financial Systems 2015*, pp. 246-252 (2015)
9. P. Jindrová, K. Seinerová, Bayesian Estimates of Health Care Costs in Slovakian Insurance Companies. *International Journal of Economics and Statistics*, **3**, pp. 174-177 (2015)
10. P. Jindrová, K. Seinerová, Comparison of Bayesian Estimates of Health Care Costs for Czech and Slovakian Insurance Companies, *Scientific Papers of the University of Pardubice, Series D*, **37** (2/2016), pp. 48-59 (2016)
11. B. Linda, J. Kubanová, Credibility Premium Calculation in Motor Third-Party Liability Insurance, WSEAS conference *Advances in Mathematical and Computational Methods*, Malta, pp. 259-263 (2012)
12. V. Pacáková, Bayesian Estimations in Insurance Theory and Practice. In *Advances in Mathematical and Computational Methods: proceedings of the 14th International Conference on Mathematical and Computational methods in Science and Engeneering*, pp. 127-131 (2012)
13. V. Pacáková, Credibility models for permanently updated estimates in insurance. *International journal of mathematical models and methods in applied sciences,* **3**(7), pp. 333-340 (2013)
14. V. Pacáková, P. Jindrová, Simulation of the highest insured catastrophe losses using quantile function. *International Journal of Mathematics and Computers in Simulation*. NAUN, **10**, pp. 242-248 (2016)
15. V. Pacáková, B. Linda, Simulation of the extreme losses in non-life insurance *E & M EKONOMIE A MANAGEMENT*, **12**(4), pp. 97-103 (2009)
16. V. Pacáková, E. Šoltés, T. Šoltésová, Credibility Estimation of the Claim frequency. *E & M EKONOMIE A MANAGEMENT*, **12**(2), pp. 122-126 (2009)
17. E. Šoltés, V. Pacáková, T. Šoltésová, Selected credibility regression models in motor hull insurance (Vybrané kredibilné regresné modely v havarijnom poistení), *Ekonomicky casopis*, **54**(2), pp. 168-182 (2006)
18. H. R. Waters, *Credibility Theory* (Edinburgh: Heriot-Watt University, 1993)
19. Natural catastrophes and man-made disasters in 2015: Asia suffers substantial losses. *Sigma No 1/2016*, Swiss Re Ltd, Economic Research & Consulting, Zurich (2016) [online] Available from: http://reliefweb.int/sites/reliefweb.int/files/resources/sigma1_2016_en.pdf
20. Sigma Swiss Re 2010-2016 [online] Available from: http://www.swissre.com/sigma/

ITM Web of Conferences **9**, 01007 (2017) DOI: 10.1051/itmconf/20170901007
AMCSE 2016

On the discrete spectrum of the Dirac operator on bent chain quantum graph

Michail Belov[1], *Igor Popov*[1,*], *Irina Blinova*[1]

[1]ITMO University, Department of Higher Mathematics, 197101 St. Petersburg, Russia

Abstract. We study Dirac operators on an infinite quantum graph of a bent chain form which consists of identical rings connected at the touching points by δ-couplings with a parameter $\alpha \in \mathbb{R}$. We are interested in the discrete spectrum of the corresponding Hamiltonian. It can be non-empty due to a local (geometrical perturbation of the corresponding infinite chain of rings. The quantum graph of analogous geometry with the Schrodinger operator on the edges was considered by Duclos, Exner and Turek in 2008. They showed that the absence of δ-couplings at vertices (i.e. the Kirchhoff condition at the vertices) lead to the absence of eigenvalues. We consider the relativistic particle (the Dirac operator instead of the Schrodinger one) but the result is analogous. Quantum graphs of such type are suitable for description of grapheme-based nanostructures. It is established that the negativity of α is the necessary and sufficient condition for the existence of eigenvalues of the Dirac operator (i.e. the discrete spectrum of the Hamiltonian in this case is not empty). The continuous spectrum of the Hamiltonian for bent chain graph coincides with that for the corresponding straight infinite chain. Conditions for appearance of more than one eigenvalue are obtained. It is related to the bending angle. The investigation is based on the transfer-matrix approach. It allows one to reduce the problem to an algebraic task. δ-couplings was introduced by the operator extensions theory method.

1 Introduction

Investigation of the physical properties of various compounds of half-crystals and nanotubes becomes more and more interesting. There are a lot of works dedicated to mathematical modeling of nanostructures, because it allows one to create diodes, transistors, amplifiers and more complex nano-scale devices (see, e.g., [1-4]). Many physical properties are related with the spectral characteristics of the system Hamiltonian. That is why the spectral problems attract additional attention. Hamiltonians of real physical systems are very complicated. Physicists usually used model Hamiltonians. Quantum graph is one of prospective models. It is simple and, at the same time, describes properties of physical systems [5,6].

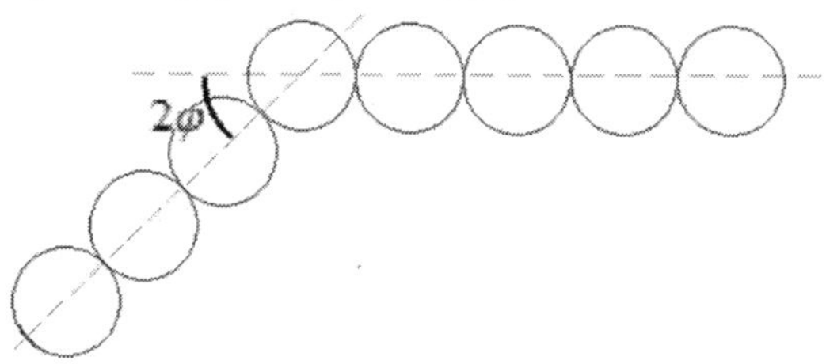

Fig. 1. A bent graph.

Mathematical models based on quantum graphs, as opposed to discrete, have a large number of parameters that allows one to make the model more realistic [7]. Usually, the Schrödinger operator is considered on the graph edges. It corresponds to non-relativistic quantum particle. As for a relativistic particle, i.e. the Dirac operator, the corresponding graph models are not so widely used. At present, due to the remarkable progress of graphene nanostructures (described better by the Dirac operator than by the Schrödinger one), the relativistic model found new applications.

In the present paper we consider bent chain graph (see Fig. 1) with the Dirac operator at the edges. A few years ago, the graph of the same geometry with the Schrödinger operator was studied. It was proved the existence of bound states for the case of delta-coupling at the graph vertices [8]. The analogous problems for branching graphs were studied in [9,10]. The mathematical background for introducing of the point-like potentials is given by the theory of self-adjoint extensions of symmetric operators [11-13]. 1D point-like potential for the Dirac operator was described in many works (see, e.g., [14-21].

2 Infinite periodic chain

The Dirac operator in one dimension for spinless particle is described by the following expression (see, e.g., [11]):

* Corresponding author: popov1955@gmail.com

DOI: 10.1051/itmconf/20170901007

$$D_0 = ic\frac{d}{dx}\otimes\sigma_1 + \frac{c^2}{2}\otimes\sigma_2,$$

$Dom(D_0) = H^1(R)\otimes C^2$, where $H^1(R)$ is the Sobolev space c, $c > 0$, is the speed of light σ_1 and σ_2 are the Pauli matrices,

$$\sigma_1 = \begin{pmatrix} 0 & 1 \\ 1 & 0 \end{pmatrix}, \sigma_2 = \begin{pmatrix} 1 & 0 \\ 0 & -1 \end{pmatrix},$$

Consider bent chain graph Γ shown in Fig. 1. It consists of two semi-infinite straight chains and a central region connecting them. At first, we will deal with the chains. Each chain is semi-periodic, i.e. it is invariant in respect to a semi-group of shifts by circular diameter. Each arc (edge of the graph) is parameterized by the polar angle ϕ, the variable is $x = R\phi$, R is the circle radius. For convenience, we take $R = 1$ (it is not a limitation). One defines the Dirac operator D on the graph by the following way. It acts in $L_2(\Gamma)\otimes C^2$ as D_0 at each edge. The operator domain consists of all 2-vector-functions

$$\begin{pmatrix} \psi_1^{(j)}(x,\lambda) \\ \psi_2^{(j)}(x,\lambda) \end{pmatrix} \in H^1(\Gamma)\otimes C^2$$

satisfying the following conditions at vertices (see Fig. 2):

$$\begin{cases} \sum_{j=1}^{4}\psi_2^{(j)}(0) = -\frac{i\alpha}{c}\psi_1(0), \\ \psi_1(0) = \psi_1^{(j)}(0), \quad \forall j = 1,..,4. \end{cases} \tag{2.0}$$

Here $-\frac{i\alpha}{c}$ characterizes the Lorentz interaction (relativistic point-like interaction).

If one consider the spectral problem, then it is necessary to solve the following Dirac system at each edge:

$$D_0\psi = \lambda\psi. \tag{2.1}$$

Let $C_{1,2}^{(j)}$ be the value of function $\psi_{1,2}^{(j)}$ at the input ($j = 1, 2$) or the output ($j = 3, 4$) of the cell (Fig. 2). Given constants $C_{1,2}^{(1)}, C_{1,2}^{(2)}$ determines the solution at the whole cell and, correspondingly, the values $C_{1,2}^{(3)}, C_{1,2}^{(4)}$. Matrix T which gives one this relation is named the transfer-matrix (or monodromy matrix) [22]. This relation gives one:

$$\begin{pmatrix} C_1^{(2n-1)} \\ C_2^{(2n-1)} \\ C_1^{(2n)} \\ C_2^{(2n)} \end{pmatrix} = T^n \begin{pmatrix} C_1^{(1)} \\ C_2^{(1)} \\ C_1^{(2)} \\ C_2^{(2)} \end{pmatrix}.$$

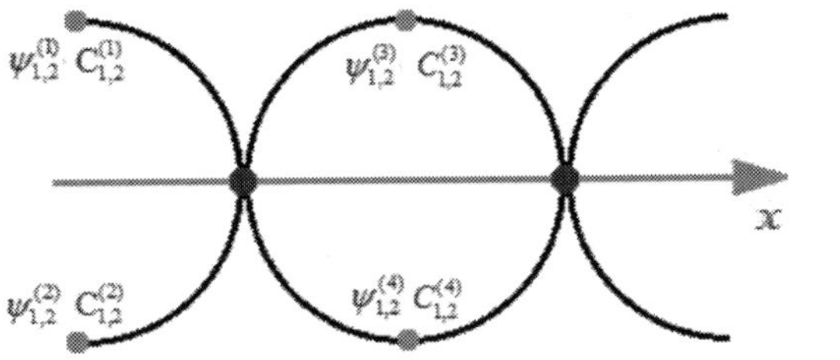

Fig. 2. The transfer-matrix definition.

To construct the transfer-matrix, one chooses directions at each edge of the basic cell (Fig. 3).

Let $\psi_{1,2}^{(1)}(x,\lambda)$, $\psi_{1,2}^{(2)}(x,\lambda)$, $\psi_{1,2}^{(3)}(x,\lambda)$ and $\psi_{1,2}^{(4)}(x,\lambda)$ be the solutions of the Dirac system (2.1) for the corresponding edge ($x \in \left[0, \frac{\pi}{2}\right]$).

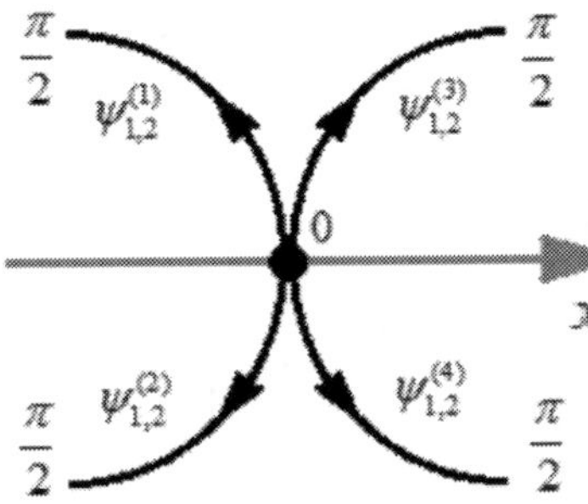

Fig. 3. The basic cell.

One starts from the given values $\psi_{1,2}^{(1)}(\frac{\pi}{2},\lambda) = C_{1,2}^{(1)}$, $\psi_{1,2}^{(2)}(\frac{\pi}{2},\lambda) = C_{1,2}^{(2)}$. The graph is symmetric. The condition at the vertex (2.0) is also symmetric. Hence, the solution should be symmetric or anti-symmetric at the cell. An anti-symmetric solution gives one zero values at the vertex, correspondingly, it is a trivial solution. We are interested in a symmetric case:

$$\begin{cases} \psi_1^{(1)}\left(\frac{\pi}{2}\right) = \psi_1^{(2)}\left(\frac{\pi}{2}\right) \\ \psi_2^{(1)}\left(\frac{\pi}{2}\right) = \psi_2^{(2)}\left(\frac{\pi}{2}\right) \end{cases}. \tag{2.2}$$

Taking into account (2.0) and (2.2), one can find the expression for the coefficients $C_{1,2}^{(3)}$ and $C_{1,2}^{(4)}$ through $C_{1,2}^{(1)}$ and $C_{1,2}^{(2)}$. Thus, one constructs the transfer-matrix.

3. Main result

Theorem. *The Dirac operator D for bent chain quantum graph has at least one bound state if and only if the parameter of the point-like potentials at the vertices is negative ($\alpha < 0$).*

Proof. Let's solve the Dirac system (2.1) at an edge. It is the homogeneous system of linear first order differential equations with constant coefficients. Its solutions can be written as:

$$\begin{cases} \psi_1 = i\sqrt{\dfrac{c^2+2\lambda}{c^2-2\lambda}}(C_1e^{kx} - C_2e^{-kx}) \\ \psi_2 = C_1e^{kx} + C_2e^{-kx} \end{cases}, \quad (3.1)$$

where $k = \dfrac{\sqrt{c^4 - 4\lambda^2}}{2c}$. Due to known spectral properties of 1D Dirac operator, it is clear that if one finds a non-zero solution corresponding to λ,

$$\lambda \in \left(-\frac{c^2}{2}, \frac{c^2}{2}\right), \quad (3.2)$$

then such λ belongs to the point spectrum of the operator. Using (2.0), (2.2), (3.1), one can express $C_{1,2}^{(3)}$ and $C_{1,2}^{(4)}$ through the $C_{1,2}^{(1)}$ and $C_{1,2}^{(2)}$:

$$\begin{pmatrix} C_1^{(3)} \\ C_2^{(3)} \end{pmatrix} = \begin{pmatrix} C_1^{(4)} \\ C_2^{(4)} \end{pmatrix} = \begin{pmatrix} \gamma-1 & -\gamma-1 \\ \gamma & -\gamma \end{pmatrix} \begin{pmatrix} C_1^{(1)} \\ C_2^{(1)} \end{pmatrix} = T\begin{pmatrix} C_1^{(1)} \\ C_2^{(1)} \end{pmatrix}, \quad \begin{pmatrix} C_1^{(1)} \\ C_2^{(1)} \end{pmatrix} = \begin{pmatrix} C_1^{(2)} \\ C_2^{(2)} \end{pmatrix},$$

where $\gamma = \dfrac{\theta}{4} = \dfrac{\alpha}{4c}\sqrt{\dfrac{c^2+2\lambda}{c^2-2\lambda}}$. The reduced (i.e. 2x2-matrix T due to symmetry) transfer-matrix is as follows:

It is simple to find the eigenvalues $\mu_{1,2}$ and the eigenvectors $e_{1,2}$ of the transfer-matrix T:

$$\mu_{1,2} = \frac{-1 \pm \sqrt{1-8\gamma}}{2}, e_{1,2} = \begin{pmatrix} a_{1,2} \\ b_{1,2} \end{pmatrix}$$

$$\frac{a_{1,2}}{b_{1,2}} = \frac{2\gamma - 1 \pm \sqrt{1-8\gamma}}{2\gamma}. \quad (3.3)$$

Relation (3.3) gives one the ratio of the coefficients $C_1^{(1)}$ and $C_2^{(1)}$:

$$\frac{C_1^{(1)}}{C_2^{(1)}} = \frac{2\gamma - 1 \pm \sqrt{1-8\gamma}}{2\gamma}. \quad (3.4)$$

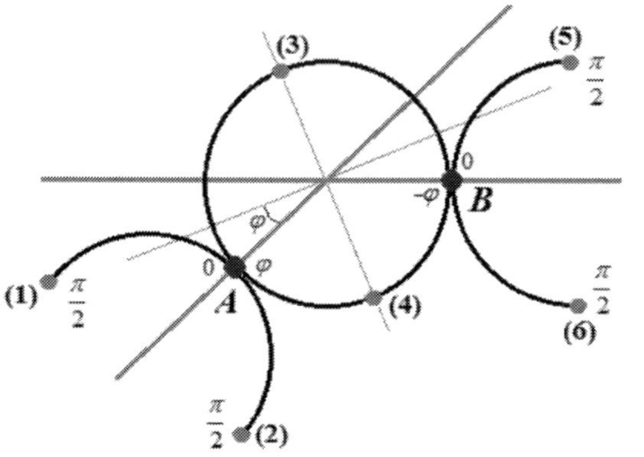

Fig. 4. The central region of the graph. φ is the angular deviation of the coupling point from the diameter.

It is well known (see, e.g., [21]) that real λ belongs to the continuous spectrum of the Dirac operator for semi-infinite chain if $|\mu_{1,2}| = 1$. In contrast, if $|\mu_1| < 1$ (or $|\mu_2| < 1$) then the corresponding λ belongs to the point spectrum of the Dirac operator for semi-infinite chain (one obtains square integrable solution at the chain). In our case, it means that $8\gamma < 1$ (the corresponding eigenvector of the transfer-matrix gives one the initial conditions for this solution). Keeping in mind this observation, we will consider the central region of the graph - chains coupling.

This part of the graph is shown in Fig. 4. There is only one difference between the central circle and any other circle of the chain – the coupling points A and B do not belong to the same diameter.

The system of equations for values of $\psi_{1,2}^{(1,2)}$, $\psi_{1,2}^{(3,4)}$ and $\psi_{1,2}^{(5,6)}$ consists of two parts corresponding to A and B respectively. It looks like (2.2), but the values of the functions $\psi_{1,2}^{(3,4)}$ are calculated at φ (for A) and $-\varphi$ (for B), i.e.:

$$\begin{cases} \psi_2^{(1)}(0) + \psi_2^{(2)}(0) + \psi_2^{(3)}(\varphi) + \psi_2^{(4)}(\varphi) = -\dfrac{i\alpha}{c}\psi_1^{(1)}(0), \\ \psi_1^{(1)}(0) = \psi_1^{(2)}(0) = \psi_1^{(3)}(\varphi) = \psi_1^{(4)}(\varphi). \end{cases}$$

$$\begin{cases} C_1^{(1)} = C_1^{(2)}, \\ C_2^{(1)} = C_2^{(2)}. \end{cases} \quad \begin{cases} C_1^{(5)} = C_1^{(6)}, \\ C_2^{(5)} = C_2^{(6)}. \end{cases}$$

$$\begin{cases} \psi_2^{(3)}(-\varphi) + \psi_2^{(4)}(-\varphi) + \psi_2^{(5)}(0) + \psi_2^{(6)}(0) = -\dfrac{i\alpha}{c}\psi_1^{(5)}(0) \\ \psi_1^{(3)}(-\varphi) = \psi_1^{(4)}(-\varphi) = \psi_1^{(5)}(0) = \psi_1^{(6)}(0). \end{cases}$$

Here we take into account the symmetry. Substituting the explicit expressions for functions ψ, one expresses the coefficients $C_{1,2}^{(5)}$, $C_{1,2}^{(6)}$ in terms of $C_{1,2}^{(1)}$, $C_{1,2}^{(2)}$:

$$\begin{pmatrix} C_1^{(5)} \\ C_2^{(5)} \end{pmatrix} = \begin{pmatrix} a_{11} & a_{12} \\ a_{21} & a_{22} \end{pmatrix} \begin{pmatrix} C_1^{(1)} \\ C_2^{(1)} \end{pmatrix}$$

$$\begin{aligned} a_{11} &= e^{-2k\varphi} + \frac{\theta^2}{16}\left(1-e^{-4k\varphi}\right), \\ a_{12} &= -\frac{\theta}{4}\left(e^{k\varphi}-e^{-3k\varphi}\right) - \frac{\theta^2}{16}\left(1-e^{-4k\varphi}\right), \\ a_{21} &= -\frac{\theta}{4}\left(e^{k\varphi}-e^{-3k\varphi}\right) + \frac{\theta^2}{16}\left(1-e^{-4k\varphi}\right), \\ a_{22} &= e^{2k\varphi} - \frac{\theta^2}{16}\left(1-e^{-4k\varphi}\right). \end{aligned} \qquad (3.5)$$

Using Eq. (3.4), (3.5), one obtains the relation between

$$\theta = \theta(\lambda) = \frac{\alpha}{c}\sqrt{\frac{c^2+2\lambda}{c^2-2\lambda}}, \quad k = k(\lambda) = \frac{\sqrt{c^4-4\lambda^2}}{2c}.$$

The result is as follows:

$$\theta = -2e^{2k\varphi} - 2e^{k\varphi}. \qquad (3.6)$$

$$\theta = 3e^{k\varphi} - e^{2k\varphi} + \sqrt{e^{2k\varphi} - 6e^{3k\varphi} + e^{4k\varphi}} \qquad (3.7)$$

$$\theta = 3e^{k\varphi} - e^{2k\varphi} - \sqrt{e^{2k\varphi} - 6e^{3k\varphi} + e^{4k\varphi}}. \qquad (3.8)$$

Equations (3.6), (3.7) and (3.8) are equations for λ. To be an eigenvalue of the Dirac operator for the bent chain, λ should be a root of at least one of these equation. To complete the proof of the main theorem, we consider the following two lemmas:

Lemma 1. *Equation (3.6) have at least one solution if and only if* $\alpha < 0$.

Lemma 2. *Equations (3.7) and (3.8), in common, have at least one solution if and only if* $\alpha < 0$ *and*

$$\varphi \ge \frac{2\ln\left(3+2\sqrt{2}\right)(r+1)}{c\sqrt{2r^2+2}}, \quad r = \left(\frac{c}{\alpha}\left(6\sqrt{2}+8\right)\right)^2.$$

Equation (3.7) have at least two another solutions if and only if $\alpha < 0$ *and*

$$\varphi \in \begin{cases} [w_0; w_1) \cup (w_2; +\infty), & \alpha \ge -4c \\ [w_0; w_2), & \alpha < -4c \end{cases}$$

where

$$w_{1,2} = \frac{2}{c}\ln\left(\frac{3\alpha \pm \alpha\sqrt{1-\frac{2\alpha}{c}}}{2(4c+\alpha)}\right),$$

$$w_0 = \frac{2}{c}\ln\left(\frac{3\sqrt{c}+\sqrt{9c-4\alpha}}{2\sqrt{c}}\right).$$

The lemmas are proved by means of analyzing the behavior of functions including in the equations. Summarizing the results of Lemma 1 and Lemma 2, we complete the proof of the main theorem.

4. Concluding remarks

The paper deals with the spectrum description of the Dirac operator for a quantum graph having the form of bent chain of rings. Such models can be applied to the description of electron behavior in grapheme-based nanostructures like grapheme nanoribbon (see, e.g., [23]). The main result of the paper is the description of the conditions ensuring the existence of bound states. As for the continuous spectrum for the bending chain, it is not so interesting because it coincides with that for the corresponding straight infinite chain (finite rank perturbation cannot lead to a change of the continuous spectrum). It is interesting to compare the result with the corresponding theorem concerning to the analogous quantum graph with the Schrodinger operator (i.e. non-relativistic particle) [8]. In the both cases, bound states appear only if there is negative delta-coupling at the graph vertices. From the physical point of view delta-coupling is related to physical characteristics of the rings connections. For non-relativistic particle, this statement was proved for more complicated cases of branched graphs too [9,10]. Naturally, the conditions depend on the bending angle. For the relativistic case, i.e. for the Dirac operator, one can use the transfer-matrix approach analogously to the non-relativistic case (for the Schrodinger operator).

This work was partially financially supported by the Government of the Russian Federation (grant 074-U01), by the Ministry of Education and Science of the Russian Federation (GOSZADANIE 2014/190, Projects No 14.Z50.31.0031 and No. 1.754.2014/K)), by grants MK-5001.2015.1 and MK-5161.2016.1 of the President of the Russian Federation, by grant 16-11-10330 of Russian Science Foundation.

References

1. D. Wei, Y. Liu, Advanced Materials, **20**, 2815-2841, (2008)
2. I.S. Lobanov, I.Yu. Popov, Nanosystems: Physics, Chemistry, Mathematics, **3**(2), 6-28 (2012)
3. E.L. Korotyaev, K.M. Schmidt, L. Reine Angew. Math., **670**, 217-248 (2012)
4. H. Santos, A. Ayuela, W. Jaskolski, M. Pelc, L. Chico., Phys. Rev. B., **80**, 035436 (2009)
5. G. Berkolaiko, P Kuchment, Introduction to Quantum Graphs, Mathematical Surveys and Monographs 186; 270 pp; AMS (2013), Providence, RI
6. P. Exner, J.P. Keating, P. Kuchment, T. Sunada, A. Teplyaev, S. Benvegn, L. Dabrowski (eds.), Analysis on Graphs and Its Applications. Proc. Symp. Pure Math. 77 (Amer. Math.Soc., Providence, RI, 2008)
7. O. Post. *Spectral Analysis on Graph-like Spaces.* (Berlin, Springer, 2012)
8. P. Duclos, P. Exner, O. Turek, J. Phys. A: Math. Theor, **41**, 415206/1-18 (2008)
9. I.Yu. Popov, P.I. Smirnov, Physics Letters A, **377**(6), 439–442 (2013)

10. I.Yu. Popov, A.N. Skorynina, I.V. Blinova, J. Math. Phys., **55**(3), 033504/1-19 (2014)
11. S. Albeverio, F. Gesztesy, R. Høegh-Krohn, H. Holden, *Solvable models in quantum mechanics,* 2nd ed., with an appendix by Pavel Exner, AMS (Chelsea, Providence, RI, 2005)
12. S. Albeverio and P. Kurasov, *Singular perturbations of differential operators,* London Math. Soc. Lecture Note Ser., 271, (Cambridge Univ. Press, Cambridge, 2000)
13. B.S. Pavlov, Uspekhi Mat. Nauk., **42**(6), 99-131 (1987)
14. S. Clark, F. Gesztesy. Trans. Am. Math. Soc., **358**, 3475 (2002).
15. W.B. Huddell III, R.J Hughes. J. Phys. A: Math. Gen., 38(22), 4803 , (2005).
16. S. Benvegnù. Letters in Math. Phys. 30(2):159-167.(1994)
17. F. Gesztesy, P. Seba, Lett. Math. Phys. **13**, 345-358 (1987)
18. K.Yoshitomi. *Michigan Math. J.* 2009, **58**, 363-384
19. R.J. Hughes, Integral Equations Operator Theory, **31**, 436–448 (1998)
20. P. Šeba, Lett. Math. Phys., **18**, 77–86 (1989)
21. E.N.Grishanov, D.A.Eremin, D.A.Ivanov, I.Y.Popov, *Chin. Phys. B*, **25** (4), 047303 (2016)
22. B.M. Levitan, I.S. Sargsyan. *Sturm-Liouville and Dirac Operators* (Moscow, Nauka, 1988)
23. K.Wakabayashi, K.Sasaki, T.Nakanishi, T.Enoki, *Sci. Technol. Adv. Mater*., **11**, 054504 (18 pp) (2010)

DOI: 10.1051/itmconf/20170901008

Model of tunnelling through periodic array of quantum dots

Dmitry Meynster[1], *Anton Popov*[1], *Igor Popov*[1,*]

[1]ITMO University, Department of Higher Mathematics, 197101, St. Petersburg, Russia

Abstract. Several explicitly solvable models of electron tunnelling in a system of single and double two-dimensional periodic arrays of quantum dots with two laterally coupled leads in a homogeneous magnetic field are constructed. First, a model of single layer formed by periodic array of zero-range potentials is described. The Landau operator (the Schrodinger operator with a magnetic field) with point-like interactions is the system Hamiltonian. We deal with two types of the layer lattices: square and honeycomb. The periodicity condition gives one an invariance property for the Hamiltonian in respect to magnetic translations group. The consideration of double quantum layer reduces to the replacement of the basic cell for the single layer by a cell including centers of different layers. Two variants of themodel for the double layer are suggested: with direct tunneling between the layers and with the connecting channels (segments in the model) between the layers. The theory of self-adjoint extensions of symmetric operators is a mathematical background of the model. The third stage of the construction is the description of leads connection. It is made by the operator extensions theory method too. Electron tunneling from input lead to the output lead through the double quantum layer is described. Energy ranges with extremely small (practically, zero) transmission were found. Dependencies of the transmission coefficient (particularly, "zero transmission bands" positions) on the magnetic field, the energy of electron and the distance between layers are investigated. The results are compared with the corresponding single-layer transmission.

1 Introduction

Electron tunnelling through periodic array of quantum dots in a homogeneous magnetic field has been intensively investigated over the last few years because it can be significant for nanotechnology applications [1-6]. Since the publication of famous Hofstadter paper [7] research interest in spectral properties of two-dimensional periodic arrays in magnetic field has greatly increased. Fractal structure of the spectrum (and "Hofstadter butterfly" type energy-flux diagrams) has been a theoretical result for long time, but later got experimental confirmation.

There are several different approaches to building models of quantum dot arrays. In this paper zero-range potential model [8-10] based on the theory of self-adjoint extensions of symmetric operators is used [11]. One of its benefits is that the model is explicitly solvable.

Because of high quality of nanostructure devices, large Fermi wave length (i.e. the de Broglie wavelength of electrons with energy close to the Fermi energy) (4×10^{-8} m) and long mean free path of electron (10^{-5} m) can be observed [12]. Therefore, one deals with the ballistic regime of electron transport. In this case the Landauer-Buttiker formalism can be used to derive the conductivity σ for the nanostructure with several leads from the transmission coefficient $T(E)|_{E=E_F}$ (here E_F is the Fermi energy).

In the simplest case (one incoming and one outgoing channel) the Landauer formula has the form $\sigma = \dfrac{e^2T}{\hbar(1-T)}$ [13]. Here e is the electron charge, $\hbar$ is the Plank's constant. The geometry of a nanostructure can influence transmission coefficient a lot.

In this article we consider tunnelling in the system consisting of double two-dimensional periodic array (with square or hexagonal (honeycomb) lattices in each layer) of quantum dots with two connected semi-infinite leads orthogonal to the plane of the array. We study the influence of the magnetic field and tunnelling electron energy on the transmission coefficient and compare our results with the tunnelling through the corresponding single-layer periodic arrays researched earlier [14-15].

2 Model of single-layer array

We start from the Hamiltonian of a single particle in constant homogeneous magnetic field B. Let us assume that particle has mass m and electric charge e. Since the space is $\mathbb{R}^3$, we choose a standard basis $\mathbf{i},\mathbf{j},\mathbf{k}$ such that $\mathbf{B}$ is collinear to $\mathbf{k}$: $\mathbf{B}=B\mathbf{k}, B\neq 0$. Then, the particle can move freely along z axis, and the state space of the model is $L^2(\mathbb{R}^2)$, where $\mathbb{R}^2$ is the plane based on vectors $\mathbf{i},\mathbf{j}$.

* Corresponding author: popov1955@gmail.com

The Hamiltonian H_0 has the following form:

$$H_0 = \frac{1}{2m}(\hat{p} - \frac{e}{c}\mathbf{A}(\mathbf{r})), \quad (1)$$

where c is the speed of light, $\hat{p} = -i\hbar\nabla$ is the two-dimensional momentum operator, $\mathbf{A}(\mathbf{r})$ is a vector potential of the field $\mathbf{B}$ ($\mathbf{B} = rot\mathbf{A}$). Vector potential is not unique, we will use the symmetric gauge ($\mathbf{A}(\mathbf{r}) = \frac{1}{2}\mathbf{B} \times \mathbf{r}$).

The following standard notations are used:

- $\omega = |eB|/cm$ is the cyclotron frequency;
- $\Phi_0 = 2\pi\hbar c/|e|$ is the quantum of the magnetic flux;
- $\xi = \pm B/\Phi_0$ --- number of the flux quanta through the unit area in $\mathbb{R}^2$.

The sign of ξ is chosen in such a way that the condition $\xi eB > 0$ holds.

We will also use the system of units in which $e = \hbar = m = c = 1$, to simplify the calculations. Then, H_0 can be rewritten in the following form:

$$H_0 = -\frac{1}{2}\left[(\frac{\partial}{\partial x} + \pi i \xi y)^2 + (\frac{\partial}{\partial y} - \pi i \xi x)^2\right]. \quad (2)$$

We will also need the Green function of H_0, which has the form ([11]):

$$G_0(\mathbf{r}, \mathbf{r}', E) = \frac{1}{2\pi}\Gamma(\frac{1}{2} - \frac{E}{\omega})\exp(-\pi i \mathbf{r} \wedge \mathbf{r}' - \frac{1}{2}\pi\xi \|\mathbf{r} - \mathbf{r}'\|^2) \quad (3)$$
$$\Phi(\frac{1}{2} - \frac{E}{\omega}, 1, \pi\xi \|\mathbf{r} - \mathbf{r}'\|^2).$$

Here $\Gamma(x)$ is the Euler Gamma-function, $\Phi(a,c,x)$ is the confluent hypergeometric function of the second kind [6].

To obtain a model for an array of quantum dots, we use the theory of self-adjoint extensions of symmetric operators. We assume that quantum dots are placed in the points of regular two-dimensional lattice Ξ. Square and hexagonal lattices are considered in the next sections. We can fix two vectors $\mathbf{a}_1, \mathbf{a}_2$ in such a way that any element $\mathbf{l}$ of Ξ has unique representation $\mathbf{l} = \lambda_1\mathbf{a}_1 + \lambda_2\mathbf{a}_2$, where λ_1 and λ_2 are integers. For the square lattice $a_1 = (1,0)$ and $a_2 = (0,1)$ can be used. Note also that the hexagonal lattice can be obtained as a superposition of two offset square lattices:

$$\Xi = \Xi_\square + K, \quad (4)$$

where $\Xi_\square$ is square lattice, $K = \{0, \mathbf{b}\}$, $\mathbf{b} = \frac{2}{3}(\mathbf{a}_1 + \mathbf{a}_2)$.

Now consider the Hamiltonian of the array of quantum dots as a perturbation of operator H_0. Then, it can be expressed in the following form:

$$H = H_0 + \sum_{\mathbf{g}_i \in \Xi} V(\mathbf{r} - \mathbf{g}_i). \quad (5)$$

Due to several physical and mathematical assumptions (namely, we consider separable potential for the lattice with short-range potentials for centers) [9,12], potentials V can be replaced by δ-like one, hence, H_0 coincides with H everywhere except points of the lattice Ξ. To obtain H, we first restrict H_0 onto the set of smooth functions vanishing at the points of the lattice (let the restriction be S_0), and then construct self-adjoint extension of S_0. The Krein's resolvent formula [8] for infinite deficiency indices describes the difference between the Green functions of H and H_0:

$$G(\mathbf{r}, \mathbf{r}'; \zeta) = G_0(\mathbf{r}, \mathbf{r}', \zeta) - \sum_{\mathbf{g}, \mathbf{g}' \in \Xi} [Q(\zeta) + A]^{-1}_{\mathbf{g},\mathbf{g}'} G_0(\mathbf{r}, \mathbf{g}; \zeta) G_0(\mathbf{g}', \mathbf{r}'; \zeta). \quad (6)$$

Here Q is the Krein Q-function and operator A parameterizes the self-adjoint extension of the operator H_0.

Finally, we can assume that non-diagonal elements of $Q(\zeta)$ are negligible, hence, it takes the form:

$$Q(\zeta)_{ij} = q(\zeta)\delta_{ij}, \quad (7)$$

where q is:

$$q = -\frac{1}{2\pi}(\psi(\frac{1}{2} - \frac{\zeta}{\omega}) + \log(\pi\xi) + 2C_E). \quad (8)$$

Here ψ is digamma function and C_E is the Euler's constant.

3 Tunnelling through double layer

3.1 Model of tunnelling between layers

First, we build the model for the square lattice. We assume that the state space of the model is a direct sum of two identical state spaces, one for each layer:

$$\mathcal{H} = L_2(\mathbb{R}^2) \oplus L_2(\mathbb{R}^2). \quad (9)$$

Then, the unperturbed Hamiltonian has the following form:

$$H_0 = H_L \oplus H_L, \quad (10)$$

where H_L is the Hamiltonian for the single layer.

To use the Krein's resolvent formula, we, first, restrict operator H_L onto the set of smooth functions vanishing at the points of the lattice, let the restricted operator be S_L. Then, the restriction of H_0 has the form:

DOI: 10.1051/itmconf/20170901008

$$S_0 = S_L \oplus S_L. \quad (11)$$

The deficiency space for S_0 is the direct sum:

$$\mathcal{G} = \mathcal{G}_L \oplus \mathcal{G}_L, \quad (12)$$

where $\mathcal{G}_L = l_2(\Xi)$ is the deficiency space for operator S_L.

In its turn, the Krein Q-function is the direct sum:

$$Q = Q_L \oplus Q_L, \quad (13)$$

where Q_L is defined as in (7), (8).

Let f be a function from $l_2(\Xi)$, then the Krein Γ-function of pair (H_L, S_L) for a single layer has the following form:

$$(\Gamma_L(\zeta)f)(\mathbf{r}) = \sum_{\mathbf{g}\in\Xi} G_0(\mathbf{r},\mathbf{g};\zeta)f(\mathbf{g}), \quad (14)$$

and the Krein Γ-function for the whole system can be defined as the direct sum:

$$\Gamma = \Gamma_L \oplus \Gamma_L. \quad (15)$$

Finally, using the Krein resolvent formula for infinite deficiency indices, we obtain an expression for the Green function of H:

$$G(\mathbf{r},\mathbf{r}';\zeta) = G_0(\mathbf{r},\mathbf{r}',\zeta) - \\ - \sum_{\mathbf{g},\mathbf{g}'\in\Xi\oplus\Xi} [Q(z)+A]^{-1}_{\mathbf{g},\mathbf{g}'} G_0(\mathbf{r},\mathbf{g};\zeta)G_0(\mathbf{g}',\mathbf{r}';\zeta) \quad (16)$$

Now, we need a self-adjoint operator. It is related with the properties of A. We assume that the probability of tunnelling between non-adjacent points of the lattice is negligible. Due to the periodicity of the system, the Hamiltonian of a single layer H_L should be invariant with respect to the magnetic translation group transformations (see, e.g., [6]), therefore, we have the following property for the terms of operator A_L of the single layer:

$$A_{L_{\mathbf{l}-\mathbf{g},\mathbf{m}-\mathbf{g}}} = \exp(i\pi\xi(\mathbf{g}\wedge(\mathbf{l}-\mathbf{m})))A_{L_{\mathbf{l},\mathbf{m}}}. \quad (17)$$

As a consequence, it is sufficient to define elements $A_{L_{\lambda,0}}$ (see [6] for more detailed explanation):

$$A_{L_{1,0}} = \alpha(\delta(\lambda_1,0)(\delta(1,\lambda_2)+\delta(-1,\lambda_2))+ \\ +\delta(0,\lambda_2)(\delta(\lambda_2,1)+\delta(\lambda_2,-1))). \quad (18)$$

Here α is some constant which characterizes the intensity of the interaction.

Now, we need to take into account tunneling between layers, hence, operator A takes the following form:

$$A = \begin{bmatrix} A_L & \alpha I \\ \alpha I & A_L \end{bmatrix}. \quad (19)$$

In the case of hexagonal lattice the main change is different magnetic translations group which lead to changes of the form of operator A_L [15]:

$$A_{L_{\mathbf{l_i},\mathbf{l_j}}} = \alpha[\delta(\mathbf{l_i}-\mathbf{a}_1,\mathbf{l_j}-\mathbf{b})+ \\ \delta(\mathbf{l_i}-\mathbf{a}_2,\mathbf{l_j}-\mathbf{b})+ \\ \delta(\mathbf{l_i}-\mathbf{a}_1-\mathbf{a}_2,\mathbf{l_j}-\mathbf{b}) \\ +\exp(\frac{2}{3}i\pi\xi(\delta(\mathbf{l_i}+\mathbf{a}_1-\mathbf{b},\mathbf{l_j})- \\ \delta(\mathbf{l_i}+\mathbf{a}_2-\mathbf{b},\mathbf{l_j})))\times \\ \times(\delta(\mathbf{l_i}+\mathbf{a}_1-\mathbf{b},\mathbf{l_j})+\delta(\mathbf{l_i}+\mathbf{a}_2-\mathbf{b},\mathbf{l_j})+ \\ \delta(\mathbf{l_i}+\mathbf{a}_1+\mathbf{a}_2-\mathbf{b},\mathbf{l_j}))] \quad (20)$$

3.2 Model of tunnelling

Now, we would like to build the general model of tunnelling through the quantum layer (single or double, it doesn't matter). To do that, we use the idea from [17]: let D be any device that is connected to a pair of semi-infinite leads at points C_- and C_+. We assume that the Green function G_D for D is already known. Let $\mathcal{H}_D$ be the state space for the device, $\mathcal{H}_\pm$ be the state spaces for leads, then, the state space for the whole system is given as follows:

$$\mathcal{H}_0 = \mathcal{H}_- \oplus \mathcal{H}_D \oplus \mathcal{H}_+. \quad (21)$$

The Hamiltonian of the system without interaction between the device and the channels is just the orthogonal sum of the corresponding operators:

$$H_0 = H_- \oplus H_D \oplus H_+, \quad (22)$$

where H_D is the Hamiltonian of a charged particle in D, $H_\pm$ is the Hamiltonian of the charged particle in space $L^2(R_\pm)$ under the Neumann condition at the edge (it has the form $-\frac{d^2}{dx^2}$). It is easy to find the Green functions for $H_\pm$:

$$G_\pm(x,x';\zeta) = \\ \frac{i}{2k}[\exp(ik|x-x'|)+\exp(\pm ik(x+x'))], \quad (23)$$

where $k^2 = \zeta$.

Using the theory of self-adjoint extensions again, we can obtain the model Hamiltonian $\mathcal{H}$. To do that, we, first, restrict H_0 onto the set of smooth functions vanishing at the points $C_\pm$, then, build its extension H.

After solving the scattering problem for H, we obtain the following formula for the transmission coefficient $T(E)$:

$$T(E) = \frac{|(\alpha_-\alpha_+Q^D_{21})^2|}{E|\det[Q(E)+A]|^2}. \quad (24)$$

Here A is the matrix of the self-adjoint operator that characterizes the extension,

$$A = \begin{bmatrix} 0 & \alpha_- & 0 & 0 \\ \alpha_- & 0 & 0 & 0 \\ 0 & 0 & 0 & \alpha_+ \\ 0 & 0 & \alpha_+ & 0 \end{bmatrix}, \quad (25)$$

$Q(E)$ is the Krein Q-function:

$$Q(E) = \begin{bmatrix} Q_-(E) & 0 & 0 & 0 \\ 0 & Q_D^{11}(E) & Q_D^{12}(E) & 0 \\ 0 & Q_D^{21}(E) & Q_D^{22}(E) & 0 \\ 0 & 0 & 0 & Q_+(E) \end{bmatrix}, \quad (26)$$

$Q_\pm(E) = G_\pm(0,0;E)$ and Q_D is 2×2 Krein Q-function for the extension, where $Q_D^{12}(E) = \overline{Q_D^{21}(E)} = G_D(\mathbf{r}_1,\mathbf{r}_2;E)$ and $Q_D^{11}(E), Q_D^{22}(E)$ are the regular parts of $G_D(r_1,r_1;E), G_D(r_2,r_2;E)$, correspondingly. Values $\alpha_\pm$ describe the quality of contacts $C_\pm$. This technique was used in [14] to construct a model of tunneling through single quantum electron layer.

4. Results and discussion

For numerical experiments in this section constants $\alpha, \alpha_-, \alpha_+$ were all selected equal to 1. Only a finite fragment of an infinite lattice is used for calculations since points which are far enough from contacts don't have much influence on transmission coefficient value. For the square lattice, the basic vectors $a_1 = (1,0), a_2 = (0,1)$ were used; the hexagonal (honeycomb) lattice is defined using (4).

The transmission coefficient T is calculated as a function of electron energy E for different values of the magnetic field B. The resulting function of two values is represented as a contour plot. We assume that the both contacts are in the same lattice cell, calculations show that putting them in different cells of the lattice leads to decreasing of T but doesn't add or remove any other significant effects.

During the investigation of tunnelling through single-layer lattices, wide zones in the resulting contour plot with very low value of T were discovered ([14], [15]). The widths of these zones are greater than the product of the Boltzmann constant and the temperature value and, thus, this phenomenon is physically measurable. Our calculations show that these zones are preserved both in the square and the hexagonal lattice cases (see Fig. (1-4)). Plots were built for the model with direct tunnelling between the layers. The model with channels between the layers gives one qualitatively analogous result such zones exist too.

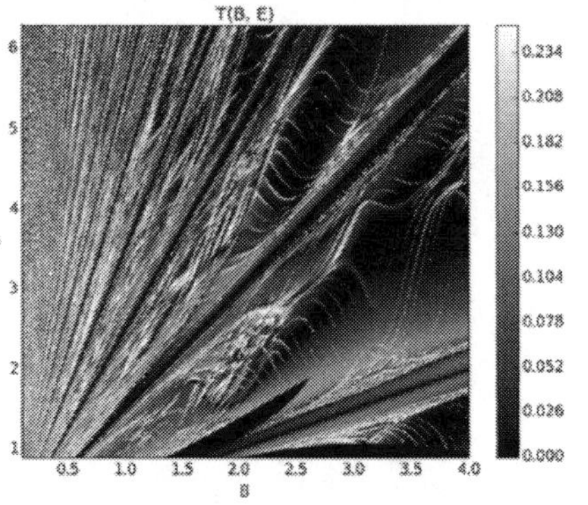

Fig.1. Dependence of transmission coefficient value T on electron energy E and magnetic field B: case of single-layer square lattice. B and E are in dimensionless units.

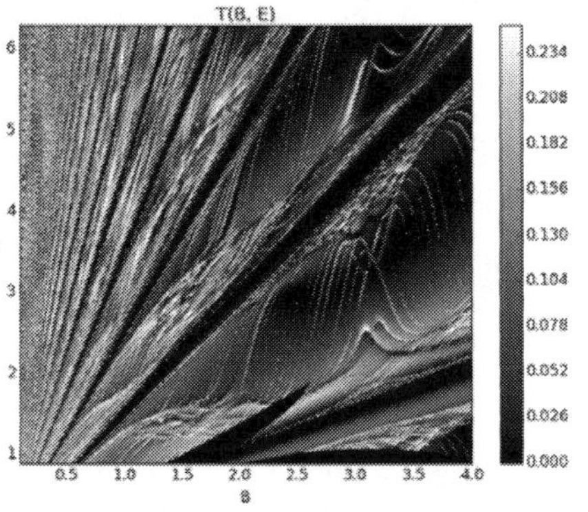

Fig. 2. Dependence of transmission coefficient value T on electron energy E and magnetic field B: case of double-layer square lattice. B and E are in dimensionless units.

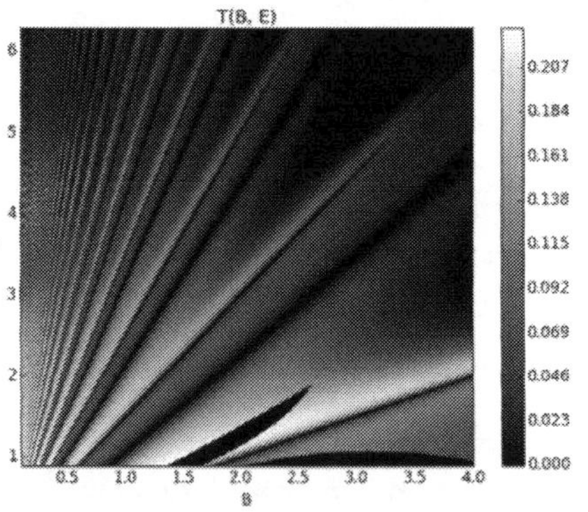

Fig. 3. Dependence of transmission coefficient value T on electron energy E and magnetic field B: case of single-layer hexagonal lattice. B and E are in dimensionless units.

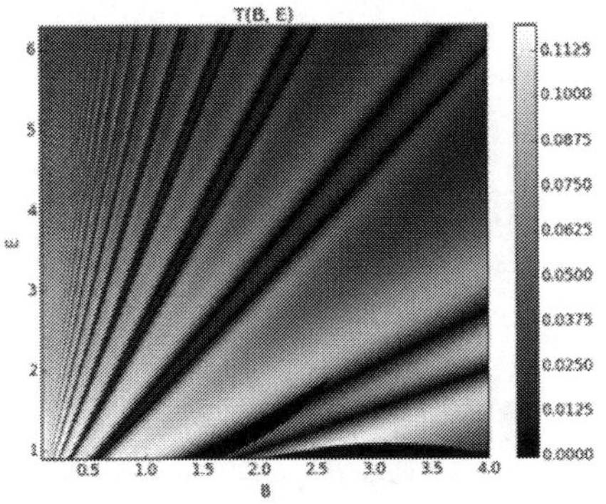

Fig. 4. Dependence of transmission coefficient value T on electron energy E and magnetic field B: case of double-layer hexagonal lattice. B and E are in dimensionless units.

It can be observed that the middle of each dark strip in Fig. (3) and Fig. (4) coincides with some singularity of $\Gamma(\frac{1}{2}-\frac{E}{\omega})$ (which is one of multipliers in (3)).

The dependence of T on B and E shown in the figures is very complicated. For the hexagonal (honeycomb) lattice, it is also complicated but local oscillations of T have less amplitude, that is why in the Fig. (5), (6) it looks like more monotone. Such effect is related to the complex structure of the spectrum of the Hamiltonian for periodic array of quantum dots (see, e.g., [9]). As for the comparison of tunnelling for single-layer and double-layer cases, one observes that in the model with double layer, dark stripes are wider and inside zones between these stripes, T value is, generally, greater. In the case of the hexagonal lattice, when considering double-layer model, dark stripes pairs are closer to each other and value of T in zones between them is less than in the single-layer case.

This work was partially financially supported by the Government of the Russian Federation (grant 074-U01), by the Ministry of Education and Science of the Russian Federation (GOSZADANIE 2014/190, Projects No 14.Z50.31.0031 and No. 1.754.2014/K)), by grants MK-5001.2015.1 and MK-5161.2016.1 of the President of the Russian Federation, by grant 16-11-10330 of Russian Science Foundation.

References

1. A. Gonzalez, R. Physica E, **10**(4), 528-534 (2001)
2. R. Brunner, R. Meisels, F. Kuchar, R. Akis, D.K. Ferry, J.P. Bird, Int. J. Mod. Phys. B., **21**, 1288-1296 (2007)
3. M.C. Roco, C.A. Mirkin, and M.C. Hersam. *Nanotechnology Research Directions for Societal Needs in 2020. Retrospective and Outlook* (Berlin, Springer, 2010)
4. I. Bloch, J. Dalibard, and W. Zwerger, Rev. Mod. Phys., **80**, 885-964 (2008)
5. V.A. Geyler, St. Petersburg Mathematical Journal, **3**(3), 489 – 532 (1992)
6. A. Mantile. Nanosystems: Phys. Chem. Math., **6**, 100-112 (2015)
7. D.R. Hofstadter, Phys. Rev.B, **14**, 2239-2249 (1976)
8. S. Albeverio and P. Kurasov. *Singular perturbations of differential operators. Solvable Schrödinger type operators* (London Mathematical Society, London, 2005)
9. V.A. Geyler, B.S. Pavlov and I.Yu. Popov. Atti Sem. Mat. Fis. Univ. Modena, **46**, 79-124 (1998)
10. V.A. Geyler, B.S. Pavlov and I.Yu. Popov. J. Math. Phys., **37**(10), 5171-5194 (1996)
11. J.P Eisenstein., et.al. Phys. Rev. Lett., **68**(9), 1383-1386 (1992)
12. V.A. Geyler, and I.Yu. Popov. Proc. Royal Soc. London A., **454**, 697-705 (1998)
13. R. Landauer. IBM J. Res. Dev. **49**(24), 17177-17184 (1957)
14. I.Yu. Popov and S.A. Osipov. Chinese Phys. B, **21**, 117306 (2012)
15. D.A Eremin, E.N. Grishanov, D.A. Ivanov, A.A. Lazutkina, E.S. Minkin, and I.Yu. Popov. Chinese J. Phys., **52**(24), 1100-1109 (2014)
16. V.A. Geyler and I.Yu. Popov, Z. Phys. B, **93**, 437-439 (1994)
17. V.A. Geyler and I.Yu. Popov. Theor. Math. Phys., **107**, 427-434 (1996)

DOI: 10.1051/itmconf/20170901009

Bound state for dielectric waveguide with high contrast inset in the core

Maria Faleeva[*], *Igor Popov*

ITMO University, Department of Higher Mathematics, 197101 St. Petersburg, Russia

Abstract. We consider cylindrical dielectric waveguide with a core having a high contrast inset. It is assumed, that for certain conditions, for a waveguide with an insertion in the core there is a bound state appears below the bottom of the continuous spectrum. In this paper we pose the problem of proving the existence of such state and obtain an upper estimate of the distance from the eigenvalue to the threshold.

1 Introduction

Optical fibers containing inclusions with high contrast in permittivity are widely used (see, e.g., [1-5]). We consider a cylindrical dielectric waveguide (permittivity $\varepsilon = \varepsilon_1$) with a core (permittivity $\varepsilon = \varepsilon_2 \gg \varepsilon_1$). The core has an insertion with width $2a$ of a material higher contrast. The waveguide is unbounded in the direction of z. We show that a bound state below the bottom of the continuous spectrum exists for the system. Variational approach is used. Propagation of electromagnetic waves along the waveguide is described by a scalar dimensionless Helmholtz equation for the electric field. For proof we need to construct the trial function, approximately satisfying the equation, and obtain the necessary estimate, which should be negative. The method is analogous to that suggested in [6]. This method is applicable to eigenvalues below the continuous spectrum. The result can be generalized to the periodic system of such insets in the core. In this case, one has an additional spectral band instead of the bound state. It is possible that such periodic system has photonic crystal property. As for eigenvalues lying in spectral gaps, the method should be essentially modified [7]. There are numerous works concerning to eigenvalue below the threshold for various waveguides systems (see, e.g., [8] and references therein). Trapped modes (i.e. eigenvalues embedded in the continuous spectrum) are of special interest. There is an extensive literature about the occurrence of trapped modes in curved closed waveguides (see, e.g., [9, 10]).

2 The method of the proof

It is assumed that we deal with monochromatic light passes along the axis. Analysis of electromagnetic waves reduces to the investigation of the scalar equation for the electric field. Mode solutions for waveguide are obtained from the dimensionless Helmholtz equation for (harmonic dependence on time t ($e^{i\omega t}$) is taken into account),

$$\frac{\partial^2 E_z}{\partial r^2}+\frac{1}{r}\frac{\partial E_z}{\partial r}+\frac{1}{r^2}\frac{\partial^2 E_z}{\partial \varphi^2}+\frac{\partial^2 E_z}{\partial z^2}+k^2E_z=0, \tag{1}$$

where $k = d\omega\sqrt{\varepsilon\mu}$, d is the radius of the waveguide. Multiply the radius of the waveguide on the wave number allows you to get the dimensionless equation, and all the variables having the dimensions of length, normalized to the d.

We take $E_z = \Psi(r,z)e^{i(\nu\varphi+\omega t)}$. Then eigenfunction Ψ satisfies the equation

$$-\frac{\partial^2 \Psi}{\partial r^2}-\frac{1}{r}\frac{\partial \Psi}{\partial r}-\frac{\partial^2 \Psi}{\partial z^2}+\frac{\nu^2}{r^2}\Psi=k^2\Psi \tag{2}$$

and can be constructed using Ψ_1, Ψ_2, and Ψ_3, where

$$\Psi=\begin{cases}\Psi_1, r\in[0,\delta]\\ \Psi_2, r\in[\delta,1]\\ \Psi_3, r\in[1,\infty]\end{cases}. \tag{3}$$

The left side of the equation is the action of the operator H, is the operator with the matching conditions at the interfaces $r = \delta$ and $r = 1$:

$$H\Psi=k^2\Psi, \tag{4}$$

$$\begin{gathered}\Psi_3\big|_{(r=1)}=\Psi_2\big|_{(r=1)}\\ \Psi_1\big|_{(r=\delta)}=\Psi_2\big|_{(r=\delta)}\\ \left.\frac{d\Psi_3}{dr}\right|_{(r=1)}=\left.\frac{d\Psi_2}{dr}\right|_{(r=1)}\\ \left.\frac{d\Psi_1}{dr}\right|_{(r=\delta)}=\left.\frac{d\Psi_2}{dr}\right|_{(r=\delta)}\end{gathered} \tag{5}$$

In accordance with the variational principles, minimum of the ratio

$$\frac{(H\Psi,\Psi)-\left\|\Psi\sqrt{k^2-\beta^2}\right\|^2}{\|\Psi\|^2}, \tag{6}$$

[*] Corresponding author: faleeva.masha@gmail.com

DOI: 10.1051/itmconf/20170901009

is reached if is an eigenfunction of the operator H, it equals to the distance from the eigenvalue to the threshold. If there exists a trial function Ψ (satisfying the matching conditions at the interfaces) such that the function is negative:

$$M(\Psi)=(H\Psi,\Psi)-\left\|\Psi\sqrt{k^2-\beta^2}\right\|^2<0, \quad (7)$$

then there exists an eigenvalue Λ^2 less than the threshold, $\Lambda^2\le\lambda^2=k^2-\beta^2$. Thus, the idea is to find a trial function, which makes the expression $M(\Psi)$ negative.

We take the trial function Ψ as $\Psi=F+G$, where $F=\alpha u(r)f(z)$, $G=\eta s(r)g(z)$,

$$u(r)=\begin{cases}N_1J_\nu(\lambda_1 r)+M_4J_\nu(\lambda_4 r), r\in[0,\delta]\\ N_2J_\nu(\lambda_2 r)+M_2N_\nu(\lambda_2 r), r\in[\delta,1],\\ N_3\,K_\nu(\gamma r), r\in[1,\infty]\end{cases} \quad (8)$$

J_ν, N_ν, K_ν are Bessel functions, $\lambda_i^2=k_i^2-\beta^2$, $k_i^2=\omega^2\varepsilon_i\mu_0 d^2$, $\lambda_3=i\gamma$, $\gamma^2=\beta^2-k_3^2$,

$$s(r)=\begin{cases}e^{\frac{\pi r}{2a}}, r\in\left[0,\frac{1}{2}\right]\\ 2(1-r)e^{\frac{\pi}{4a}}, r\in\left[\frac{1}{2},1\right],\\ 0, r\in[0,\infty]\end{cases} \quad (9)$$

$$f(z)=\min\left\{1,e^{-q(|z|-a)}\right\}=\begin{cases}1, z\in[-a,a]\\ e^{-q(|z|-a)}, z\in[-\infty,-a]\cup[a,\infty]\end{cases}, \quad (10)$$

$$g(z)=\begin{cases}\cos\left(\frac{\pi z}{2a}\right), z\in[-a,a]\\ e^{-q(|z|-a)}, z\in[-\infty,-a]\cup[a,\infty]\end{cases}, \quad (11)$$

where α,η,q are free parameters of the trial function, d is the radius of the waveguide, δ is the radius of the core $(0\le\delta\ll 1)$, 2a is a width of the inset in the core $(0\le a\ll 1)$.

The functions $f(z)$ and $g(z)$ are z symmetrical.

$$f(-\infty)=f(\infty)=0,\ g(-\infty)=g(\infty)=0. \quad (12)$$

Functions $u(r)$ and $s(r)$ satisfy the condition:

$$\int_0^\infty ru(r)dr=\int_0^\infty rs(r)dr=0. \quad (13)$$

Lemma 1: The trial function Ψ defined in (8)-(11) satisfies the following condition:

$$(H\Psi,\Psi)=\|\Psi_r\|^2+\|\Psi_z\|^2+\int_{-\infty}^{\infty}\int_{-\infty}^{\infty}\frac{\nu^2}{r}\Psi^2 drdz. \quad (14)$$

Proof:

$$(H\Psi,\Psi)=\int_{-\infty}^{\infty}\int_{-\infty}^{\infty}r\left(-\frac{\partial^2\Psi}{\partial r^2}-\frac{1}{r}\frac{\partial\Psi}{\partial r}-\frac{\partial^2\Psi}{\partial z^2}+\frac{\nu^2}{r^2}\Psi\right)\Psi drdz=$$
$$=-\int_{-\infty}^{\infty}\int_{-\infty}^{\infty}\left(\frac{\partial}{\partial r}\left(r\frac{\partial\Psi}{\partial r}\right)+r\frac{\partial^2\Psi}{\partial z^2}-\frac{\nu^2}{r}\Psi\right)\Psi drdz \quad (15)$$

Using integration by parts, one can write:

$$(H\Psi,\Psi)=-\int_{-\infty}^{\infty}\left(r\frac{\partial\Psi}{\partial r}\Psi\right)\Bigg|_0^\infty dz+\int_{-\infty}^{\infty}\int_0^{\infty}r\left(\frac{\partial\Psi}{\partial r}\right)^2 drdz- \quad (16)$$
$$-\int_0^\infty r\left(\frac{\partial\Psi}{\partial z}\Psi\right)\Bigg|_{-\infty}^\infty dr+\int_0^\infty r\int_{-\infty}^{\infty}\left(\frac{\partial\Psi}{\partial z}\right)^2 dz+\int_{-\infty}^{\infty}\int_0^{\infty}\frac{\nu^2}{r}\Psi^2 drdz$$

Then, from conditions (12)-(13), one has:

$$(H\Psi,\Psi)=\int_{-\infty}^{\infty}\int_0^{\infty}r\left(\frac{\partial\Psi}{\partial r}\right)^2 drdz+\int_0^\infty r\int_{-\infty}^{\infty}\left(\frac{\partial\Psi}{\partial z}\right)^2 dzdr \quad (17)$$
$$+\int_{-\infty}^{\infty}\int_0^{\infty}\frac{\nu^2}{r}\Psi^2 drdz=\|\Psi_r\|^2+\|\Psi_z\|^2+\int_{-\infty}^{\infty}\int_0^{\infty}\frac{\nu^2}{r}\Psi^2 drdz$$

Lemma 2: The form $M(\Psi)$ for the trial function Ψ defined in (8)-(11) can be represented as follows:

$$M(\Psi)=\|F_z\|^2+\|G_z\|^2+2\int_{-\infty}^{\infty}\int_0^{\infty}rF_zG_z drdz \quad (18)$$

Proof:

$$M(\Psi)=\|\Psi_r\|^2+\|\Psi_z\|^2+\int_{-\infty}^{\infty}\int_{-\infty}^{\infty}\frac{\nu^2}{r}\Psi^2 drdz-\left\|\Psi\sqrt{k^2-\beta^2}\right\|^2=$$
$$=\int_{-\infty}^{\infty}\int_0^{\infty}r\left(\frac{\partial(F+G)}{\partial r}\right)^2 drdz+\int_0^\infty r\int_{-\infty}^{\infty}\left(\frac{\partial(F+G)}{\partial z}\right)^2 dzdr-$$
$$-\int_{-\infty}^{\infty}\int_{-\infty}^{\infty}r\left(k^2-\beta^2-\frac{\nu^2}{r^2}\right)(F+G)^2 drdz \quad (19)$$

We take some integrals by parts and after some calculations, one has

$$M(\Psi)=-\int_{-\infty}^{\infty}\int_0^{\infty}rF\left((F_{rr}+G_{rr})+\frac{1}{r}(F_r+G_r)\right)drdz-$$
$$-\int_{-\infty}^{\infty}\int_0^{\infty}rF(F+G)\left(k^2-\beta^2-\frac{\nu^2}{r^2}\right)drdz-$$
$$-\int_{-\infty}^{\infty}\int_0^{\infty}rG\left((F_{rr}+G_{rr})+\frac{1}{r}(F_r+G_r)\right)drdz+$$
$$+\int_{-\infty}^{\infty}\int_0^{\infty}rG\left((F+G)\left(k^2-\beta^2-\frac{\nu^2}{r^2}\right)\right)drdz+$$
$$+\int_{-\infty}^{\infty}\int_{-\infty}^{\infty}(rF_z^2+rG_z^2+2rF_zG_z)drdz \quad (20)$$

The first and the second integrals in the previous formula are equal to zero and we obtain our lemma. Next, we need to estimate the resulting expression. Write out its as:

$$M(\Psi)=\alpha^2\int_{-\infty}^{\infty}\left(\frac{df(z)}{dz}\right)^2 dz\int_0^\infty r(u(r))^2 dr+\eta^2\int_{-\infty}^{\infty}\left(\frac{dg(z)}{dz}\right)^2 dz$$
$$\int_0^\infty r(s(r))^2 dr+2\alpha\eta\int_{-\infty}^{\infty}\frac{df(z)}{dz}\frac{dg(z)}{dz}dz\int_0^\infty ru(r)s(r)dr \quad (21)$$

The square of the norm of the trial function has the form:

$$\|\Psi\|^2 = \alpha^2 \int_{-\infty}^{\infty} (f(z))^2 dz \int_0^{\infty} r(\mathrm{u}(r))^2 dr + \eta^2 \int_{-\infty}^{\infty} (g(z))^2 dz$$

$$\int_0^{\infty} r(\mathrm{s}(r))^2 dr + 2\alpha\eta \int_{-\infty}^{\infty} \mathrm{a}(z) g(z) dz \int_0^{\infty} r u(r) \mathrm{s}(r) dr \tag{22}$$

Integrals of the functions $f(z)$ and $g(z)$ can be calculated. To estimate the integrals containing Bessel functions use their asymptotic representations:

$$J_\nu(\lambda r) \approx \sqrt{\frac{2}{\pi\lambda r}} \cos\left(\lambda r - \frac{\nu\pi}{2} - \frac{\pi}{4}\right)$$

$$N_\nu(\lambda r) \approx \sqrt{\frac{2}{\pi\lambda r}} \sin\left(\lambda r - \frac{\nu\pi}{2} - \frac{\pi}{4}\right)$$

$$\mathrm{K}_\nu(\gamma r) \approx \sqrt{\frac{\pi}{2\gamma r}} e^{-\gamma r}$$

(23)

For the trial function Ψ defined in (8)-(11) the estimation of the ratio has the form:

$$\frac{\mathrm{M}(\Psi)}{\|\Psi\|^2} \le \left(\frac{2\alpha^2 q}{\pi} A + qB + \eta^2 \left(\frac{a}{4} + \frac{qa^2}{\pi^2}\right) C\right) \tag{24}$$

$$\left(\frac{2\alpha^2}{\pi}\left(2a + \frac{1}{q}\right) A + \frac{4a}{\pi} B + \frac{\eta^2 a^3}{\pi^2} C\right)^{-1},$$

where

$$A = \left(\left(\frac{N_1}{\sqrt{\lambda_1}} + \frac{M_4}{\sqrt{\lambda_4}}\right)^2 \delta + \frac{(N_2 + M_2)^2}{\lambda_2}(1-\delta) + \frac{\pi^2 N_3^2}{4\gamma^2}\right), -$$

$$B = \frac{4}{3}\sqrt{\frac{2}{\pi}}\alpha\eta\left(\left(\frac{N_1}{\sqrt{\lambda_1}} + \frac{M_4}{\sqrt{\lambda_4}}\right)\delta^{\frac{3}{2}} + \frac{(N_2 + M_2)}{\sqrt{\lambda_2}}\left(2^{-\frac{3}{2}} - \delta^{\frac{3}{2}} + \right.\right.$$

$$\left.\left. + 2e^{-\frac{\pi}{4a}} \frac{(8\sqrt{2} - 7)}{20\sqrt{2}}\right)\right),$$

$$C = \left(e^{-\frac{\pi}{2a}}\left(\frac{5\pi^2}{48a^2} - \frac{\pi}{2a} - 1\right) + 1\right).$$

We take $N_3 = 1$. The coefficients N_2, M_2, N_1, M_4 are obtained from the system of boundary conditions:

$$\begin{cases} N_1 J_\nu(\lambda_1\delta) + M_4 J_\nu(\lambda_4\delta) = N_2 J_\nu(\lambda_2\delta) + M_2 N_\nu(\lambda_2\delta), \\ N_2 J_\nu(\lambda_2) + M_2 N_\nu(\lambda_2) = K_\nu(\gamma), \\ N_1\left(\frac{d}{dr} J_\nu(\lambda_1 r)\right)\Big|_{r=\delta} + M_4\left(\frac{d}{dr} J_\nu(\lambda_4 r)\right)\Big|_{r=\delta} = \\ = N_2\left(\frac{d}{dr} J_\nu(\lambda_2 r)\right)\Big|_{r=\delta} + M_2\left(\frac{d}{dr} N_\nu(\lambda_2 r)\right)\Big|_{r=\delta}, \\ N_2\left(\frac{d}{dr} J_\nu(\lambda_2 r)\right)\Big|_{r=1} + M_2\left(\frac{d}{dr} N_\nu(\lambda_2 r)\right)\Big|_{r=1} = \\ = \left(\frac{d}{dr} K_\nu(\gamma r)\right)\Big|_{r=1}. \end{cases}$$

This coefficients obtained numerically for some values of the parameters of the trial function and for fixed dependencies of geometrical characteristics of the waveguide. The results are summarized in Table 1.

The tables show that there are many different combinations of parameter sets the trial function for which there is a value of the permittivity of the inset in the core of the waveguide that the ratio $\mathrm{M}(\Psi)/\|\Psi\|^2$ will be negative. This means that there is a spectral point below the threshold of the continuous spectrum.

3 Conclusions

Thus, we show that for a cylindrical dielectric waveguide having a core with a high contrast inset, there is a bound state below the bottom of the continuous spectrum and we obtain an upper estimate of the distance from the eigenvalue to the threshold. This result allows us to assume that in the case of the periodic system of inserts in the core, system can obey the photonic crystal property which can find many applications.

This work was partially financially supported by the Government of the Russian Federation (grant 074-U01), by Ministry of Science and Education of the Russian Federation (GOSZADANIE 2014/190, Projects No 14.Z50.31.0031 and No. 1.754.2014/K)), by grants MK-5001.2015.1, MK-5161.2016.1 of the President of the Russian Federation, by grant 16-11-10330 of Russian Science Foundation

References

1. S.G. Johnson, J.D. Joannopoulos, *Photonic Crystals, the Road from Theory to Practice* (Boston, Kluwer, 2002)
2. W. Axmann, P. Kuchment, L. Kunyansky, J. Light. Tech., **17**, 1996–2007 (1999)
3. A. Tip, A. Moroz, J.M. Combes, J. Phys. A: Math. Gen., **33**, 6223–6252 (2000)
4. I.Y. Popov, A.I. Trifanov, E.S. Trifanova, Comp. Math Math. Phys., **50**(11), 1830-1836 (2010)
5. A.D. Kiselev, D.O. Plutenko, Nanosystems: Physics, Chemistry, Mathematics, **7**(2), 349-370 (2016)
6. P. Exner, S. Vugalter, Ann. Inst. Henri Poincare, **65**, 109–123 (1996)
7. Langer, M., Strauss, M., 2016, Triple variational principles for self-adjoint operator functions, J. Func. Anal., Vol. 270, pp. 2019-2047
8. Linton, C. M., McIver, P., 2007, Embedded trapped modes in water waves and acoustics, Wave Motion, Vol. 45, pp. 16-29.
9. Dauge, M., Raymond, N., 2012, Plane waveguides with corners in the small angle limit, J.Math. Phys., Vol. 53, p. 123529.
10. Duclos, P., Exner, P., 1995, Curvature-induced bound states in quantum waveguides in two and three dimensions, Rev. Math. Phys., Vol. 7, pp. 73-102.

DOI: 10.1051/itmconf/20170901009

Table 1. Parameters of the trial function which give negative values of the ratio

№	α	η	ν	q	λ_1	λ_2	λ_4	γ	a	$\delta(a)$
1	0.001	1000	2	1	50	10	26	0.1	0.01	$a^{0.5}$
2	0.001	1000	2	1	50	10	46	0.1	0.01	$a^{0.1}$
3	0.01	1000	2	1	100	1	82	1	0.001	$a^{0.01}$
4	0.001	1000	2	1	10	1	130	0.1	0.001	$a^{0.01}$
5	0.5	500	0	1	100	1	202	0.1	0.001	$a^{0.01}$
6	0.01	100	0	10	100	1	162	0.1	0.001	$a^{0.01}$

ITM Web of Conferences 9, 01010 (2017) DOI: 10.1051/itmconf/20170901010
AMCSE 2016

Modeling of helix molecules formation on a surface of nanotube and inside it

Anna Belolipetskaya[1], *Nikita Lisitsa*[1], *Igor Popov*[1], *Maria Faleeva*[1*]

[1]ITMO University, Department of Higher Mathematics, 197101 St. Petersburg, Russia

Abstract. "On-surface" synthesis of large molecules and assemblies is a great challenge in molecular electronics, biomedical devices, sensors, energy harnessing and catalysis. Usually, plane surfaces are used for these purposes. We study a possibility of macro molecules formation on the surface of nanotube and inside it. It can be used for synthesis of helix molecules, particularly, organic and biological. Nanotubes of different structures were considered. Point-like approximation is used for atoms. The Lennard-Jones 6-12 potential is taken as the interaction potential. Possibilities of helix molecules and ring-like molecules formations were shown. A hypothesis of such processes influence on the organic molecules formation during the early Earth history is suggested and discussed.

1 Introduction

In the fields of systems biology and systems chemistry, attaining compositional and topological diversity of macromolecules is fundamental for realizing complex functions [1]. "On-surface" synthesis with prefunctionalized molecules is a major research challenge in molecular electronics, biomedical devices, sensors, energy harnessing and catalysis [2-4]. Recently, it has led to specific one- and two-component products with remarkable efficiency [5-7].
There are four strategies now followed in the synthesis of large molecules and assemblies [8]: (i) controlled formation of covalent bonds, (ii) covalent polymerization, (iii) self-organization, (iv) molecular self-assembly. The third synthetic strategy abandons the covalent bond as a required connection between atoms and relies instead on weaker bonds, such as ionic bonds, hydrogen bonds, van der Waals interaction, to organize atoms, ions, or molecules into structures. Such method is used, e.g., for obtaining molecular crystals [9] or self-assembled monolayers [10]. We will discuss an approach close to the last two methods. Molecular self-assembly is the spontaneous association of molecules under equilibrium conditions into stable, structurally well-defined aggregates joined by noncovalent bonds A promising route to construction of functional systems with nanometre dimensions is the autonomous ordering and assembly of atoms and molecules on atomically well-defined surfaces [11]. Site-specific on-surface synthesis and on-surface formation of nanostructures were studied in many works (see, e.g., [12,13] and references therein).
In the present paper we investigate a possibility of helix structure (or long helix molecule) formation on the surface of nanotube and inside the nanotube. Nanotubes of different structures are considered. The most simple model for nanotube is based on zero-range potentials [14-16]. The mathematical background of the model is given by the theory of self-adjoint extensions of symmetric operators. We use more realistic model : a point-like approximation for atoms and the Lennard-Jones 6-12 interaction potential:

$$V(r) = 4\varepsilon\left(\left(\frac{\sigma}{r}\right)^{12} - \left(\frac{\sigma}{r}\right)^{6}\right).$$

Here ε is the depth of potential well, σ is the distance for which the potential becomes zero, r is the distance from the center.

2 Model construction

Different models are used for the description of nanotubes (see, e.g., [7] and references therein). We use the model of point–like potentials similar to that in [8] but with another interaction potential. We compute an energy surface which allows us to find local minima corresponding to preferable position of an atom disposed on the nanotube surface or inside it. The formula for the energy surface contains infinite series over all point-like potentials. This sum can be estimated by using of asymptotic approaches. Minima were found by numerical calculations.
The standard Lennard-Jones 6-12 potential with parameters σ=2.28 and ε=51.2 [1] without a cutoff distance was used to approximate interactions between a nanotube and a foreign particle. The potential isosurfaces were generated using the marching tetrahedra algorithm [2] on a tetrahedral grid, formed by tesellating each cell of a 3-dimensional cubic grid into six tetrahedra, with

* Corresponding author: popov1955@gmail.com

DOI: 10.1051/itmconf/20170901010

the cell's diagonal being a common edge. The potential's values on the grid vertices were precomputed and cach ed to speedup generation of multiple isosurfaces on the same data.

The most interesting pictures are observed when one considers the internal part of the nanotube. Figures show the position of energy minima as a dark area. The nanotube diameter and the structure of the nanotube (it is shown in the Figures) vary. A helix structure of the potential minimum set is observed in many cases. if one puts an atom in the nanotube, It will follow this helix way. Correspondingly, it is a place which is preferable for molecules formation. Thus, the nanotube plays a role of specific catalyst, ensuring the creation of helix-like molecular structure.

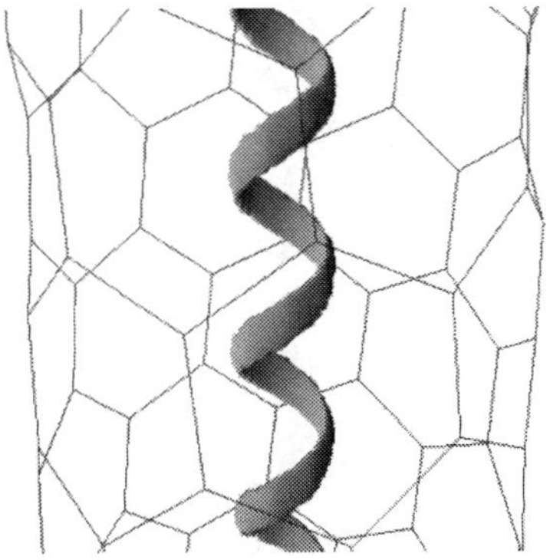

Fig. 1. Single helix structure of the potential minimum set..

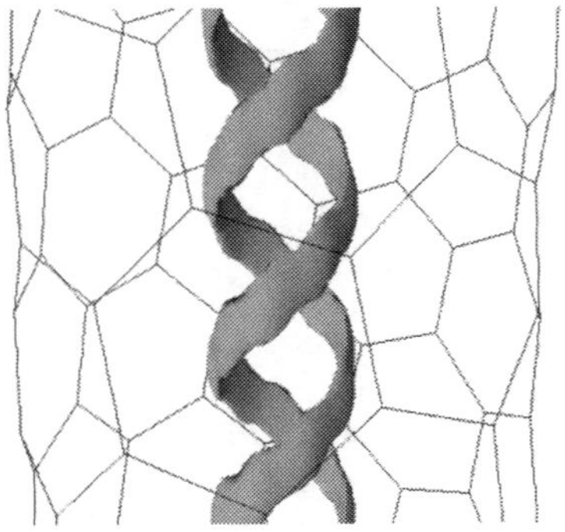

Fig. 2. Double helix structure of the potential minimum set.

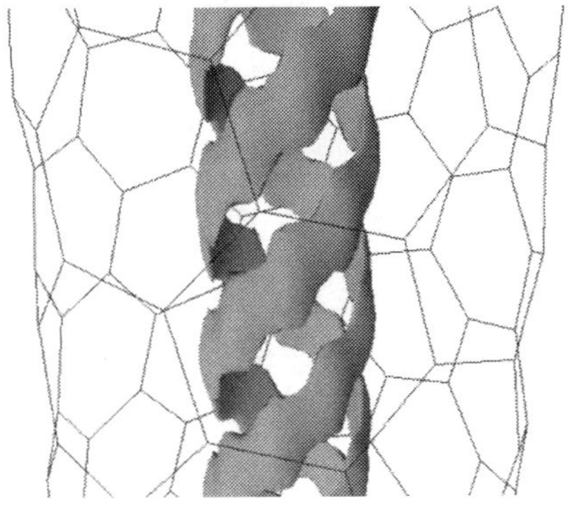

Fig. 3. Triple helix structure of the potential minimum set.

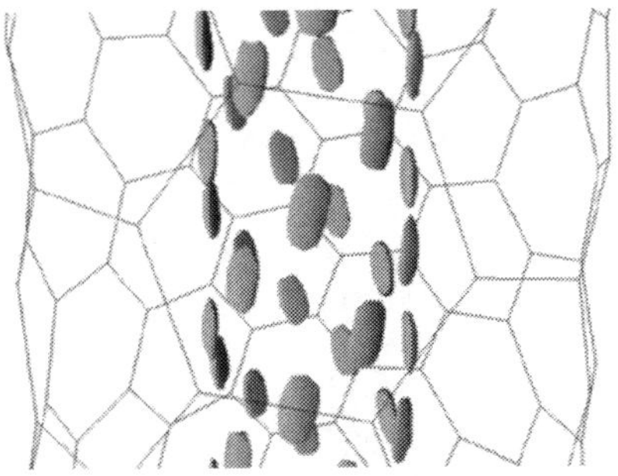

Fig. 4. Single helix structure of the potential minimum set.

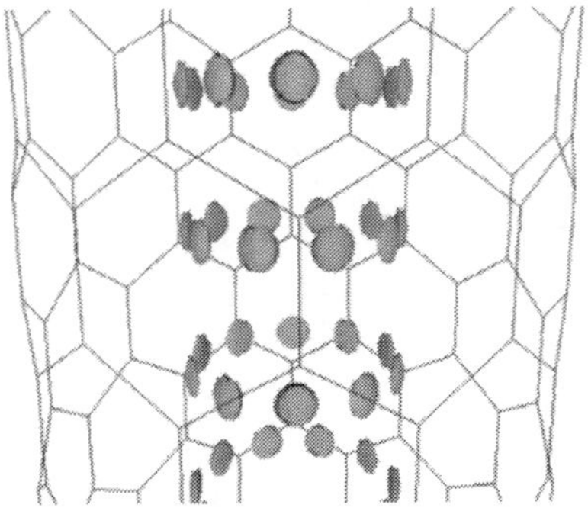

Fig. 5. Ring-like structure of the potential minimum set

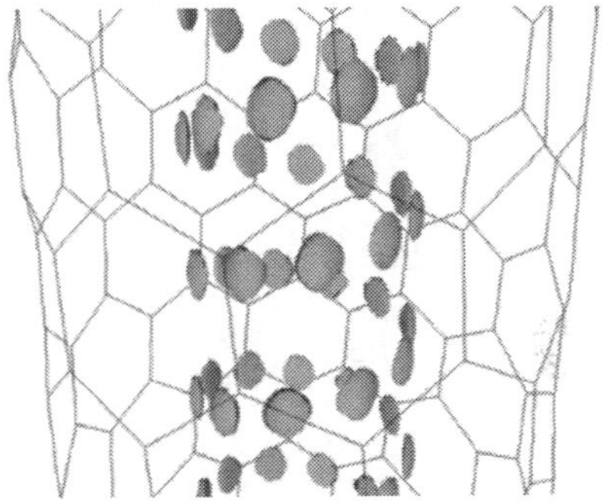

Fig. 6. Helix structure of the minima set.

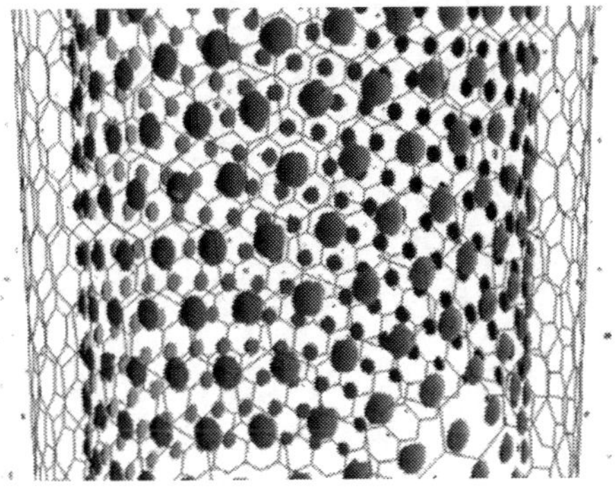

Fig. 7. . Helix structure of the minima set for large nanotube radius.

For the internal nanotube domain, wall confinement is an additional stabilizing factor stimulating macromolecules formation. As for the preferable positions for particles (atoms) at the external surface of the nanotube, one observes an analogous situation, but energy minima are not so deep as in the interior domain. The pictures are shown in the following figures (for various nanotube types).

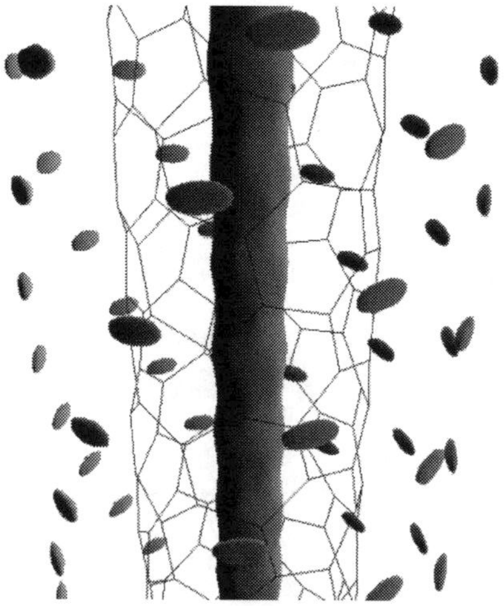

Fig. 8. . Helix structure of the external minima set for (6,2) nanotube.

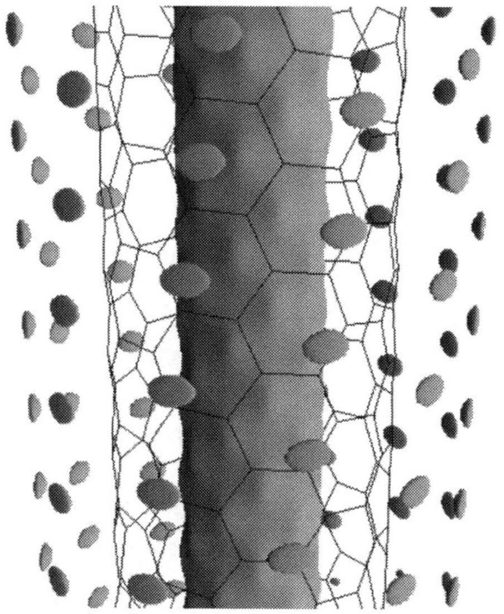

Fig. 9. . Helix structure of the external minima set for (7,5) nanotube.

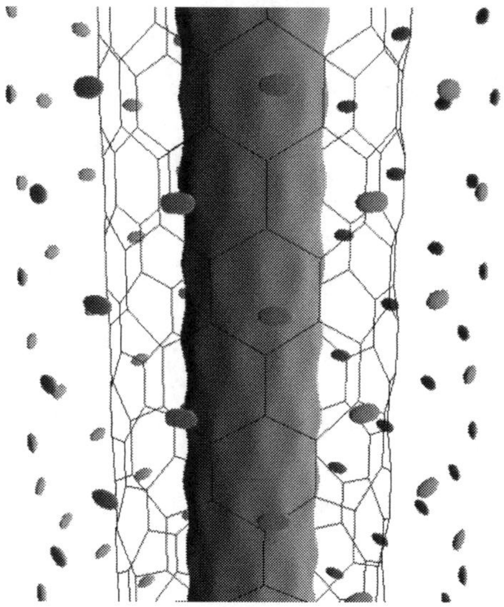

Fig. 10. . Helix (ring) structure of the external minima set for (10,0) nanotube.

4. Concluding remarks

The obtained results show that one can use nanotubes for the process of helix structure macromolecules synthesis. In nanoelectronics, such molecules are useful for nano devices based on using of a magnetic field. Obviously, helix-like and ring-like molecules are interesting for pharmacy and bio engineering. We can also put a hypothesis concerning to the problem of life origin on the Earth, more precisely, to the process of organic molecules formation on the Earth. Namely, the process described above, can lead to formation (inside the nanotube or on its external surface) of a helix-like and ring-like molecules having biological nature. As for the nanotubes playing the role of a background, they appeared on the Earth essentially earlier than organic (biological) molecules. Correspondingly, such nanotubes can play a catalytic role for the organic molecules formation processes.

This work was partially financially supported by the Government of the Russian Federation (grant 074-U01), by the Ministry of Education and Science of the Russian Federation (GOSZADANIE 2014/190, Projects No 14.Z50.31.0031 and No. 1.754.2014/K)), by grants MK-5001.2015.1 and MK-5161.2016.1 of the President of the Russian Federation, by grant 16-11-10330 of Russian Science Foundation.

References

1. R.F.. Ludlow, S. Otto. Chem. Soc. Rev **37**, 101–108 (2008)
2. E. C. P.Smits, S. G. J. Mathijssen, , P. A. van Hal, S. Setayesh, T. C. T Geuns,.; K. A. H. A. Mutsaers, E. Cantatore,.; H. J Wondergem.; O.Werzer, R Resel, Nature, **455**, 956–959 (2008)

DOI: 10.1051/itmconf/20170901010

3. T.Aida, E.W. Meijer, S.I. Stupp. . Science, **335**, 813–817 (2012)
4. M.Koepf, F. Cherioux, J.A.Wytko, J. Weiss. Coord. Chem. Rev. **256**, 2872–2892 (2012)
5. L.Bartels. Nat. Chem. **2**, 87–95 (2010)
6. J. Greenwood, H.A. Fruechtl, C.J. Baddeley. J. Phys. Chem. C, **117**, 4515–4520 (2013)
7. Lackinger, M.; Heckl, W. M. J. Phys. D: Appl. Phys., **44**, 464011 (2011).
8. G.M.Wintersides, J.P.Mathias, C.T.Seto. Science, **254**, 1312-1319, (1991)
9. J.D.Wright. Molecular crystals. {Cambriudge Univ. Press, Cambridge, 1987)
10. G.M.Wintersides, et.al. *J. Am. Chem. Soc.* **111**, 321 (1989)
11. J. Barth, K Constantini, K.Kernth. Nature, **437**, 671-67 (2005)
12. C.J.Villagomez, T. Sasaki, J.M. Tour, L. Grill. *J. Am. Chem. Soc.* **111**, 321 (1989)
13. S. Schlogl, V.M.Heckl, M.Lackinger. Surf. Sci. **606**, 999-1004 (2012).
14. S. Albeverio and P. Kurasov. *Singular perturbations of differential operators. Solvable Schrödinger type operators* (London Mathematical Society, London, 2005
15. I.S. Lobanov, I.Yu. Popov, Nanosystems: Physics, Chemistry, Mathematics, **3**(2), 6-28 (2012)
16. E.N.Grishanov, I.Y. Popov, Nanosystems: Physics, Chemistry, Mathematics, **7**(5), 880-883 (2016)

ITM Web of Conferences 9, 01011 (2017) DOI: 10.1051/itmconf/20170901011
AMCSE 2016

ANALYTICAL AND NUMERICAL SOLUTION OF CONJUGATE HEAT AND MASS TRANSFER IN FALLING FILM ABSORPTION PROCESS

Vladimir E. Nakoryakov, and *Maria V.* Bartashevich*

Kutateladze Institute of Thermophysics SB RAS, 1, Acad. Lavrentiev Avenue, 630090 Novosibirsk, Russia

Abstract. Investigation of heat and mass transfer at nonisothermal absorption on the falling liquid film surrounded by pure vapour is presented. The analytical method for entrance region and numerical method for solving the problem are proposed. The analytical solution is obtained for the entry region, where temperature and concentration perturbations are near the interface. Then, the analytical solution is used as an inlet condition for further numerical solution by the finite-difference method. The key dimensionless criteria characterizing heat and mass transfer at absorption on the falling liquid film are defined.

1 Introduction

Thin liquid films are used in various technological processes more often to intensify the heat and mass transfer processes. The numerous theoretical and experimental studies of heat and mass transfer in liquid films are explained by the applied significance of film flows. The classical problem of heat transfer in a smooth laminar film flowing down the wall with a fixed temperature was firstly described in the work of Nusselt [1]. Graetz was the first, who considered a similar problem of liquid flow heating in a cylindrical pipe with isothermal wall [2]. One of the important applications of non-isothermal film flow is the absorption thermotransformers. So, the absorption heat pump is a unique device consisted of the desorber, condenser, evaporator, and absorber. In such devices, vapor absorption in liquid goes along with heat release, so the heat and mass transfer processes are conjugated. The first theoretical description of the processes of simultaneous heat and mass transfer at vapor absorption in a laminar falling liquid film was made in [3-6] with some simplifications of the problem formulation. The authors of [7] investigated the process of vapor absorption by the liquid film for two limiting cases: constant wall temperature and thermally insulated wall. Equations of heat transfer and diffusion were solved jointly throughout the entire flow region by numerical method and analytical method (for the thermal entry region). The profiles of temperature and concentration were presented in the form of power series of the transverse coordinate with coefficients dependent on the longitudinal coordinate. In subsequent theoretical studies [8-19] the process of simultaneous heat and mass transfer at gas absorption by the liquid film was also examined under various simplifying assumptions. In [8] the problem of the vapour absorption into a falling film, where the concentration levels of absorbate and absorbent are comparable was studied. The conjugated heat and mass transfer processes were analysed. The adiabatic and isothermal wall conditions were considered. In [11] the falling film absorption in aqueous LiBr solution was investigated numerically using package Fluent. The convective boundary condition at the cooling water side was considered. In [14] the comparison of absorption and desorption has been conducted using a detailed model describing heat and mass transfer. In [15] the conjugate heat and mass transfer process taking place during absorption of a refrigerant vapor into a falling liquid film has been analyzed using the Laplace transform method. In [16] the numerical and experimental investigations of a vertical LiBr falling film absorber considering wave regimes have been performed. In [17] a model was developed for analysis of the combined heat and mass transfer processes for laminar liquid film evaporation. The model was developed in cylindrical coordinate. The energy and diffusion equations are solved using the Laplace transform. The comparison between theoretical calculation and experimental data has been carried out.

This work offers the conjugation of the analytical method for solving the problem of heat and mass transfer in nonisothermal absorption on the falling liquid film surrounded by pure vapour at the entrance region, with numerical method, using the analytical solution as an inlet condition for further numerical calculations.

2 Problem statement

It is assumed that the laminar liquid film with constant thickness h flows under the action of gravity on the wall, inclined at angle θ to the horizon, with uniform velocity u. In the case of falling film absorption, when the vapour

* Corresponding author: bartashevichmv@gmail.com

 DOI: 10.1051/itmconf/20170901011

or gas phase is stationary, coupled heat and mass transfer is considered under the assumptions [18, 19], which are used in absorption heat pumps and refrigerators. Introducing the Cartesian coordinate system with axis *Ox*, directed to the film flow, and axis *Oy*, directed from the interface to the film depth. At $x = 0$ we set inlet conditions T_0, C_0. The wall temperature is also equal T_0. Liquid density ρ, dynamic viscosity μ, heat conductivity λ, thermal diffusivity a and diffusivity D are considered constant. The uniform velocity profile in the liquid film takes the form:

$$u = \frac{gh^2 \sin\theta}{3\nu}.$$

The processes of heat and mass transfer in the solution film are described by the system of heat conduction and diffusion equations:

$$u\frac{\partial T}{\partial x} = a\frac{\partial^2 T}{\partial y^2}, \quad (1)$$

$$u\frac{\partial C}{\partial x} = D\frac{\partial^2 C}{\partial y^2}, \quad (2)$$

with corresponding conditions at the inlet

$$x=0 \quad T = T_0,\ C = C_0, \quad (3)$$

with conjugate conditions at the interphase boundary

$$y=0 \quad C_i = k_1 - k_2 \cdot T_i,\ \lambda\frac{\partial T}{\partial y} = \frac{r_a \rho D}{1-C_0}\frac{\partial C}{\partial y}, \quad (4)$$

and with the conditions on the wall

$$y = h \quad T = T_0,\ \frac{\partial C}{\partial y} = 0. \quad (5)$$

We introduce dimensionless variables $\bar{x} = \frac{x}{\text{Pe}h}$, $\text{Pe} = \frac{uh}{a}$, $\bar{y} = \frac{y}{h}$, $\theta = \frac{(T-T_0)}{(T_e-T_0)}$, $\gamma = \frac{(C-C_0)}{(C_e-C_0)}$. Here $C_e = k_1 - k_2 T_0$ is equilibrium concentration corresponding to the given initial temperature of solution T_0; T_e is the equilibrium temperature corresponding to the given initial concentration of solution C_0 ($C_0 = k_1 - k_2 T_e$). The the coefficients k_1 and k_2 are determined by the pressure. In dimensionless variables, Eq.(1), Eq.(2) and the boundary conditions (3-5) take the form:

$$\frac{\partial\theta}{\partial\bar{x}} = \frac{\partial^2\theta}{\partial\bar{y}^2}, \quad (1')$$

$$\frac{\partial\gamma}{\partial\bar{x}} = \text{Le}\frac{\partial^2\gamma}{\partial\bar{y}^2}, \quad (2')$$

$$\bar{x}=0 \quad \theta = 0,\ \gamma = 0, \quad (3')$$

$$\bar{y}=0 \quad \theta_i + \gamma_i = 1,\ \frac{\partial\theta}{\partial\bar{y}} = \text{KaLe}\frac{\partial\gamma}{\partial\bar{y}} \quad (4')$$

$$\bar{y} = 1 \quad \theta = 0,\ \frac{\partial\gamma}{\partial\bar{y}} = 0. \quad (5')$$

where $\text{Ka} = \frac{r_a}{C_p(T_e-T_0)}\frac{C_e-C_0}{1-C_0}$, $\text{Le} = \frac{D}{a}$.

The parameters for calculations were chosen close to the characteristic values of the two-phase system: water solution of lithium bromide – water vapor, under the assumptions corresponding to the conditions of working absorber of a heat pump.

3 Analytical solution for entrance region

By using self-similar variable $\xi = \frac{\bar{y}}{\bar{x}^{1/2}}$ the ordinary differential equations can be obtained

$$\frac{d^2\theta}{d\xi^2} + \frac{\xi}{2}\frac{d\theta}{d\xi} = 0, \quad (6)$$

$$\frac{d^2\gamma}{d\xi^2} + \text{Le}^{-1}\frac{\xi}{2}\frac{d\gamma}{d\xi} = 0, \quad (7)$$

with boundary conditions

$$\xi=0 \quad \theta_i + \gamma_i = 1,\ \frac{\partial\theta}{\partial\xi} = \text{KaLe}\frac{\partial\gamma}{\partial\xi}, \quad (8)$$

$$\xi \to \infty \quad \theta = 0,\ \gamma = 0. \quad (9)$$

The self-similar solutions of Eq. (6), (7) with boundary conditions (8) and (9) take the form:

$$\theta = \theta_i\left(1 - erf(\xi/2)\right), \quad (10)$$

$$\gamma = \gamma_i\left(1 - erf\left(\text{Le}^{-1/2}\xi/2\right)\right), \quad (11)$$

where $\theta_i = \frac{\text{KaLe}^{1/2}}{\text{KaLe}^{1/2}+1}, \gamma_i = \frac{1}{\text{KaLe}^{1/2}+1}$. Note that the formulas for temperature and concentration in the entrance region of the free surface of the film θ_i, γ_i agree with the formulas obtained for film absorption at other velocity profiles [18,19].

4 Numerical solution

The numerical calculations by finite-difference method are performed using the self-similar temperature and concentration profiles (10), (11) at the inlet (see Fig.1) at $x = 0.01$. Further y^1 is measured from the wall and varied from 0 to h ($y^1 = 1 - \bar{y}$). Equations (1') and (2') in the difference form with approximation of order $O\left(\Delta\overline{x}, \Delta y^{1^2}\right)$ are as follows:

$$\frac{\theta_{i,j}-\theta_{i-1,j}}{\Delta\overline{x}}=\frac{\theta_{i,j+1}-2\theta_{i,j}+\theta_{i,j-1}}{\Delta(y^1)^2}, \quad (12)$$

$$\frac{\gamma_{i,j}-\gamma_{i-1,j}}{\Delta\overline{x}}=\frac{\gamma_{i,j+1}-2\gamma_{i,j}+\gamma_{i,j-1}}{\Delta(y^1)^2}. \quad (13)$$

Here $\theta_{i,j}$ is the temperature value in node $(\overline{x}_i, y^1{}_j)$. The values of the grid step $\Delta\overline{x}=10^{-3}$ and $\Delta y^1=2.\cdot 10^{-3}$ are rather small, so the results of calculations practically do not change with a further decrease in the grid step. For everyone x_i Eq. (12), (13) have been solved by a sweep method on coordinate y^l using boundary conditions (3'-5'). The interfacial values of the temperature and concentration have been found from the conjugated calculations of the sweep coefficients.

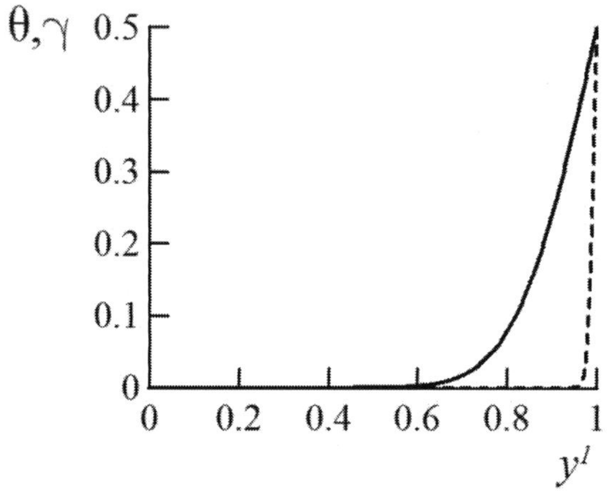

Fig. 1. The self-similar temperature and concentration profiles (10), (11) at the inlet at $x = 0.01$, Ka = 10, Le = 0.01. Solid line – initial temperature profile, dashed line – initial concentration profile.

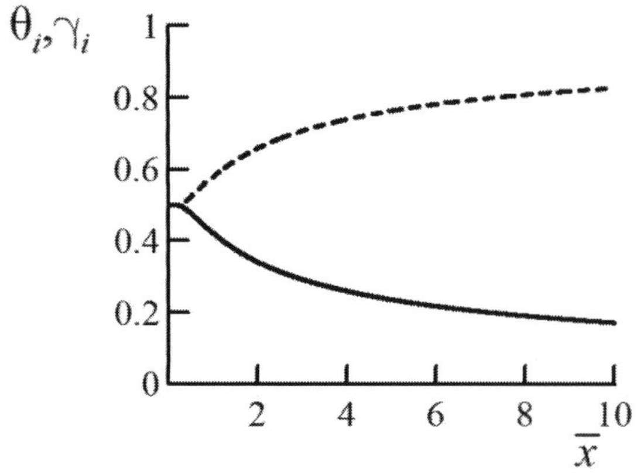

Fig. 2. Film surface temperature and concentration depending on the distance in the case Ka = 10, Le = 0.01. Solid line – temperature profile, dashed line –concentration profile.

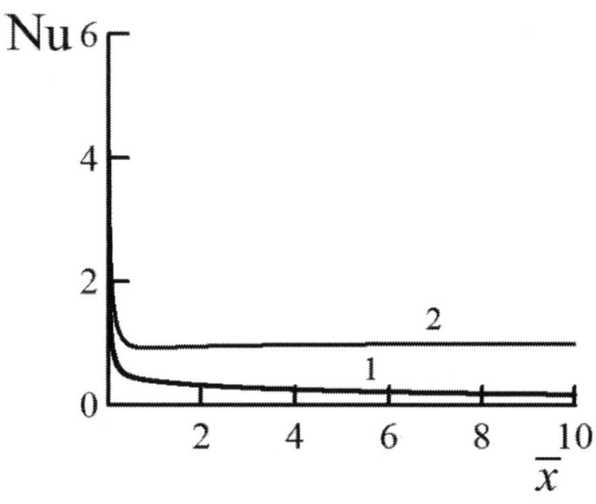

Fig. 3. Nusselt numbers Nu_1 and Nu_2 along the film-flow direction. 1 – Nu_1, 2 – Nu_2.

Fig. 2 shows the interfacial temperature and concentration profiles depending on the distance. The Nusselt numbers $Nu_1=\frac{h}{\Delta T_1}\frac{\partial T}{\partial y}\Big|_{y=0}$, $Nu_2=\frac{h}{\Delta T_2}\frac{\partial T}{\partial y}\Big|_{y=0}$ are shown in Fig.3 depending on the longitudinal coordinate. Here $\Delta T_1 = T_e - T_0$, $\Delta T_2 = T_i - T_0$.

5 Conclusion

The following conclusions can be drawn from the above study. The self-similar solutions for temperature and concentration appropriated only for the entrance region have been obtained. Although, the self-similar solutions have a limited applicability, these solutions are simple and give a clear interrelation between physical quantities. Another important aspect of the present study is using the self-similar solutions as the initial conditions for solving the problem beyond the entrance region. So, in the following numerical calculations we can vary the thermal regime conditions. Using these self-similar solutions allow us to avoid the difficulties associated with the inconsistency of the boundary conditions. Also, the key criteria characterizing heat and mass transfer in the film absorption with uniform velocity profile and with a constant thickness (Ka – the absorption analog of phase transition criterion, Lewis, Peclet, and Nusselt complexes) have been determined in the present study. The problem was solved for lithium bromide solutions with Lewis number smaller than unity, Le = *D*/*a*. It was shown that with increasing distance, the value of Nusselt number significantly decreases to the equilibrium value for the isothermal wall.

The work was financially supported by the grant MK-6384.2016.8.

References

1. W. Nusselt, Zeitschrift des Vereines Deutscher Ingenieure, **60** (1916).
2. L. Graetz, Annalen der Physik, **25** (1885).
3. S.H. Chiang, H.L. Toor, AIChE J., **10** (1964).
4. R. B. Bird, W. E. Stewart, E. N. Lightfoot, *Transport Phenomena* (John Wiley & Sons, 2002).
5. V. E. Nakoryakov, N. I. Grigorjeva, Theor. Found. Chem. Eng., **14** (1980).
6. N. I. Grigorieva, V. E. Nakoryakov, J. Engineering Physics and Thermophysics, **33** (1977).
7. G. Grossman, Int. J. Heat Mass Transfer, **26** (1983).
8. N. Brauner, Int. J. Heat Mass Transfer, **34** (1991).
9. A. T. Conlisk, Chem. Eng. Sci., **50** (1995).
10. W. Chen, R. N. Christensen, Int. J. Heat and Mass Transfer, **43** (2000).
11. S. Bo, X. Ma, Zh. Lan, J. Chen, H. Chen, Chem. Eng. Journal, **156** (2010)
12. T. Meyer, F. Ziegler, Int. J. of Heat and Mass Transfer, **73** (2014).
13. M. Mittermaier, P. Schulze, F. Ziegler, Int. J. Heat and Mass Transfer, **70** (2014).
14. M.Mittermaier, F.Ziegler, Int. J. Refrig., **59** (2015).
15. Mortazavi M., Moghaddam S., Int. J.Refrig., **66** (2016).
16. E. Garcia-Rivera, J. Castro, J. Farnos, A. Oliva, Int. J. Thermal Sciences, **109** (2016).
17. Y. Wu, Int. J. Heat and Mass Transfer, **93** (2016).
18. V. E. Nakoryakov, N. I. Grigoryeva, *Non-isotermal absorption in thermotransformers* (Nauka, Novosibirsk, 2010).
19. Nakoryakov V.E., Grigoryeva N.I., Bartashevich M.V., Int. J. Heat and Mass Transfer, **54** (2011).

ITM Web of Conferences **9**, 01012 (2017) DOI: 10.1051/itmconf/20170901012
AMCSE 2016

A ternary relation for structuring the digital plane

Josef Šlapal[1,*]

[1] *IT4Innovations Centre of Excellence, Brno University of Technology, 616 69 Brno, Czech Republic*

Abstract. We discuss certain ternary relations, called plain, and show that each of them induces a connectedness on its underlying set. This connectedness allows for definitions of concepts of simple closed and Jordan curves. We introduce a particular plain ternary relation on the digital plane $\mathbb{Z}^2$ and, as the main result, we prove a digital analogue of the Jordan curve theorem for the connectedness induced by this relation. It follows that the ternary relation introduced may be used as a convenient structure on the digital plane for the study of the geometric properties of digital images that are related to boundaries because boundaries of objects in digital images are represented by digital Jordan curves. An advantage of this structure over the Khalimsky topology is that it allows Jordan curves to turn at the acute angle $\frac{\pi}{4}$ at some points.

1 Introduction

Digital topology is a theory that was founded for the study of geometric and topological properties of digital images. The classical, graph theoretic approach to digital topology is based on using the 4-adjacency and 8-adjacency graphs for structuring $\mathbb{Z}^2$ (cf. [6] and [7]). Unfortunately, neither 4-adjacency nor 8-adjacency itself allows for an analogue of the Jordan curve theorem (cf. [4]) so that a combination of the two adjacency graphs has to be used. To eliminate this deficiency, a new, purely topological approach to the problem was proposed in [2] which utilizes a convenient topology for structuring the digital plane, namely the Khalimsky topology. The convenience of the Khalimsky topology for structuring the digital plane was shown in [2] by proving an analogue of the Jordan curve theorem for the topology (recall that the classical Jordan curve theorem states that a simple closed curve in the Euclidean plane separates this plane into exactly two connected components). The topological approach was then developed by many authors - see, e.g., [3]-[5] and [9].

Since the Khalimsky topological space is an Alexandroff T_0-space, it corresponds to a partial order on $\mathbb{Z}^2$, the so-called specialization order. The connectedness in the Khalimsky space then coincides with the connectedness in the underlying (simple) graph of the specialization order. Thus, when studying the connectedness of digital images, this graph, rather than the Khalimsky topology itself, may be used for structuring the digital plane. A disadvantage of this approach is that Jordan curves in the (specialization order of the) Khalimsky topology may never turn at the acute angle $\frac{\pi}{4}$. It would, therefore, be useful to find some new, more convenient structures on $\mathbb{Z}^2$ that would allow Jordan curves to turn, at some points, to form the acute angle $\frac{\pi}{4}$. In the present note, to obtain such a convenient structure, we replace the specialization order of the Khalimsky topology, hence a binary relation on $\mathbb{Z}^2$, with a ternary relation on $\mathbb{Z}^2$. We will define a connectedness provided by this ternary relation and will prove a digital Jordan curve theorem for this connectedness thus showing that the ternary relation provides a convenient structure on the digital plane for the study of digital images.

*e-mail: slapal@fme.vutbr.cz

2 Preliminaries

For every point $(x, y) \in \mathbb{Z}^2$, we denote by $A_4(x, y)$ and $A_8(x, y)$ the sets of all points that are 4-adjacent and 8-adjacent to (x, y), respectively. Thus, $A_4(x, y) = \{(x + i, y + j);\ i, j \in \{-1, 0, 1\},\ ij = 0,\ i + j \neq 0\}$ and $A_8(x, y) = A_4(x, y) \cup \{(x + i, y + j);\ i, j \in \{-1, 1\}\}$. The simple graphs $(\mathbb{Z}^2, A_4)$ and $(\mathbb{Z}^2, A_8)$ are called the *4-adjacency graph* and *8-adjacency graph*, respectively (for the basic graph-theoretic concepts used see [1]).

In digital image processing, the 4-adjacency and 8-adjacency graphs are the most frequently used structures on the digital plane. But, since the late 1980's, another structure on $\mathbb{Z}^2$ has been used too, namely the Khalimsky topology [2]. The specialization order of the Khalimsky topology (see Introduction) is the binary relation $\leq$ on $\mathbb{Z}^2$ given as follows:

For any $(x, y), (z, t) \in \mathbb{Z}^2$, $(x, y) \leq (z, t)$ if and only if

- $(x, y) = (z, t)$ or
- x, y are even and $(z, t) \in A_8(x, y)$ or
- x is even, y is odd, $z = x + i$ where $i \in \{-1, 1\}$, and $t = y$ or
- x is odd, y is even, $z = x$, and $t = y + i$ where $i \in \{-1, 1\}$.

A portion of the specialization order $\leq$ of the Khalimsky topology is demonstrated in Figure 1 by a directed graph with the vertex set $\mathbb{Z}^2$ where an oriented edge from a point p to a point q means that $p \leq q$.

 DOI: 10.1051/itmconf/20170901012

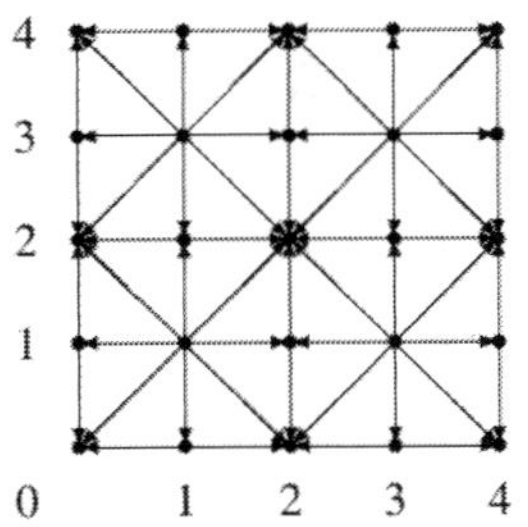

Figure 1. A portion of the specialization order of the Khalimsky topology.

Recall that, given a directed graph (i.e., a set with a binary relation), its *underlying* (undirected) simple graph is obtained by just ignoring the direction of the edges. A circle in a simple graph is said to be a simple closed curve if, with each of its vertices, it contains precisely two vertices adjacent to it. A simple closed curve J in a simple graph with the vertex set V is called a *Jordan curve* if it separates the set V into precisely two components, i.e., if the induced subgraph $V - J$ has exactly two components.

The famous Jordan curve theorem proved for the Khalimsky topology in [2] may be formulated as follows:

Theorem 1 *In the underlying graph of the specialization order of the Khalimsky topology, every simple closed curve with at least four points is a Jordan curve.*

It is readily verified that a simple closed curve (and thus also a Jordan curve) in the underlying graph of the specialization order of the Khalimsky topology may never turn at the acute angle $\frac{\pi}{4}$. It could therefore be useful to replace the specialization order of the Khalimsky topology with some more convenient structure (relation on $\mathbb{Z}^2$) that would allow Jordan curves to turn at the acute angle $\frac{\pi}{4}$ at some points. And this is what we will do in the next section.

3 Plain ternary relations and induced connectedness

Recall that, given a positive in integer n and a set X, an *n-ary relation* on X is a subset $R \subseteq X^n$. Thus, the elements of R are finite sequences (ordered n-touples) $(x_0, x_1, ..., x_n) = (x_i | i < n)$ consisting of elements of X (for the basic properties of n-ary relations see [8]). In the sequel, we will restrict our considerations to $n = 3$, i.e., to ternary relations.

Definition 1 A ternary relation R on a set X is said to be *plain* if, for any $f, g \in R$, $f \neq g$ implies card$(f \cap g) \leq 1$.

Definition 2 Let R be a plain ternary relation on a set X and n a nonnegative integer. A sequence $C = (c_i | i \leq n)$ of elements of X is called an *R-walk* if the following two conditions are satisfied:

I. For every positive integer $i < n$, there exists $(x_0, x_1, x_2) \in R$ such that

$$\{c_i, c_{i+1}\} = \{x_0, x_1\} \text{ or } \{c_i, c_{i+1}\} = \{x_1, x_2\},$$

II. Every $(a_0, a_1, a_2) \in R$ satisfies the following two conditions:

(i) if there exists $i \in \{0, 1, ..., n-1\}$ such that $c_i = a_1$ and $c_{i+1} = a_2$, then $i > 0$ and $c_{i-1} = a_0$,

(ii) f there exists $i \in \{1, 2, ..., n\}$ such that $c_{i-1} = a_2$ and $c_i = a_1$, then $i < n$ and $c_{i+1} = a_0$.

An R-walk $(c_i | i \leq n)$ with the property that $n \geq 2$ and $c_i = c_j \Leftrightarrow \{i, j\} = \{0, n\}$ is said to be an *R-circle*.

Observe that, if $(x_0, x_1, ..., x_n)$ is an R-walk, then $(x_n, x_{n-1}, ..., x_0)$ is an R-walk, too (so that R-walks are closed under reversion) and, if $(x_i | i \leq m)$ and $(y_i | i \leq p)$ are R-walks with $x_m = y_0$, then, putting $z_i = x_i$ for all $i \leq m$ and $z_i = y_{i-m}$ for all i with $m \leq i \leq m + p$, we get an R-walk $(z_i | i \leq m + p)$ (so that R-walks are closed under composition).

Given a plain ternary relation R on a set X, a subset $Y \subseteq X$ is said to be *$\mathcal{R}$-connected* if, for every pair $a, b \in Y$, there is an $\mathcal{R}$-walk $(c_i | i \leq n)$ such that $c_0 = a$, $c_n = b$ and $c_i \in Y$ for all $i \in \{0, 1, ..., n\}$. A maximal (with respect to set inclusion) R-connected subset of X is called an *R-component* of X.

Definition 3 Let R be a plain ternary relation on a set X. A nonempty, finite and R-connected subset J of X is said to be an *R-simple closed curve* if every element $(a_0, a_1, a_2) \in R$ with $\{a_0, a_1\} \subseteq J$ satisfies $a_2 \in J$ and every $z \in J$ fulfills one of the following two conditions:

(1) There are exactly two elements $(a_0, a_1, a_2) \in R$ satisfying both $\{a_0, a_1, a_2\} \subseteq J$ and $z \in \{a_0, a_2\}$ and there is no element $(b_0, b_1, b_2) \in R$ satisfying both $\{b_0, b_1, b_2\} \subseteq J$ and $z = b_1$.

(2) There is exactly one element $(b_0, b_1, b_2) \in R$ satisfying both $\{b_0, b_1, b_2\} \subseteq J$ and $z = b_1$ and there is no element $(a_0, a_1, a_2) \in R$ satisfying both $\{a_0, a_1, a_2\} \subseteq J$ and $z \in \{a_0, a_2\}$.

Clearly, every R-simple closed curve is an R-circle.

Definition 4 Let R be a plain ternary relation on a set X. An R-simple closed curve J is called an *R-Jordan curve* if the subset $X - J \subseteq X$ consists (i.e., is the union) of precisely two R-components.

From now on, R will denote the plain ternary relation on $\mathbb{Z}^2$ given as follows: For every $((x_i, y_i) | i < 3)$ such that $(x_i, y_i) \in \mathbb{Z}^2$ for every $i < 3$, $((x_i, y_i) | i < 3) \in R$ if and only if one of the following eight conditions is satisfied:

(1) $x_0 = x_1 = x_2$ and there is $k \in \mathbb{Z}$ such that $y_i = 4k + i$ for all $i < 3$,

(2) $x_0 = x_1 = x_2$ and there is $k \in \mathbb{Z}$ such that $y_i = 4k - i$ for all $i < 3$,

DOI: 10.1051/itmconf/20170901012

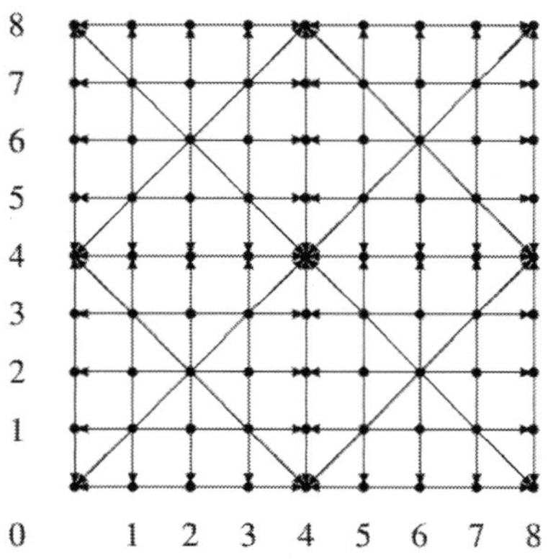

Figure 2. A portion of the relation R.

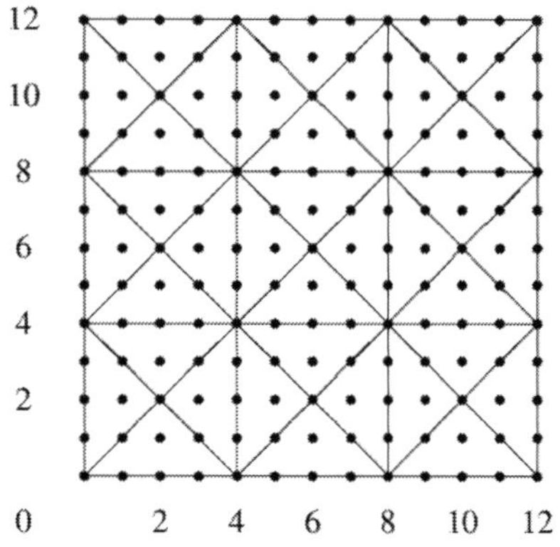

Figure 3. R-Jordan curves.

(3) $y_0 = y_1 = y_2$ and there is $k \in \mathbb{Z}$ such that $x_i = 4k + i$ for all $i < 3$,

(4) $y_0 = y_1 = y_2$ and there is $k \in \mathbb{Z}$ such that $x_i = 4k - i$ for all $i < 3$,

(5) there is $k \in \mathbb{Z}$ such that $x_i = 4k + i$ for all $i < 3$ and there is $l \in \mathbb{Z}$ such that $y_i = 4l + i$ for all $i < 3$,

(6) there is $k \in \mathbb{Z}$ such that $x_i = 4k + i$ for all $i < 3$ and there is $l \in \mathbb{Z}$ such that $y_i = 4l - i$ for all $i < 3$,

(7) there is $k \in \mathbb{Z}$ such that $x_i = 4k - i$ for all $i < 3$ and there is $l \in \mathbb{Z}$ such that $y_i = 4l + i$ for all $i < 3$,

(8) there is $k \in \mathbb{Z}$ such that $x_i = 4k - i$ for all $i < 3$ and there is $l \in \mathbb{Z}$ such that $y_i = 4l - i$ for all $i < 3$.

A portion of R is demonstrated in Figure 2. The ordered triples belonging to R are represented by arrows oriented from first to last terms.

Theorem 2 *Every circle in the graph demonstrated in Figure 3 that does not turn at any point $(4k + 2, 4l + 2)$, $k, l \in \mathbb{Z}$, is an R-Jordan curve.*

Proof. Clearly, every circle in the graph demonstrated in Fig. 3 that does not turn at any point $(4k + 2, 4l + 2)$, $k, l \in \mathbb{Z}$, is an R-simple closed curve. Let $z = (x, y) \in \mathbb{Z}^2$ be a point such that $x = 4k + p$ and $y = 4l + q$ for some $k, l, p, q \in \mathbb{Z}$ with $pq = \pm 1$. Then we define the *fundamental triangle $T(z)$* to be the fifteen-point subset of $\mathbb{Z}^2$ given as follows:

$$T(z) = \begin{cases} \{(r,s) \in \mathbb{Z}^2; 4k \leq r \leq 4k + 14,\ 4l \leq s \leq 4l + 4k + 4 - r\} \\ \text{if } x = 4k + 1 \text{ and } y = 4l + 1 \text{ for some } k, l \in \mathbb{Z}, \\ \{(r,s) \in \mathbb{Z}^2;\ 4k \leq r \leq 4k + 4,\ 4l \leq s \leq 4l + r - 4k\} \\ \text{if } x = 4k + 3 \text{ and } y = 4l + 1 \text{ for some } k, l \in \mathbb{Z}, \\ \{(r,s) \in \mathbb{Z}^2;\ 4k \leq r \leq 4l,\ 4l + 4k + 4 - r \leq s \leq 4l + 4\} \\ \text{if } x = 4k + 3 \text{ and } y = 4l + 3 \text{ for some } k, l \in \mathbb{Z}, \\ \{(r,s) \in \mathbb{Z}^2;\ 4k \leq r \leq 4k + 4,\ 4l + r - 4k \leq s \leq 4l + 4\} \\ \text{if } x = 4k + 1 \text{ and } y = 4l + 3 \text{ for some } k, l \in \mathbb{Z}. \end{cases}$$

Graphically, every fundamental triangle $T(z)$ consists of fifteen points and forms a rectangular triangle obtained from a 4×4-square by dividing it by a diagonal. More precisely, each of the two diagonals divides the square into just two fundamental triangles having a common hypotenuse coinciding with the diagonal. In every fundamental triangle $T(z)$, the point z is one of the three inside points of the triangle. The (four types of) fundamental triangles are demonstrated in the following figure:

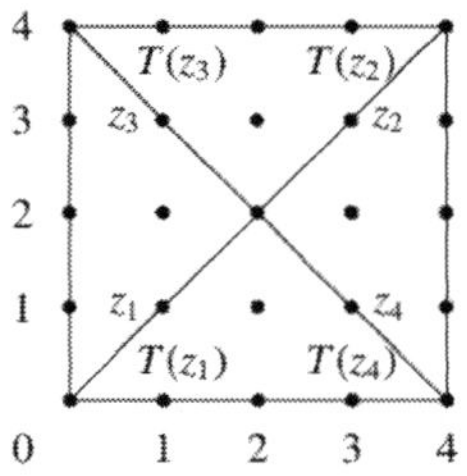

Given a fundamental triangle, we speak about its sides - it is clear from the above picture which sets are understood to be the sides (note that each side consists of five points and that two different fundamental triangles may have at most one side in common).

Now, one can easily see that:

(1) Every fundamental triangle is R-connected (so that the union of two fundamental triangles having a common side is connected).

(2) If we subtract from a fundamental triangle some of its sides, then the resulting set is still R-connected.

(3) If S_1, S_2 are fundamental triangles having a common side D, then the set $(S_1 \cup S_2) - M$ is R-connected whenever M is the union of some sides of S_1 or S_2 different from D.

(4) Every R-connected subset of $\mathbb{Z}^2$ with at most two points is a subset of a fundamental triangle.

DOI: 10.1051/itmconf/20170901012

We will show that the following is also true:

(5) For every circle C in the graph demonstrated in Fig. 3 that does not turn at any point $(4k+2, 4l+2)$, $k, l \in \mathbb{Z}$, there are sequences $\mathcal{S}_F, \mathcal{S}_I$ of fundamental triangles, $\mathcal{S}_F$ finite and $\mathcal{S}_I$ infinite, such that, whenever $\mathcal{S} \in \{\mathcal{S}_F, \mathcal{S}_I\}$, the following two conditions are satisfied:

(a) Each member of $\mathcal{S}$, excluding the first one, has a common side with at least one of its predecessors.

(b) C is the union of those sides of fundamental triangles in $\mathcal{S}$ that are not shared by two different fundamental triangles from $\mathcal{S}$.

Put $C_1 = C$ and let S_1^1 be an arbitrary fundamental triangle with $S_1^1 \cap C_1 \neq \emptyset$. For every $k \in \mathbb{Z}$, $1 \leq k$, if $S_1^1, S_2^1, ..., S_k^1$ are defined, let S_{k+1}^1 be a fundamental triangle with the following properties: $S_{k+1}^1 \cap C_1 \neq \emptyset$, S_{k+1}^1 has a side in common with S_k^1 which is not a subset of C_1 and $S_{k+1}^1 \neq S_i^1$ for all i, $1 \leq i \leq k$. Clearly, there will always be a (smallest) number $k \geq 1$ for which no such fundamental triangle S_{k+1}^1 exists. Denoting by k_1 this number, we have defined a sequence $(S_1^1, S_2^1, ..., S_{k_1}^1)$ of fundamental triangles. Let C_2 be the union of those sides of fundamental triangles in $(S_1^1, S_2^1, ..., S_{k_1}^1)$ that are disjoint from C_1 and not shared by two different fundamental triangles in $(S_1^1, S_2^1, ..., S_{k_1}^1)$. If $C_2 \neq \emptyset$, we construct a sequence $(S_1^2, S_2^2, ..., S_{k_2}^2)$ of fundamental triangles in an analogous way to $(S_1^1, S_2^1, ..., S_{k_1}^1)$ by taking C_2 instead of C_1 (and obtaining k_2 analogously to k_1). Repeating this construction, we get sequences $(S_1^3, S_2^3, ..., S_{k_3}^3)$, $(S_1^4, S_2^4, ..., S_{k_4}^1)$, etc. We put $\mathcal{S} = (S_1^1, S_2^1, ..., S_{k_1}^1, S_1^2, S_2^2, ..., S_{k_2}^2, S_1^3, S_2^3, ..., S_{k_3}^3, ...)$ if $C_i \neq \emptyset$ for all $i \geq 1$ and $\mathcal{S} = (S_1^1, S_2^1, ..., S_{k_1}^1, S_1^2, S_2^2, ..., S_{k_2}^2, ..., S_1^l, S_2^l, ..., S_{k_l}^l)$ if $C_i \neq \emptyset$ for all i with $1 \leq i \leq l$ and $C_i = \emptyset$ for $i = l+1$.

Further, let $S_1' = T(z)$ be a fundamental triangle such that $z \notin S$ whenever S is a member of $\mathcal{S}$. Having defined S_1', let $\mathcal{S}' = (S_1', S_2', ...)$ be a sequence of fundamental triangles defined analogously to $\mathcal{S}$ (by taking S_1' instead of S_1^1). Then one of the sequences $\mathcal{S}, \mathcal{S}'$ is finite and the other is infinite. Indeed, $\mathcal{S}$ is finite (infinite) if and only if its first member equals such a fundamental triangle $T(z)$ for which $z = (k, l) \in \mathbb{Z}^2$ has the property that the cardinality of the set $\{(x, l) \in \mathbb{Z}^2;\ x > k\} \cap C$ is odd (even). The same is true for $\mathcal{S}'$. If we put $\{\mathcal{S}_F, \mathcal{S}_I\} = \{\mathcal{S}, \mathcal{S}'\}$ where $\mathcal{S}_F$ is finite and $\mathcal{S}_I$ is infinite, then the conditions (a) and (b) are clearly satisfied.

Given a circle C in the graph demonstrated in Fig. 3 that does not turn at any point $(4k+2, 4l+2)$, $k, l \in \mathbb{Z}$, let S_F and S_I denote the union of all members of $\mathcal{S}_F$ and $\mathcal{S}_I$, respectively. Then $S_F \cup S_I = \mathbb{Z}^2$ and $S_F \cap S_I = C$. Let $\mathcal{S}_F^*$ and $\mathcal{S}_I^*$ be the sequences obtained from $\mathcal{S}_F$ and $\mathcal{S}_I$ by subtracting C from each member of $\mathcal{S}_F$ and $\mathcal{S}_I$, respectively. Let S_F^* and S_I^* denote the union of all members of $\mathcal{S}_F^*$ and $\mathcal{S}_I^*$, respectively. Then S_F^* and S_I^* are connected by (1), (2) and (3) and it is clear that $S_F^* = S_F - C$ and $S_I^* = S_I - C$. So, S_F^* and S_I^* are the two components of $\mathbb{Z}^2 - C$ by (4) ($S_F - C$ is called the *inside* component and $S_I - C$ is called the *outside* component). The proof is complete.

The circles in the graph demonstrated in Fig. 3 that do not turn at any point $(4k+2, 4l+2)$, $k, l \in \mathbb{Z}$, which are R-Jordan curves by the previous Theorem, provide a rich enough variety of circles to be used for representing borders of objects in digital images. The advantage of the circles over the Jordan curves in the Khalimsky topology is that they may turn at at acute angle $\frac{\pi}{4}$ at some points.

Example 1 Consider the following (digital picture of a) triangle:

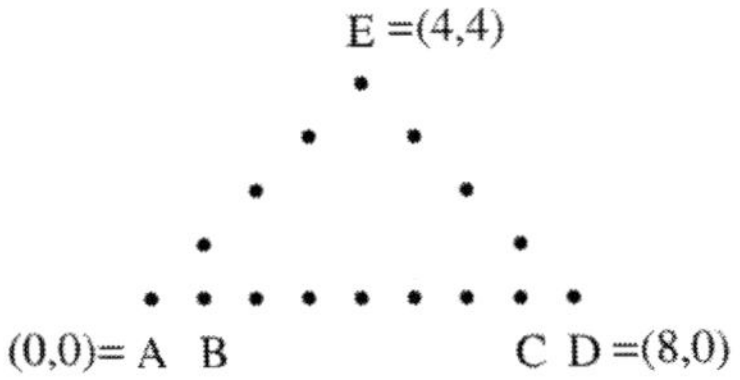

While the triangle ADE is a R-Jordan curve, it is not a Jordan curve in the (underlying graph of the specialization order of) Khalimsky topology. For this triangle to be a Jordan curve in the Khalimsky topology, we have to delete the points A,B,C and D. But this will lead to a considerable deformation of the triangle.

4 Conclusions

We have shown that every plain ternary relation induces connectedness on its underlying set. This connectedness was used to define concepts of simple closed and Jordan curves in the underlying set of a given plain ternary relation. We then introduced and discussed a particular plain ternary relation on the digital plane $\mathbb{Z}^2$ and showed that the connectedness induced by this relation allows for a digital analogue of the Jordan curve theorem. Thus, we have shown that the ternary relation introduced provides a convenient structure on the digital plane for the study of digital images. An advantage of this structure over the Khalimsky topology is that it allows the Jordan curves to turn at the acute angle $\frac{\pi}{4}$ at some points. Since Jordan curves represent borders of objects in digital images, the structure on $\mathbb{Z}^2$ provided by the ternary relation introduced may be used in digital image processing for solving problems related to boundaries such as pattern recognition, boundary detection, contour filling, data compression, etc.

This work was supported by The Ministry of Education, Youth and Sports of the Czech Republic from the National Programme of Sustainability (NPU II) project "IT4Innovations excellence in science - LQ1602".

References

[1] J.A. Bondy, U.S.R. Murty, *Graph Theory* (Springer, 2008)

[2] E.D. Khalimsky, R. Kopperman, P.R. Meyer, Topology Appl. **36**, 1–17 (1990).

[3] E.D. Khalimsky, R. Kopperman, P.R. Meyer, Jour. of Appl. Math. and Stoch. Anal. **3**, 27–55 (1990)

DOI: 10.1051/itmconf/20170901012

[4] T.Y. Kong, R. Kopperman, P.R. Meyer, Amer. Math. Monthly **98**, 902–917 (1991)

[5] R. Kopperman, P.R. Meyer, R.G. Wilson, Discr. and Comput. Geom. **6**, 155–161 (1991)

[6] A. Rosenfeld, Amer. Math. Monthly **86**, 621–630 (1979)

[7] A. Rosenfeld, *Picture Languages* (Academic Press, New York, 1979)

[8] J. Šlapal, Publ. Math. Debrecen **38**, 39–48 (1991)

[9] J. Šlapal, Lect. Notes in Comput. Sci. **5852**, 425-436 (2009)

Mathematical model of an integrated circuit cooling through cylindrical rods

Luis Antonio Beltrán-Prieto[1,*], *Juan Carlos* Beltrán-Prieto[2], *Zuzana* Komínková-Oplatková[1]

[1]Department of Informatics and Artificial Intelligence, Faculty of Applied Informatics, Tomas Bata University in Zlín, nám. T. G. Masaryka 5555, 760 01 Zlín, Czech Republic
[2]Department of Automation and Control Engineering, Faculty of Applied Informatics, Tomas Bata University in Zlín, nám. T. G. Masaryka 5555, 760 01 Zlín, Czech Republic

Abstract. One of the main challenges in integrated circuits development is to propose alternatives to handle the extreme heat generated by high frequency of electrons moving in a reduced space that cause overheating and reduce the lifespan of the device. The use of cooling fins offers an alternative to enhance the heat transfer using combined a conduction-convection systems. Mathematical model of such process is important for parametric design and also to gain information about temperature distribution along the surface of the transistor. In this paper, we aim to obtain the equations for heat transfer along the chip and the fin by performing energy balance and heat transfer by conduction from the chip to the rod, followed by dissipation to the surrounding by convection. Newton's law of cooling and Fourier law were used to obtain the equations that describe the profile temperature in the rod and the surface of the chip. Ordinary differential equations were obtained and the respective analytical solutions were derived after consideration of boundary conditions. The temperature along the rod decreased considerably from the initial temperature (in contatct with the chip surface). This indicates the benefit of using a cilindrical rod to distribute the heat generated in the chip.

1 Introduction

Integrated circuits are integral part of a diversity of electronics and electro-mechanical instruments. [1] The development of electronic devices is currently pushing the time rate of energy transfer per unit volume. Transistors are widely used as integral part of several electronic components consuming high amounts of electric power. In general, the failure rate of these devices is halved for each 10°C reduction in the junction operating temperature, consequently, the operating temperature at which each electronic devices work has to be kept below specific levels to minimize the risk of failure.As a result, new adjustments in element design are performed to facilitate the dissipation of energy, reduction of temperature and to provide alternatives for cooling [2]. Alternatively, schemes including single-phase high heat flux cooling along with hybrid microchannel/jet-impingement modules are also considered [3, 4]. Therefore it is important to take into account not only the electrical parameters but also thermal. Both of these parameters have equivalent laws that are used to perform calculations and develop dynamic models and numerical simulation. One of the most important parameters in thermal management of an electronic device and cooling is the temperature at the transistor chip surface. Precise information of this parameter is critical because functional properties of the device are diminished when care of temperature control is not considered [5]. Many integrated circuits are constituted from silicon. Technologies for embedding thinned integrated circuits in polyimide sheets are being performed to increase thermal resistance of chips; however it is inevitable that the operational life of the product is diminished when the devise is exposed to temperature increments. Integrated circuits are normaly attached to materials including copper, ceramics or metal matrix composites. Due to differences in dilatation coeficients of the elements or compounds, differential thermal expansion occurs when the chip reaches higher temperatures, affecting the performance of the transistor. For these reasons, thermal management solutions are intensively sought [6], including engineering of fin geometry [7], i.e. optimization of heat transfer density is extensively investigated for the design of thin-film electronic circuits. Accordingly, natural and forced convection are commonly used to achieve a device's cooling requirements [8]. In this work, we analyzed the cooling of a transistor surface studying the mathematical model of the heat transfer by conduction from the surface to the finand further dissipation to the surrounding by convection.

*Corresponding author:beltran_prieto@fai.utb.cz

2 Description of the mathematical model

The analysis of Newton's law of cooling gives important information to increase the rate of heat transfer. In general, this can be achieved by two different methods. The first one is to raise the coefficient of convection heat transfer by means, for example, of a fan. The second possibility is to increase the surface area, which can be achieved by connecting the surface with extended surfaces made of highly conductive materials, which are usually named fins. [9]

An alternative to achieve convective removal of heat from a surface is by using extensions on the surface of the transistor to increase the area [10, 11]. There exist several designs of fin, namely triangular, rectangular, cylindrical, circular, trapezoidal, concave parabolic, radial or even in the form of spiral tubes [12]. A fan is used to supply air to allow uniform dissipation of the heat that goes to the integrated circuits. Figure 1 shows a general scheme of the system under study. For simplification purposes, only one rod is represented, but in real practice, the surface is covered by several cylindrical rods. Cooling fins are generally used to improve heat transfer by increasing the surface area available for convection. Many geometric shapes can be used but the most common are rectangular, cylindrical and triangular. Initially, the surface of the chip is at temperature T_s, which will be normally greater than the surrounding temperature T_0. The length of the rod (L) is cooled by a fan [11]. The heat it conducted through the surface and dissipated to the surroundings by convection.

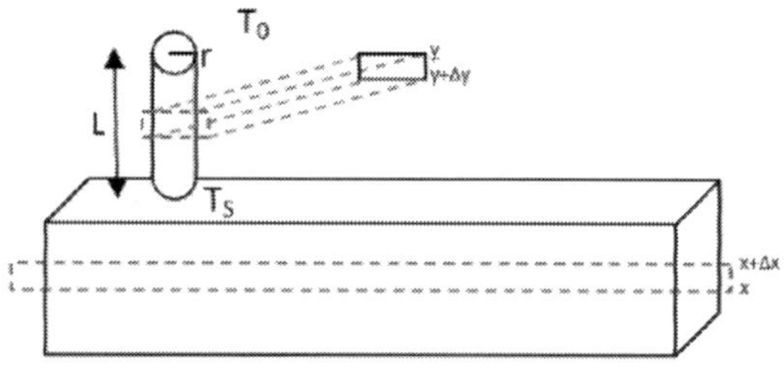

Fig. 1. Schematic diagram of the physical system under study

Fundamental assumptions are contemplated to propose the mathematical model. We consider an isotropic material with three constant parameters, namely the psychical properties of the material (conduction of heat, density, specific heat), the generation of heat in the dispositive, and finally the rate of heat energy transfer through the surface of contact per unit time and temperature difference[13].

We proceed to model the transfer of heat removal around the chip at position y in the rod by considering Newton's law of cooling as represented in the steady state heat balance equations (1) and (2) describing the heat in the chip and the heat dissipated by the rod to the surroundings. [14]

$$Sq|_y - Sq|_{y+\Delta y} - q_L S' = 0 \quad (1)$$

$$\pi r^2 q|_y - \pi r^2 q|_{y+\Delta y} - (2\pi r \Delta y)h(T_1 - T_0) = 0 \quad (2)$$

where S is the cross sectional area, S' is the lateral external area, r is the radius of the rod, q corresponds to the heat flux in y direction (W/m^2), q_L is the convection heat transfer from the rod to the surrounding h the convection heat transfer coefficient, T_r and T_0 are the temperature in the rod and the surrounding respectively. Dividing equation (2) by $\pi r^2 \Delta y$ and taking the limit when $\Delta y \to 0$ leads to equations (3) and (4) respectively. Accordingly, considering the Fourier law, equation (4) is transformed into (5)

$$\frac{q|_y - q|_{y+\Delta y}}{\Delta y} - \frac{2h(T_r - T_0)}{r} = 0 \quad (3)$$

$$-\frac{dq}{dy} - \frac{2h(T_r - T_0)}{r} = 0 \quad (4)$$

$$k_r \frac{d^2 T_r}{dy^2} - \frac{2h(T_r - T_0)}{r} = 0 \quad (5)$$

Additionally, we can transform the previous equation into dimensionless variables of temperature, distance and heat transfer coefficient. Introducing the dimensionless variables represented in (6), equation (7) is obtained:

$$\xi = \frac{y}{L}, \text{for } \epsilon\ (0,1)\ ;\ \alpha = \sqrt{\frac{2hL^2}{rk_r}} \quad (6)$$

$$\frac{d^2 T}{d\xi^2} - \alpha(T_r - T_0) = 0 \quad (7)$$

We can change variables according to the temperature difference $T_r - T_0$ to have $d^2\theta/d(\xi)^2 - \alpha\theta = 0$.

Solving the previous second ordinary differential equation by the method of characteristics, the characteristic roots for the heat balance equation are $m = \pm\alpha$. Therefore, the solution of the equation leads to $\theta = Ae^{\alpha\xi} + Be^{-\alpha\xi}$. Boundary conditions are required to evaluate the arbitrary constants A and B. the heat loss occurs at the rod's lateral surface, and the flux from the top surface is small as expressed in boundary condition (8), whereas, the surface temperature at lenght corresponds to the exterior hot surface as expressed in (9)

$$\frac{d\theta}{d\xi}(0) = 0 \quad (8)$$

$$\theta(1) = T_h \quad (9)$$

To evaluate the arbitrary constants A and B, the boundary conditions (8) and (9) is used. However, for the purpose of applying boundary conditions, it is often convenient to represent the solution of equation (7) using hyperbolic functions, i.e. $\theta = A\cosh\alpha\xi + B\sinh\alpha\xi$.As a result $B=0$, and hence $\theta = A\cosh\alpha\xi$. The value of A

can be obtained using boundary condition (9) which corresponds to $A = T_h/\cosh\alpha$. Substitution of the previous values of A and B obtained leads to equation (10)

$$\theta = \frac{T_h}{\cosh\alpha}\cosh\alpha\xi \qquad (10)$$

And returning to original variables,

$$T_r - T_0 = \frac{T_h}{\cosh\sqrt{\frac{2hL^2}{rk_r}}}\cosh\sqrt{\frac{2hL^2}{rk_r}}\frac{y}{L} \qquad (11)$$

As a result, the temperature in the rod can be derived from previous equation (11). The next step, is to describe the transfer of heat around the computer chip at the position x with a thickness Δx. This is expressed in equation (12):

$$Sq|_x - Sq|_{x+\Delta x} + (S\Delta x)Q = 0 \qquad (12)$$

where S is cross-sectional surface area of the chip, Q is the heat generated per unit volume and Δx the thickness. Dividing equation (12) by $S\Delta x$ we obtain:

$$\frac{q|_x - q|_{x+\Delta x}}{\Delta x} + Q = 0 \qquad (13)$$

Taking the limit as $\Delta x \to 0$ gives the differential equation that represents the heat flux.

$$-\frac{dq}{dx} + Q = 0 \qquad (14)$$

If the chip is homogenous, we can consider the vector form of Fourier's law of heat conduction which describes that the heat flux along the axis is proportional to the gradient in temperature, $q = -k\nabla T$. Therefore, equation (14) yields (15)

$$k_c\frac{d^2T_c}{dx^2} + Q = 0 \qquad (15)$$

where k_c and T_c corresponds to the thermal conductivity and temperature of the chip respectively.

The boundary conditions are set in equations (16) and (17)

$$\frac{dT_c}{dx}(0) = 0 \qquad (16)$$

$$T_c(\delta) = T_h \qquad (17)$$

Solution of equation (15) is represented in (18),

$$T_c = \frac{1}{k_c}\left(Cx + D - \frac{1}{2}Qx^2\right) \qquad (18)$$

After application of the boundary condition (16) we obtain $C=0$ and hence $T_c = (1/k_c)(D - Qx^2/2)$. The value of D can be obtained using boundary condition (17) which corresponds to $D = k_cT_h + Q\delta^2/2$. Substitution of the previous values of C and D leads to equation (19)

$$T_c = \frac{1}{k_c}\left(k_cT_h + \frac{Q\delta^2}{2} - \frac{1}{2}Qx^2\right) \qquad (19)$$

And simplifying (19), the temperature in the chip surface can be expressed as :

$$T_c = T_h + \frac{Q}{2k_c}(\delta^2 - x^2) \qquad (20)$$

From equation (20) it is evident that for the case of $x=\delta$, the temperature in the chip equals the exterior hot surface ($T_c=T_h$).

4 Discussion

Thermal management in electronic devices is important to protect the correct functioning and operation of the dispositive. Accordingly, the knowledge of temperature distribution across the chip surface is a key factor also for operational safety reasons. In Fig. 1 we present two cases of temperature distribution. The first case is when the temperature at the surface of the chip is 75°C and the second is 50°C. In both examples, the final temperature (near the surface that is in contact with the surrounding at y=0) approximates to 30°C. The information obtained from the plot is also important for the design and determination of rod geometry, i.e. to know the required length that is necessary to achieve a specific temperature. In the present example, if the temperature level is acceptable at 40°C, the required dimensionless length is 0.4 when the temperature in the chip surface is 75°C, while in the second case (temperature chip surface of 50°C) is 0.2. This represents a numerical and graphical example for which the formula obtained for the temperature profile in the rod was found to be useful. As a result, the use of cylindrical rods provided to be useful to aids in heat transfer by increasing the surface of the chip and allowing better heat distribution

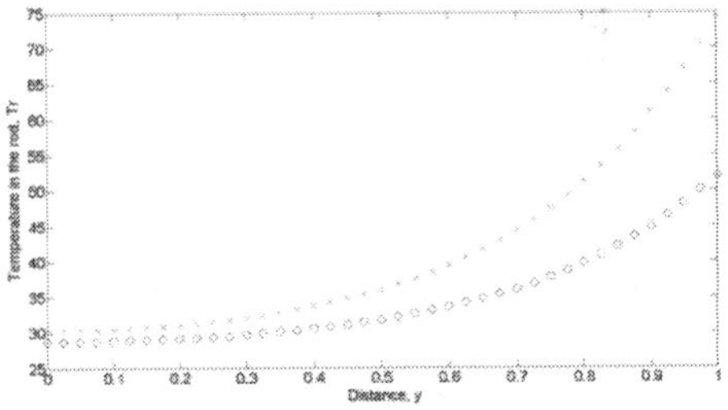

Fig.1 Temperature profile in the rod according to distance from the upper part (contact with surrounding). T_h=75°C(×), T_h=50°C(O).

3 Conclusions

The development of microelectronic devices that are advanced in technology and smaller in size confront considerable challenges in thermal design aiming to dissipate the heat generated inside the device that could reduce reliability of an equipment, malfunctioning and problem of operation. An analysis of combined conduction-convection system was performed in the

DOI: 10.1051/itmconf/20170901013

present study to describe the cooling of a transistor surface by the use of a cooling rod. Mathematical model of these phenomena was performed to obtain the profile of temperature in the rod and the transistor surface.

This work was supported by the Ministry of Education, Youth and Sports of the Czech Republic within the National Sustainability Programme project No. LO1303 (MSMT-7778/2014) and also by the European Regional Development Fund under the project CEBIA-Tech No. CZ.1.05/2.1.00/03.0089, further it was supported by Grant Agency of the Czech Republic—GACR 588 P103/15/06700S and by Internal Grant Agency of Tomas Bata University in Zlín under the project No. IGA/CebiaTech/2016/007. L.A.B.P author also thanks the doctoral scholarship provided by the National Council for Science and Technology (CONACYT) and the Council for Science and Technology of the State of Guanajuato (CONCYTEG) in Mexico.

References

1. A.R.R Adli, K.M.B Jansen. *Microelectron Reliab* 2016; **62**: 26–38.
2. Y. Cengel, R. Turner. *Fundamentals of Thermal-Fluid Sciences*. (McGraw-Hill Higher Education, Singapore, 2001).
3. D.B. Tuckerman, R.F.W. Pease. *IEEE Electron Device Lett* 1981; **2**: 126–9.
4. C-B. Kim, C. Leng, X-D. Wang, T-H. Wang, W-M. Yan. *Int J Heat Mass Transf* 2015; **89**: 838–45.
5. S.P. Jang, S.J. Kim, K.W. Paik. *Sensors Actuators A* 2003; **105**: 211–24.
6. L.T. Yeh. *J Electron Packag* 1995; **117**: 333.
7. R. Baby, C. Balaji *Appl Therm Eng* 2013; **54**: 65–77.
8. R. Palanichamy, P. Nagaraj, S.M. Somasundaram. *IUP J Mech Eng* 2010; **3**: 35–47.
9. R. Kalbasi, M.R. Salimpour. *Appl Therm Eng* 2015; **84**: 339–49.
10. F.P. Incropera, D.P. DeWitt, T.L. Bergman, S. Lavine Adrienne, *Fundamentals of Heat and Mass Transfer* (John Wiley & Sons Inc, United States, 2007).
11. J.P. Holman. *Heat transfer*. (McGraw-Hill, London, 1992).
12. J.H. Lienhard IV, J.H. Lienhard V. *A Heat Transfer Textbook*. (Philogiston Press, Cambridge, 2004).
13. H. Nguyen, A. Aziz *Wärme- und Stoffübertragung* 1992; **27**: 67–72.
14. R.G. Rice, D.D. Do. *Applied mathematics and modeling for chemical engineers*. (John Wiley & Sons, Inc, New Jersey, 2012).
15. R.B. Bird, E.W. Stewart, N.E. Lightfoot. *Transport Phenomena*. (John Wiley & Sons, Inc, New York, 2002).

ITM Web of Conferences **9**, 02001 (2017) DOI: 10.1051/itmconf/20170902001
AMCSE 2016

A variant of genetic algorithm for non-homogeneous population

Najmeh Alibabaie[1], *Mohammad Ghasemzadeh* [2], and *Christoph Meinel* [3]

[1] Computer Department, Engineering Campus, Yazd University, Yazd, Iran
[2] Assoc. Prof. at Yazd University in Iran and Guest Researcher at HPI, Potsdam, Germany
[3] President and CEO of Hasso Plattner Institute (HPI), at Potsdam University, Potsdam, Germany

Abstract. Selection of initial points, the number of clusters and finding proper clusters centers are still the main challenge in clustering processes. In this paper, we suggest genetic algorithm based method which searches several solution spaces simultaneously. The solution spaces are population groups consisting of elements with similar structure. Elements in a group have the same size, while elements in different groups are of different sizes. The proposed algorithm processes the population in groups of chromosomes with one gene, two genes to k genes. These genes hold corresponding information about the cluster centers. In the proposed method, the crossover and mutation operators can accept parents with different sizes; this can lead to versatility in population and information transfer among sub-populations. We implemented the proposed method and evaluated its performance against some random datasets and the Ruspini dataset as well. The experimental results show that the proposed method could effectively determine the appropriate number of clusters and recognize their centers. Overall this research implies that using heterogeneous population in the genetic algorithm can lead to better results.

1 Introduction

Optimization is one of the most important topics in various fields of science and technology. In computer science, this issue is concerned with finding optimal solutions for some problems. There are exact and efficient algorithms for a subset of optimization problems, but for lot of them we need to do a comprehensive search in an exponentially large solution space, which most of times is impractical. In order to find a solution for these problems, we usually use heuristic and approximation methods. These problems include wide range of applications such as data mining, computer vision, knowledge discovering and decision support systems. The heuristic methods do not necessarily find the optimal solutions, but often they manage to find a solution which is close to the optimal solution. This can be accomplished by consuming reasonable time and processing resources. Many problems involve complex search spaces with conflicting objectives. In these cases, optimization of an objective may prevent to achieve the other objectives. Here, sometimes there are a set of optimal solutions which have no clear superiority on each other. In multi objective optimization problems we try to find a trade-off among the requested objectives.

Clustering is a data mining technique, which is used widely in solving related problems [1]. The main issue in clustering is to partition the given data, with a specified number of clusters such that minimize the total within cluster variation (TWCV) and maximize the variance between clusters. The k-means algorithm is the simplest and the most popular clustering method [2].

The initial cluster centers usually have a high efficacy on performance of the k-means algorithm. If these centers are selected randomly, the process may converge to suboptimal partitions [3]. Regardless of the applied clustering method, one of the other most important parameters is number of clusters. In non-hierarchical clustering methods like k-means, this issue corresponds to finding a proper value for k. This parameter must be determined before the clustering process starts. There is no straightforward or easy way to determine this value, because it depends on different issues like the form and distribution of data, as well as the clustering resolution expected by the user.

The ultimate goal of this research is to find the best cluster centers and the best number of clusters simultaneously. The proposed method takes advantage of a genetic algorithm in which the population consists of some sub-populations. Since each point in the k-gene solution space can be a solution candidate, we allow these points to stay in the population at the same time,

DOI: 10.1051/itmconf/20170902001

and in their corresponding sub-populations. Furthermore we let them contribute in evolution of the final solution by applying genetic operators on them. In this regard, the genetic operators are implemented in such a way that could combine individuals of different sub-populations to create new varied individuals for the sub-population groups. Individuals in a subpopulation compete with each other, and also the best individuals of each subpopulation compete with each other as well. The main idea is that we want the genetic algorithm select the best solution from non-homogeneous solutions which are being evaluated by the same criteria.

2 Literature Review

In K-Means (KM), the initial cluster centers usually have a high impact on the clustering process. If these centers are selected randomly from the dataset, then each run of KM on the same dataset may lead to a different result. In other words choosing different centroid points every time gives different clusters. Furthermore the process may converge to suboptimal partitions [3]. In order to cope this dilemma, Sawant [4] in his research, proposed to calculate the neighbourhood distance between the first data item and the other data items, and then sort the items according to the obtained distances. Afterwards he divides the arranged data items into K equal portions. As an initial candidate, the first data item of each partitioning takes part in the KM process. The above method, manages to find appropriate results with higher degree of accuracy in fewer number of steps, but since it requires computing neighbourhood distance between data items and rearranging them, it would be very slow in clustering large datasets.

The proposed method by Karegowda et al. [5], also the proposed method by Al-Shboul and Myaeng [3] use genetic algorithm (GA) to determine the initial values of the clusters. In both of them a population is generated by using a GA process. They apply classic KM algorithm to evaluate fitness of each candidate center. When the terminating condition of GA goes true, the chromosome with highest fitness value decides which samples will be the k-means initial cluster centers. Since KM algorithm must be run for many times, these methods are usually computationally expensive.

The research carried out by Krishna and Murty [2], and another investigation performed by Yi Lu et al. [6] are also based on using GA. These methods compared to the already mentioned methods[5,3] are faster and always converge to the global optimum. In addition they are not sensitive to the initial values. These methods maintain a population of some coded solutions and apply a single step K-means operator (KMO) to evaluate the chromosomes. The common point among all the existing methods which try to find better initial centers, is that they need to start with a given number of clusters.

Another challenge in clustering algorithms is related to deciding about number of clusters, in other words, determination of the K parameter. The algorithm proposed by Hamerly and Elkan [7] is based on a statistical test. It relies on the hypothesis that a subset of data follows a Gaussian distribution. The algorithm starts with a small number of clusters, and then increases them when needed. In each iteration of the algorithm, if any cluster appear whose data dispersion do not follow the Gaussian distribution then that cluster would be divided into two clusters. In order to determine the number of clusters, some of the recent research works rely on using GA. In this regard, Llet′ et al. [8] use the silhouette to find the appropriate number of clusters. The common point among all the existing methods which are developed to find an appropriate number of clusters is that they evaluate the concerned criteria separately for different values of k. In other words, they need to run the KM algorithm for different values of K; this process is usually very time consuming.

The research works reported by Li and Chang [9] and also Chittu and Sumathi [10] are based on introduction of modified genetic algorithms. They divide the existing population into a number of subpopulations. The chromosomes of a population have the same structure, but they use different parameters for the subpopulations. This could let information be transferred between subpopulations. In fact, migration of individuals between subpopulations along with application of genetic operators would lead to generation of new individuals. The purpose of using this modified genetic algorithm with multiple subpopulations and dynamic parameters is to make GA run faster.

3 The proposed method

One of the most important topics and applications of computer science is concerned with information clustering. The K-means algorithm is one of the most popular methods used widely for this purpose.

This algorithm gets the data set and the desired number of clusters, and then it finds the clusters and their centers. In order to find the best number of clusters, typically the user must run this algorithm for different number of clusters and evaluate the obtained results. It can be shown that this parameter may be determined automatically by using genetic algorithms. Genetic algorithm is a search algorithm inspired from nature and genetics. Unlike many other search algorithms, which perform local and greedy search, GA performs a stochastic universal search. It is mainly composed of three operators: reproduction, crossover and mutation.

In many clustering problems, the classical genetic algorithm can be applied easily and get acceptable results. But still there are some disadvantages that need special consideration. One of these disadvantages is concerned with deciding about the suitable number of clusters and their centers. In order to cope with these disadvantages we proposed a modified structure for the genetic algorithm. In fact, in this paper, we show how we can organize the population in several sub-populations and mend the relevant operators to attain an improved genetic algorithm. We accomplish this by considering a population consisting of several set of chromosomes. The structure of these sets is almost similar because these genes hold corresponding

DOI: 10.1051/itmconf/20170902001

5 Experimental results

In order to evaluate the proposed method, we examined its performance against four different datasets [8, 3]. The first three datasets are generated by mathematical models. These datasets are shown in Fig. 4. They include some samples consisting of three or four clusters with different distributions, the other database is known as Ruspini data [11].

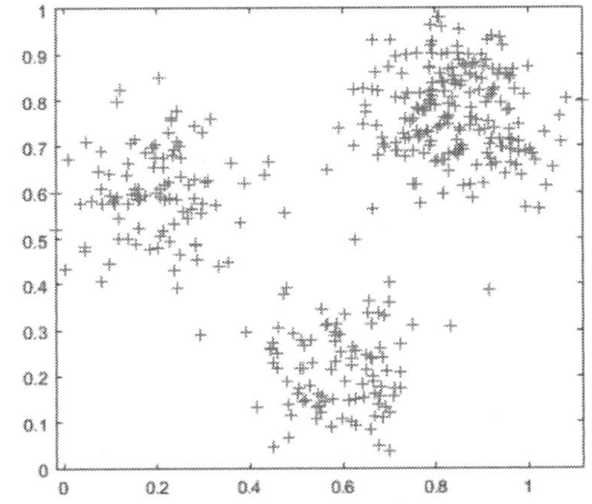

a. Dataset 1

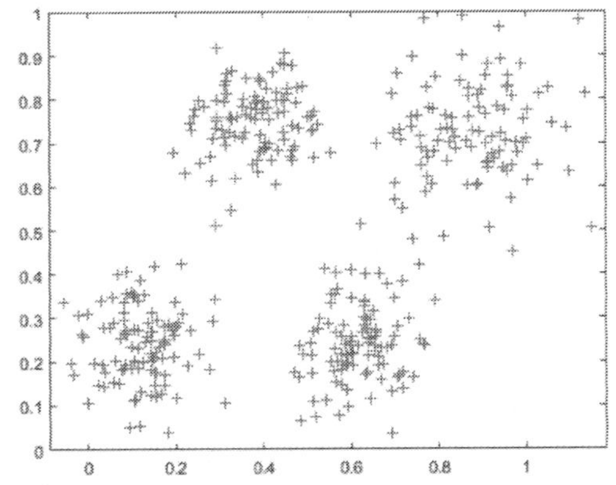

b. Dataset 2

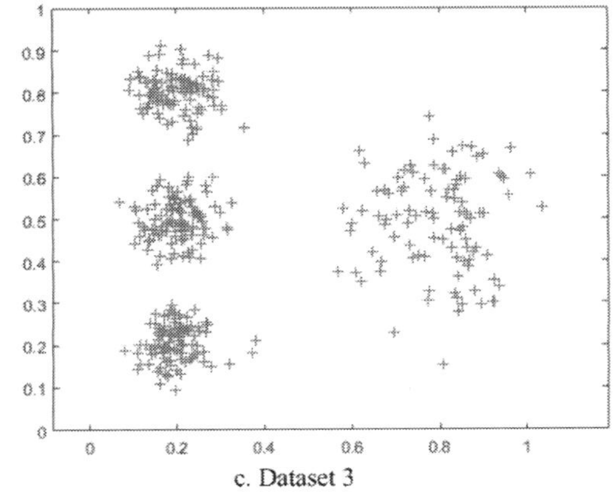

c. Dataset 3

Fig. 4. Random dataset distribution.

Already experimental results and official analysis have shown that silhouette value can also be used efficiently for higher dimensions [8]. We evaluate our proposed method against these data sets which consist of two-dimensional data, but since we have applied the silhouette value, the obtained results would also apply to dataset with higher dimensions. In fact, we chose the two-dimensional datasets for capability of easier presentation and spatial perception.

Dataset 1 consists of 300 points gathered around 3 clusters. The points are scattered with a radius of 0.2 around 3 specific points (0.2, 0.6), (0.6, 0.2) and (0.8, 0.8). They have same values in their boundary points.

Dataset 2 Consists of 400 points scattered around 4 specific points with a radius of 0.2, the points are (0.125, 0.25), (0.625, 0.25), (0.375, 0.75) and (0.875, 0.75). The points two and four have horizontal interleaving on the boundary. This dataset is to be clustered into 4 clusters.

Dataset 3 Consists of 400 points scattered around 4 points with a radius of 0.3 for the first 3 centers and radius 0.4 for the last center. Points are (0.2, 0.2), (0.2, 0.5), (0.2, 0.8) and (0.8, 0.5). The point of the last cluster seems to be isolated except for the points between the cluster center and the other three cluster; the first three clusters have common boundary points. This dataset is to be clustered into 4 clusters.

Dataset 4 is the well-known Ruspini data [11]. This dataset consists of 75 samples, with two features for each sample. This dataset is frequently used in evaluation of clustering methods. Fig. 5-a shows the initial raw dataset. It is clear that this dataset consists of four clusters (Fig. 5-b). If we consider number of clusters k=5, then k-means would give the clustering shown in Fig. 5-c. Depending on the initial centers, for k=5, we may get another form of clustering like the one shown in Fig. 5-d; this time the bold circles are the objects which have been separated to form the fifth cluster.

Usually, in order to find the best number of clusters and corresponding centers for a clustering problem using k-means, it is usually required to run it for different number of clusters and for different initial centers.

In addition, if these best values are being found by a classic genetic algorithm, we would presumably need to make some changes in the encoding: The proposed algorithm looks simultaneously for these goals and it doesn't require any alternations in the coding of solutions. In the experiments, we set population size to be 100 and the number of generations to be 30. The termination condition is met when we reach the predefined number of generations. The crossover and mutation operators were implemented as mentioned earlier. Maximum number of clusters set to be 6 to avoid too much calculation. Table 1 shows these parameters along with some other related parameters.

By considering a different structure for the genetic algorithm we managed to find the suitable centroids and the appropriate number of clusters, in a reasonable time. Fig. 6 shows the experimental results of running the proposed algorithm on datasets 2.

DOI: 10.1051/itmconf/20170902001

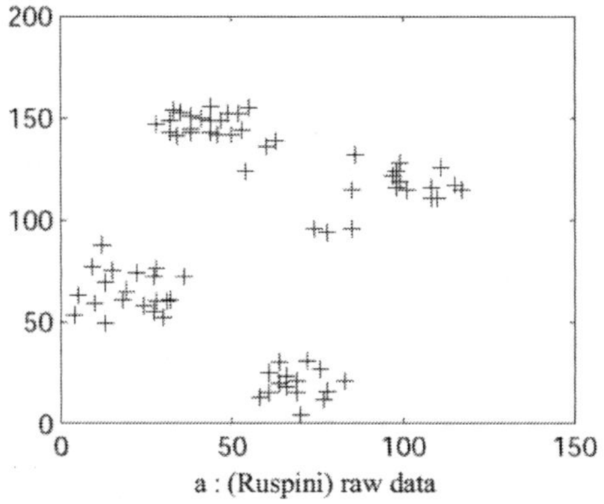

a : (Ruspini) raw data

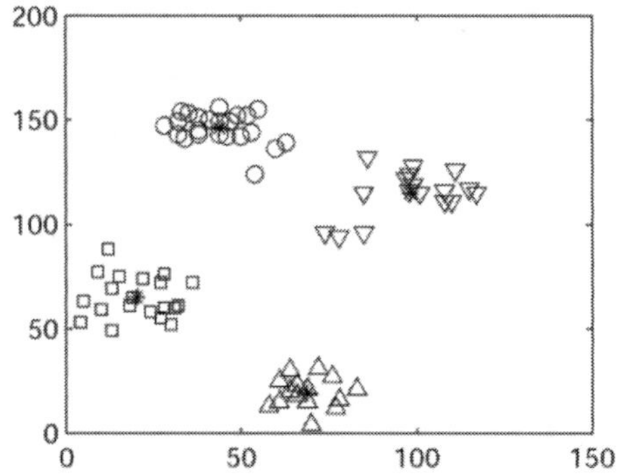

b: Clustering with k=4.

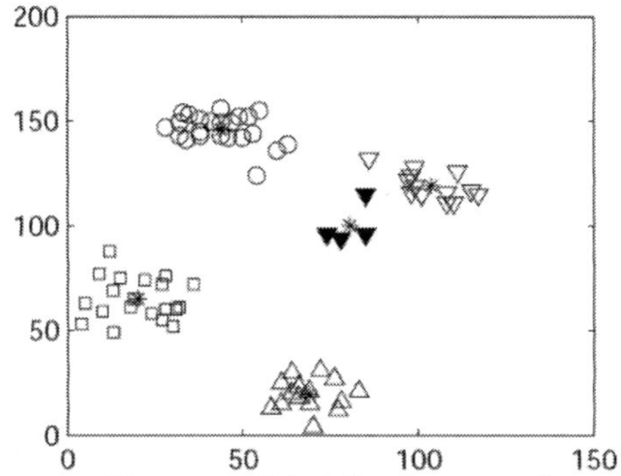

c: Clustering with k=5 (Random centers1).

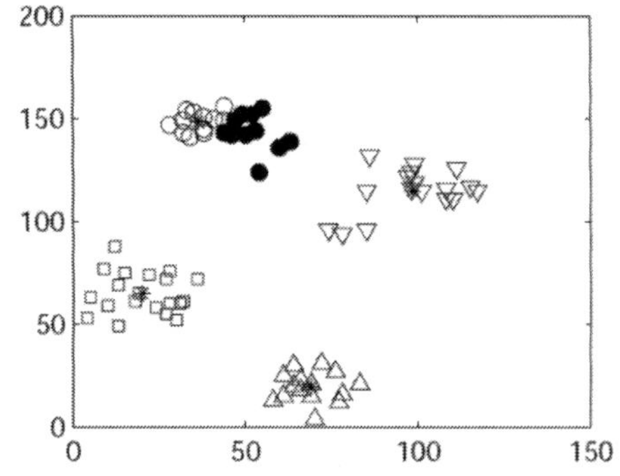

d : Clustering with k=5 (Random centers2)

Fig. 5. The Ruspini dataset [8].

Table 1. Configuration parameters for the proposed algorithm.

Parameter	**Value**
Population size	100
Sub-populations size	Randomly
Initial crossover probability	0.8
Initial mutation probability	0.1
Probability of changing Nr. of genes	*f(iteration)*
Probability of changing genes values	
Maximum Nr. of clusters (Kmax)	6
Number of generations	30

Fig. 6-a Shows how the proposed method managed to determine the clusters centroids truly. Fig. 6-b shows how the proposed algorithm could converge after only 15 iterations. Fig. 6-c shows (average) number of clusters of the fittest chromosome in every generation. These information were obtained from averaging results of ten independent runs of the algorithm during the first 30 generations. In early iterations, some points of the search space with k=3, compared to the others gain higher fitness (Fig. 6-c) but mutation and transformation of information between subpopulations could eventually lead the process to converge toward the better solution k=4.

Table 2 shows the clusters centroids and number of clusters found by the proposed method. In fact it shows how the proposed algorithm manages to find clusters centers and number of clusters correctly.

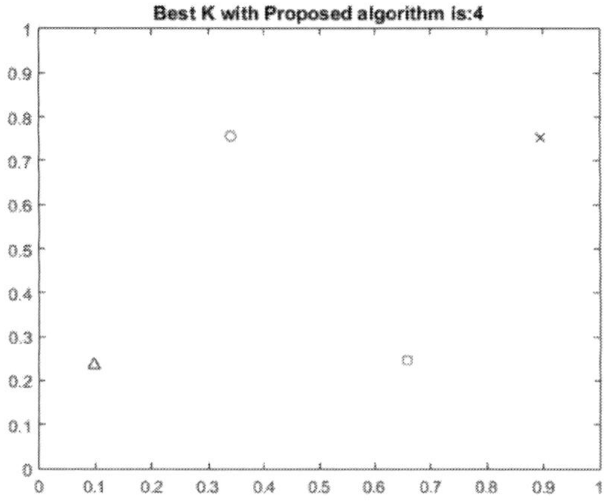

6-a. The proposed centroids by the proposed method

DOI: 10.1051/itmconf/20170902001

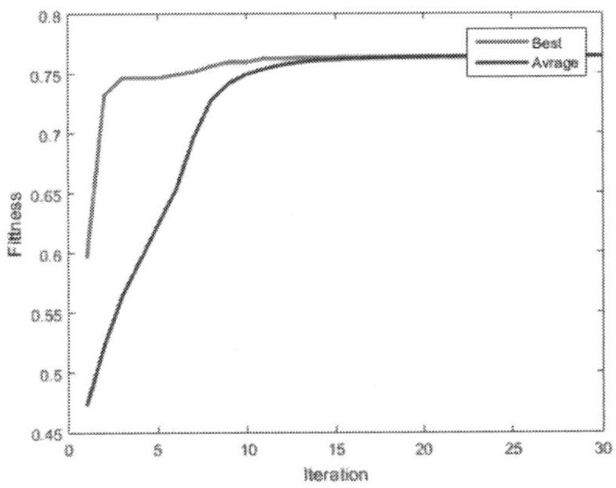

6-b. Fitness of fittest chromosome and average fitness of population in every generation.

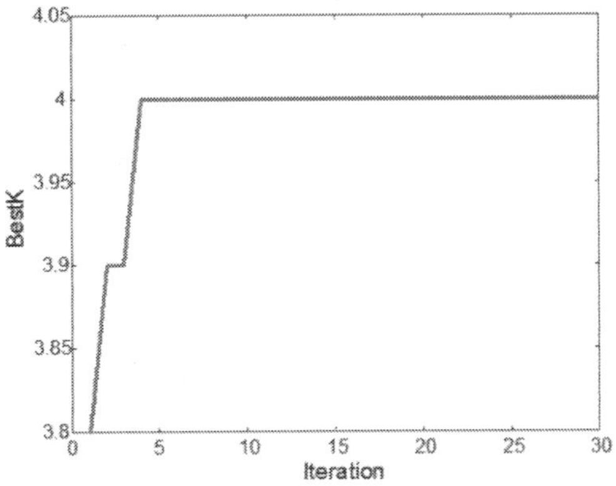

6-c. Nr. of clusters in the chromosome with the best fitness in every iteration (Average of 10 run)

Fig. 6. The proposed method output for dataset2.

Table 2. The clusters centroids and the number of clusters found by the proposed algorithm.

	Nr. of clusters	The clusters centroids obtained by the proposed method			
Dataset 1	3	(.79 , .79)	(.61 , .17)	(.18 , .60)	---
Dataset 2	4	(.30 , .77)	(.10 , .24)	(.62 , .25)	(.84 , .76)
Dataset 3	4	(.79 , .50)	(.18 , .49)	(.20 , .79)	(.19 , .20)
Ruspini	4	(65.22 , 18.95)	(16.36 , 69.26)	(148.37 , 40.18)	(115.70 , 101.94)

Considering a different structure for the population and chromosome patterns could let the procedure not require a separate run for every value of k, this means reducing runtime. It is obvious that we would get better results when the objective function could establish a better trade-off between minimize the total within cluster variation (TWCV) and maximize the variance between clusters. Although the fitness function we adopted from Llet′ et al [8] can also work efficiency for high dimension datasets, but combining it with other measures can improve it to work efficiently for overlapping data as well. In the testing phase, each dataset was used 20 times and then we calculated the average running time. The obtained results are shown in Table 3.

Table 3. Performance evaluation

	Proposed algorithm	K-means algorithm	GA+KM (constant k)
Dataset 1	2.60	0.01	18.51
Dataset 2	1.03	0.013	15.32
Dataset 3	2.61	0.02	11.12
Ruspini	0.87	0.02	8.84

If we try to use the classic genetic algorithm for solving problems which involve several search spaces, then for each search space we need to apply a different encoding and run the algorithm from the beginning; while the proposed approach doesn't require any change in the encoding and it performs the corresponding operations in parallel.

Since the proposed model for population consists of some subpopulations with different structures and we use improved genetic operators for recombining them, it is expected that offsprings will inherit appropriate information from their parent, leading to find an appropriate solution for the concerned problem. This cooperation and information exchange among subpopulations could also reduce the required processing time. In the implemented algorithm, despite presence of individuals from different search spaces, because of using the modified mutation operator, during consecutive generation, the best search space dominates and most of individuals in the last generation belong to the best solution space. When we want to apply our idea in solving optimization problems involving solutions from different search spaces, if we fix the number of individuals in subpopulations and allocate a certain proportion of the population to every subpopulation in successive generations, the algorithm can converge to the best solutions of different spaces.

DOI: 10.1051/itmconf/20170902001

5 Conclusion and future work

In this research a different structure for the genetic algorithm was introduced. This method can be applied to problems that several groups of different solutions are evaluated according to the same criteria. It shows that a heterogeneous population in genetic algorithm could accelerate the search process and improve the final results. In a further research, the fitness function can be improved by considering other criteria such as entropy for overlapping data or by considering a constant proportion for each subpopulation, apply it in solving problems that require to find best solutions from several search spaces.

References

1. A. Amirkhanyan, F. Cheng, C. Meinel. Real-time clustering of massive geodata for online maps to improve visual analysis. *11th International Conference on Innovations in Information Technology (IIT)*. 308-313. Dubai. (2015).
2. K. Krishna, M. N. Murty. Genetic K-means algorithm. *IEEE Transactions on Systems, Man, and Cybernetics,*, **29**(3): 433-439. (1999).
3. B. Al-Shboul, S. H. Myaeng. Initializing KMeans using genetic algorithms. *The International Journal of Computer, Electrical, Automation, Control and Information Engineering*, **3**(6): 1481-1485. (2009).
4. K. B. Sawant. Efficient Determination of Clusters in K-Mean Algorithm Using Neighborhood Distance. *The International Journal of Emerging Engineering Research and Technology* **3**(1): 22-27. (2015).
5. A. G. Karegowda, V. T. Shama, M. A. Jayaram, A. S. Manjunath. Improving Performance of K-Means Clustering by Initializing Cluster Centers Using Genetic Algorithm and Entropy Based Fuzzy Clustering for Categorization of Diabetic Patients. , *In Proceedings of International Conference on Advances in Computing*, 899-904. MSRIT, Bangalore: Springer India. (2013).
6. Y. Lu, S. Lu, F. Fotouhi, Y. Deng, S. J. Brown. Incremental Genetic K-Means Algorithm And Its Application In Gene Expression Data Analysis. *BMC Bioinformatics* **5**(1):172. (2004).
7. G. Hamerly, C. Elkan. Learning the k in k-means. *Advances in Neural Information Processing Systems* 17:281–288. (2004).
8. R. Llet′, M.C. Ortiz, L.A. Sarabia, M.S. Sanchez. Selecting variables for k-means cluster analysis by using a genetic algorithm that optimises the silhouettes. *Analytica Chimica Acta* **515**(1): 87-100. (2004).
9. R. Li, X. Chang. A Modified Genetic Algorithm with Multiple Subpopulations and Dynamic Parameters Applied in CVaR Model. *International Conference on Computational Intelligence for Modelling Control and Automation*, 151. Washington, DC: IEEE Computer Society. (2006).
10. V. Chittu, N. Sumathi. A Modified Genetic Algorithm Initializing K-Means Clustering. *Global Journal of Computer Science and Technology* **11**(2): 54-62. (2011).
11. E. H. Ruspini. Numerical methods for fuzzy clustering. *Information Sciences* **2**(3): 319–350. (1970).

Time Series Analysis of Air Pollutants for Karabük Province

Ali *CAN*

Karabuk University, Engineering Faculty, Mechanical Engineering Department, Balıklar Kayası Mevkii, Karabük, 78100, Turkey

Abstract. The air is polluted and the atmospheric pollution has becoming a great problem for human beings due to the high rate consumption of energy. Most of the scientists dealing with the air pollution state that the first reason of all environmental problems is the population increase. Depending on population, the high consumption of energy (including heating purposes, electricity production, transportation and industries) agricultural activities, waste disposal and the pressure for destroying forest areas create air pollution problem. The atmosphere is polluted and the concentrations of pollutants have increased tremendously. Karabük Province is one of the famous provinces of heavy industries. There are 3 air quality measurement stations. In this study, the last 2 years hourly and daily air pollutants (PM10, SO_2, NO, NO_x, NO_2, O_3 and CO) were studied as a time series by using graphical and statistical approaches. The results were considered seasonally and annually. The meteorological condition is also very effective in atmospheric pollution. Moreover, the industrial production and consumption of energy are also high pressure on Karabük province. The concentrations in each station are also showing different characteristics during two years' period. The highest concentrations (such as PM10, SO_2, NO, NO_x) are, as usual, seen in winter season due to industrial usage of energy for production and household consumption of energy for heating. However, for O_3, the highest concentration is observed in summer season, because the atmospheric ozone trend is supposed to show an opposite trend compared with NO due to photochemical reaction with these gases. For CO, the maximum concentration is recorded as 4779 µg/m3 in august in 2015. The analyses of data have shown that, the atmosphere is polluted highly with PM10, SO_2, NO, NO_x, NO_2, and CO during winter season. The households are very effective using high coal for heating purposes in Karabük Province.
Keywords: Air Pollutant Concentration, Time Series Analyses, Atmospheric Pollution

1 INTRODUCTION

The air pollution data are always evaluated as time series. Time series are mainly used to estimate missing data or used to determine trend for future estimation [1]. In the air pollution, the future scenarios are always important to tackle with this global scale problem [2]. The statistical, mathematical or graphical approach can be used for data analyses [6, 11]. The biggest problem in time series analyses is the missing data. In a study resulted by Junger and Leon [7], the complete data analysis will yielding satisfactory results regardless of the generating mechanism of the missing data if the amount of missing data is equal or lower than 5%, whereas the validity began to decline when the proportion of missing values are exceeding 10%. The increases of missing data over 10% are always resulted in some assumptions in the analyses. And the assumptions mean that the decrease in the precision [5].

The air pollutant in the atmosphere is increasing and the local studies have shown that the effect of long-range air transport is varying due to the different aspect in the regions. The meteorology, geographical aspect, persistence in the air, climate and seasons have big effect in the air pollution [9, 12, 13]. In another words, the time series are the sequence of many segments and environment [4, 8].

The main aim of this analysis is to put the characteristic of the data. To emphasize the trend and statistical analyses are mainly the complementary of this kind of studies. According to the results, the ordering of the data is natural. Time is selected as the depending variable. Finding the changes and fluctuations of air pollutants at local state is aimed. The main difficulty is the missing data due to equipment failure for some days during the study years.

2 Materials and Methods

Study Area: Karabük Province is one of the famous provinces of heavy industries. The map is prepared by using a GIS programme (Fig.1). In this study, the last 2 years air pollutants were studied. There are 3 air quality measurement stations. Because of private confidentiality, the places of air quality measurement stations are not given in this study. In one station, only SO_2 and PM10 are measured, the other pollutants are not measured for the study period. Therefore, the daily averages of pollutants for Karabük were calculated according to measured parameters.

* Corresponding author: alican@karabuk.edu.tr

DOI: 10.1051/itmconf/20170902002

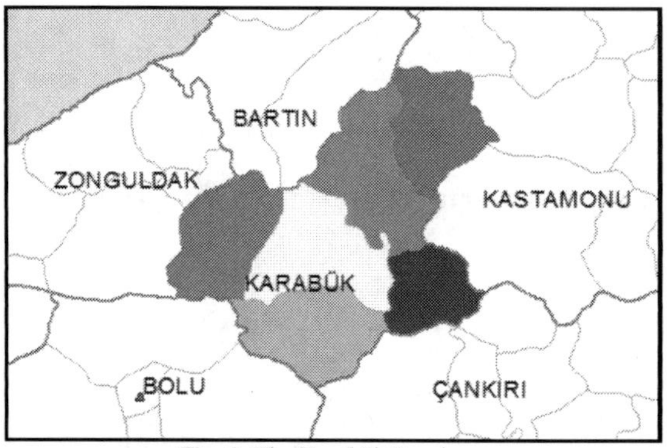

Fig. 1. The Karabük Province

Data: The data is gathered from the Ministry of Environment and Urbanization Air Quality Monitoring Network (AQMN) web page [10]. The last 2 years hourly air pollutants (PM10, SO_2, NO, NO_x, NO_2, O_3 and CO) were collected and analysed.

Method: The air pollutants are generally passed two evolutionary stages. The first one is the quantification and second one is qualification according to the air standards [2, 3]. Within this study, the pollutant data is analysed by graphical and statistical approaches. The daily average data of three stations was used. In the first part of study (graphical approach), the annual trend of data was determined. In order to see the trend, a moving averages were also calculated and inserted in the graph. In statistical analyses, the annual minimum, maximum and averages of the pollutants were calculated as supplementary information.

3 Results and Discussion

The time series analyses have shown that the pollutant is fluctuating throughout the years. In winter season the concentrations of air pollutant is increasing and during the spring and summer time, the average concentration is lower except for ozone. The situation is just opposite for atmospheric ozone concentration, which is higher during the summer time. By the way, the highest missing data was seen for 2015 data for NO, NO_x, NO_2, O_3 and CO, but even the lack of many measurements in 2015, the missing ratio is not higher than 10%. Therefore the data is accepted as precise and it reflects the situation of Karabük Province truly.

If the PM10 graph (Fig.2) is examined, the maximum concentrations are determined in the winter season with values 307 μg/m³ in 2014 and 276 μg/m³ in 2015. These two values are higher than EPA criteria. It means it is very hazardous for human beings. On the contrary, the PM10 concentration in summer is always lower than 100 μg/m³.

According to the moving averages, there is a decline trend between the winter and autumn seasons. it can be stated that the atmospheric temperature is very affective for the pollutant PM10.

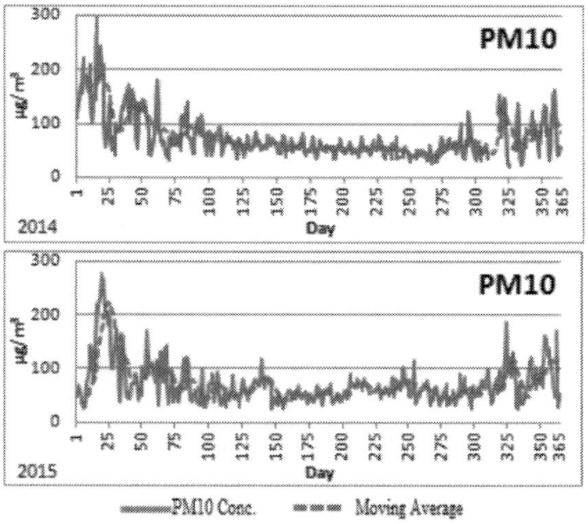

Fig. 2. The PM10 concentration in 2014 and 2015

The trend of SO_2 concentration is also very similar to the PM10 graph in 2014. However, it is not the case for 2015, especially in December 2015, the SO_2 concentration is increasing considerably and reach the highest value, which is 298 μg/m³ (Fig.3), in this month.

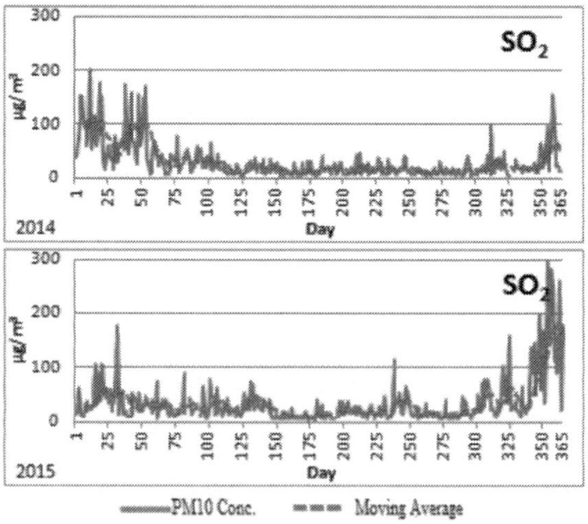

Fig. 3. The SO_2 concentration in 2014 and 2015

The figures of NO, NO_x and NO_2 concentrations are showing different characteristics. For NO and NO_x, the concentrations are showing the same trend during study period. There is a sharp increase after late august in 2015.

The highest NO and NO_x values are determined during this period with respective values 673 and 745 μg/m³ which are also very high according to the EPA criteria (Fig.4 and Fig.5).

DOI: 10.1051/itmconf/20170902002

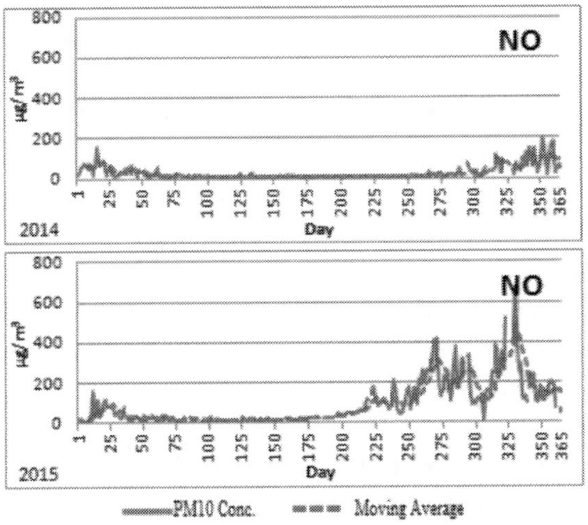

Fig. 4. The NO concentration in 2014 and 2015

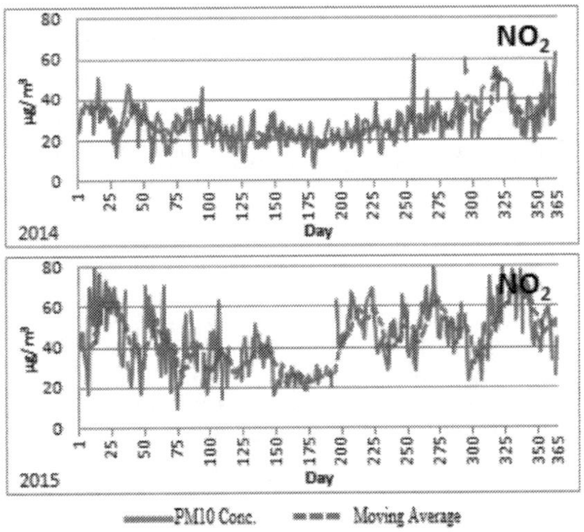

Fig. 6. The NO_2 concentration in 2014 and 2015

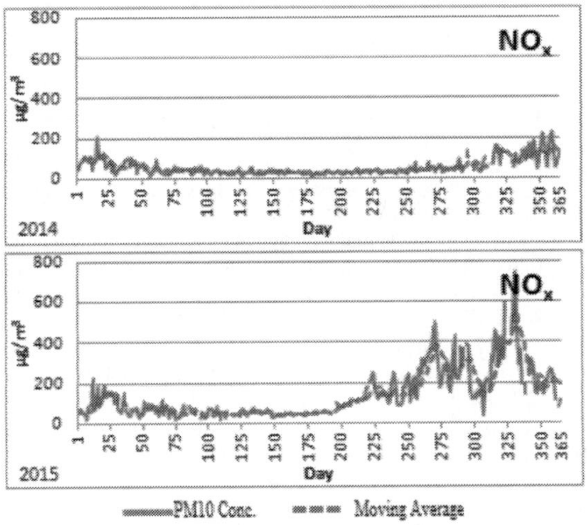

Fig. 5. The NO_X concentration in 2014 and 2015

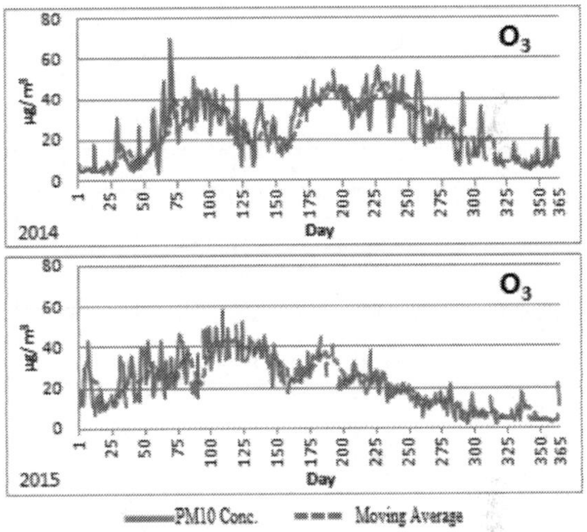

Fig. 7. The O_3 concentration in 2014 and 2015

The $N0_2$ concentration in 2015 is increasing and the concentration increases on the figure is considerable. The $N0_2$ concentrations in 2015 are about twice that in 2014 (Fig. 6).

During day time, photochemical reaction NO into NO_2 is always decreasing ozone concentrations in the atmosphere. Therefore, the atmospheric ozone trend is supposed to show an opposite fluctuation compared with NO. On the other hand, the photolysis of NO_2 by a photons forms NO and O_3. It means, the NO-NO_2 cycle starts again [2].

In figure 7, there is an opposite trend compared to the NO graph. Then, the O_3 concentration graph is showing a sharp increase after winter in 2015.

The air quality level for CO is varying tremendously between 74 μg/m³ to 4779 μg/m³. The inefficient burning, which results CO emission, is very dangerous and it will cause death if CO exposure is continues due to increasing concentration in atmosphere.

The EPA criteria are also exceeded for this pollutant on some days. The concentration of CO is increasing during winter season according to the figure. The annual average for this gases is around 900-950 μg/m³. Moreover, in 2014, the concentration of CO is higher than that in 2015 (Fig.8).

In statistical analyses, the annual minimum, maximum and averages of the pollutants were also calculated for series.

DOI: 10.1051/itmconf/20170902002

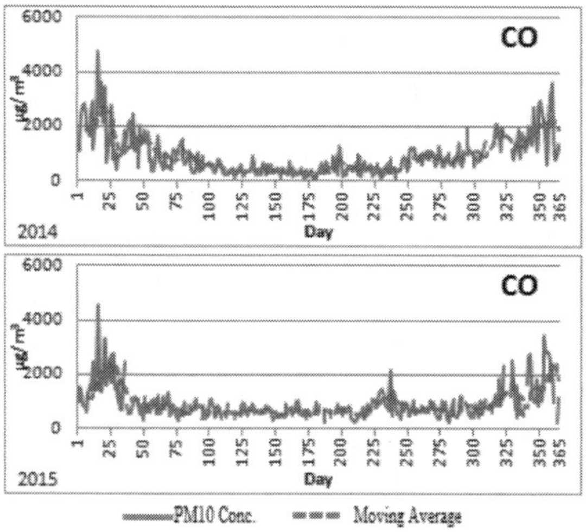

Fig. 8. The CO concentration in 2014 and 2015

The main difficulty is the missing data due to equipment failure for short period. Moreover, the complementary table (Tab. 1) is given below to support the findings.

Tab. 1. The data analysis in 2014 and 2015

2014	PM10	SO_2	NO	NO_x	NO_2	O_3	CO
Missing Data (%)	2,5	0,5	3,6	3,6	3,6	3,6	3,6
Average	75	31	26	54	28	25	941
Mimimum	21	4	3	10	7	4	74
Maximum	307	203	196	225	63	70	4779

2015	PM10	SO_2	NO	NO_x	NO_2	O_3	CO
Missing Data (%)	0,3	2,2	8,8	8,8	8,8	8,2	8,2
Average	73	34	95	139	44	22	923
Mimimum	25	4	5	18	10	2	223
Maximum	276	298	673	745	86	58	4587

The pollutants mentioned above are mostly created by the people living in Karabük province. Therefore, the solution for this problem will also be founded by the same people living in this city. With small actions such as; changing coal into natural gas, controlled used of fertilizers ending with atmospheric deposition of nitrogen, and use electrical furnaces instead of coal furnace in industries will comparable change the atmosphere in a short time. The geological aspect is also not permitting the pollutants transfer from local area of Karabük. Moreover, the wind is not powerful for the good dispersion in the atmosphere. Shortly, the air quality data of Karabük province is showing that the air is polluted as in crowded cities and the pollution level is not considered as healthful.

3 Conclusion

The highest concentrations (such as PM10, SO_2, NO, NO_x NO_2, CO) are, as usual, seen in winter season. It means household burning is very affective on the atmosphere. However, for O_3, the highest concentration is observed in summer season, because the atmospheric ozone trend is supposed to show an opposite fluctuation compared with NO due to photochemical reaction. For CO, the maximum concentration is recorded as 4779 $\mu g/m^3$ in august in 2015. The atmosphere is polluted and the high anthropogenic emissions are created in Karabük province. Therefore, the solution for this problem will also be founded by the same people living in this city. With applicable actions such as; changing coal into natural gas and use electrical furnaces instead of coal furnace in industries will change the atmosphere in this city comparably in a short time.

References

1. Alier M., Tauler R., 2013. **Multivariate Curve Resolution of incomplete data multisets**. Chemometrics and Intelligent Laboratory Systems 127 (2013).p.17–28.
2. Boubel R.A., Fox D.L., Turner D.B., Stern A.C.,1994. **Fundamentals of Air Pollution** – Academic Press, Third Addition, P.41-59,72-74.
3. Cooper C. D., Alley F.C, 2002. **Air Pollution Control**-A design Approach.p.58.63.
4. Dirgawati M, Heyworth J. S., Wheeler A.J., McCaul K.A., D.Black., Boeyen J., Cope M., Yeap B.B., Nieuwenhuijsen M., Brunekreef B., Hinwood A., 2016, **Development of Land Use Regression models for particulate matter and associated components in a low air pollutant concentration airshed**. Atmospheric Environment 144.p.69-78.
5. Greenland S., Finkle W.D., 1995. **A critical look at methods for handling missing covariates in epidemiologic regression analyses**. J. Epidemiol. 142 (12), p.1255-1264.
6. Guyet T., Nicolas H., 2016. **Long term analysis of time series of satellite images**. Pattern Recognition Letters 70 (2016).p.17–23.
7. Junger W.L., Leon A. P., 2014. **Imputation of missing data in time series for air pollutants**- Atmospheric Environment 102 (2015) p.96-104.
8. Juodis L., Filistovic V., Maceika E., Remeikis V., 2016. **Analytical dispersion model for the chain of primary and secondary air pollutants released from point source**. Atmospheric Environment 128.p.216-226.
9. Meraz M., Rodriguez E., Femat R., Echeverria J.C., Alvarez-Ramirez J., 2015. **Statistical persistence of air pollutants (O_3, SO_2, NO_2 and PM10) in Mexico City.** Physica A 427. (2015).p.202-217.
10. MEU. Ministry of Environment and Urbanization. Air Quality Monitoring Network. web address: http://www.havaizleme.gov.tr/Default.ltr.aspx date: 11.02.2016.
11. Perepu S. K., Tangirala A. K., 2016. **Reconstruction of missing data using compressed sensing techniques with adaptive dictionary**-Journal of Process Control 47 (2016)p.175–190.
12. Poberznik M., Strumbelj E., 2016.**The effects of air mass transport, seasonality, and meteorology on pollutant levels at the Iskrba regional background station (1996-2014)**. Atmospheric Environment 134 (2016)p.138-146.
13. Qu Y., An J., He Y., Zheng J. 2016. **An overview of emissions of SO_2 and NO_x and the long-range transport of oxidized sulfur and nitrogen pollutants in East Asia**. Journal of Environmental Science p.13–25.

DOI: 10.1051/itmconf/20170902003

The application of an artificial immune system for solving the identification problem

Irina Astachova[1], *Stanislav* Ushakov[1], *Andrei* Selemenev[1], and *Juliya* Hitskova[2]

[1]Department of Applied Mathematics and Informatics, Voronezh State University, Voronezh 394018, Russia
[2]Department of Information Systems Design in Civil Engineering, Voronezh State Technical University, Voronezh 394006, Russia

Abstact. Ecological prognosis sets the identification task, which is to find the capacity of pollution sources based on the available experimental data. This problem is an inverse problem, for the solution of which the method of symbolic regression is considered. The distributed artificial immune system is used as an algorithm for the problem solving. The artificial immune system (AIS) is a model that allows solving various problems of identification, its concept was borrowed from biology. The solution is sought using a distributed version of the artificial immune system, which is implemented through a network. This distributed network can operate in any heterogeneous environment, which is achieved through the use of cross-platform Python programming language. AIS demonstrates the ability to restore the original function in the problem of identification. The obtained solution for the test data is represented by the graph.

1 Introduction

Symbolic regression is a method of construction of regression models by enumeration of various arbitrary superpositions of functions in a given set. The superposition of functions is called "program" and the stochastic optimization algorithm for constructing of such superpositions is called genetic programming. However, instead of commonly used genetic algorithms the artificial immune system will be used.

The problem of finding of the optimal regression structure model will be formulated further. Thus, the input data for the problem stated is the collection of points of n-dimensional space and values of the function at a point. The problem solution is a function, which depends on n variables, that best approximates the original (at fixed points). To solve the problem of symbolic regression the algorithms of genetic programming are used [17]. Genetic programming is a way to create programs with the use of genetic algorithms [18]. In this case, the program will be presented by the function in the form of a tree, which is allegorically represented a chromosome itself. Crossover operators and mutations occur over the tree of function representation [17], the adaptation function is a measure of how well the program (function) solves the necessary task.

Let us consider an artificial immune system, which will be used for solving symbolic regression problems, and let take a look which of the basic properties of natural immune system will be useful for solving this problem. At the second step, after the creation of a simple model of the immune system, we use the fact that the immune system is inherently distributed, thus we modify the algorithm in such a way, so that it uses distributed computing that can speed it up and improve.

1) The first step is lymphocytes representation. The first task is to determine how lymphocytes will be presented. Since the solution is the function (superposition of simple defined functions), then lymphocyte should represent this function encoded in some way. A tree of expression can be considered a convenient representation method (for implementation and for the further algorithm operation). As the number of variables is known in advance, lymphocyte also keeps a list of valid variables. Lymphocytes should support the following operations:
• Calculation of the value of the provided function at a point.
• Calculation of the affinity measure (discussed below) of the given lymphocyte.
• Simplification of the corresponding expression.
• Return of the string representation of an expression (with properly spaced brackets).

2) The second step is affinity (the objective function). Let us introduce the concept of affinity. In biology the affinity of antibodies is the binding strength of active sites of antibody molecule with determinant antigen groups. The antigen is a set of input data (function values of at the given points). In this model affinity characterizes the degree of approximation "success", i.e. the value of the objective function

3) The third step- is the definition of network structure. The network is homogeneous and consists only of B-lymphocytes described above.

Let us consider the use of described symbolic regression algorithm for solving classification problem, which emerge by ecological forecasting [1]; the foundation of which is the diffusion equation in the moving media: let the concentration of a pollutant in a certain region in a certain time is known. Let assume that instead of the pollution sources capacity referred to in the problem formulation, we know the experimental observations data derived from a set of detectors. The problem of identification is to restore the function of the source, along this parallel is the approximate solution of the problem, i.e. solving function.

2 Problem formulation

Let us state the brief formulation of the symbolic regression problem [5]. There are many values of free variables $\{x_1,\ldots,x_r\}$, where $x_i \in R$, and corresponding values of the function $\{y_1,\ldots,y_r\}$. These two sets form a set of baseline data *D*. Also, a set of functions is given that will be used in the construction of superposition. We will consider only continuously differentiable functions: $g: R^n \to R$, such as sin, cos, polynomials.

Let us consider the arbitrary superposition f, consisting of no more than m functions g. It is required to find such a superposition, which would provide the maximum (minimum) of the given functional $p(f,D)$.

This functional defines the objective function. It is chosen in such a way that it shows the degree of approximation of the built functions to the desired one.

Model of an artificial immune system consists of the lymphocytes representation and the algorithm of artificial immune system functioning. Lymphocytes are presented in the form of the expression trees. The solution - is the function (superposition of simple functions), therefore a lymphocyte should represent itself as a function encoded. A tree of expression can be considered a convenient representation for the realization, as well as for the subsequent operation of the algorithm. Since only binary and unary operations will be used, this tree will be binary. Function represented by a lymphocyte, can be written as

$$F = f_1(f_2(\ldots f_m(x_1, x_2, \ldots x_n))),$$

where $f_1,\ldots f_m$ are functions from the given set, m is the number of functions, less than or equal to the maximum permissible height of the expression tree, and x_i are free variables.

As the number of variables is known in advance, a lymphocyte also stores the list of feasible variables.

Lymphocytes should support the following operations:

- Calculation of value of the provided function at a point.
- Calculation of affinity measure of the given lymphocyte.
- Simplification of the corresponding expression.
- Return of a string representation of expression.

As the affinity (the objective function), let us use the following function: $\sqrt{\sum_{i=1}^{r}(f(x_i)-y_i)^2}$, where $\{x_1,\ldots,x_r\}$, $x_i \in R^n$ is a given set of values of free variables; $\{y_1,\ldots,y_r\}$ is a given set of values of the unknown function at the corresponding points x_i, and $f(x_i)$, $i=\overline{1,r}$ is the range of function values represented by the given lymphocyte at the corresponding points x_i. The form of the function *U* after 200 iterations of an artificial immune system is shown in Fig. 2

The first task is to define how the lymphocytes will be presented. Since we have a function as a solution of the problem, i.e. superposition of the defined simple functions, then the lymphocyte should be this function, encoded in a certain way. A convenient presentation of the lymphocytes for direct implementation, as well as for the further work of the algorithm, could be the expression tree. This tree will be binary by means of binary and unary operations. For example, for the expression $x-2*(1/x+x/3)$ the tree will look like the following (Fig. 1) [4-7]:

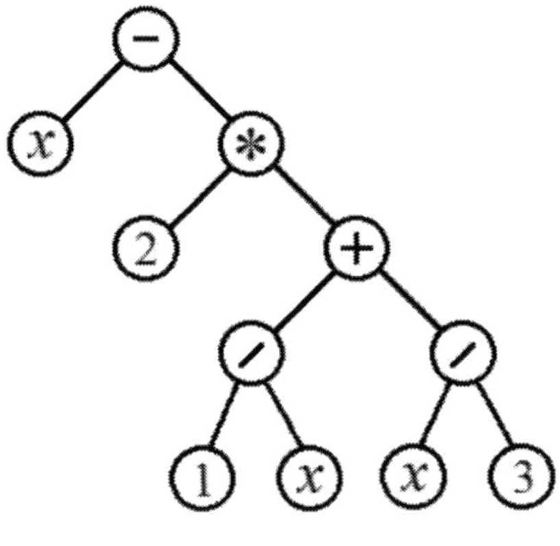

Fig. 1. The expression tree sample.

3 Algorithms of AIS realization

Let us describe a distributed immune system. The model p2p-interection [10] was chosen as a form of architecture of interaction of computing units. Peer-to-peer decentralized or peering (P2P - peer to peer) network is a computer network based on the equality of its participants. In this network there are no dedicated servers, and each node (peer) is both a client and a server. Unlike client-server architecture, this organization allows to maintain network performance with any number and any combination of available nodes.

In a distributed artificial immune system, each compute node will store some lymphocytes and at

DOI: 10.1051/itmconf/20170902003

certain times share best lymphocytes with the neighboring nodes. Let us consider the application of the described algorithm of symbolic regression for the solution of classification problems arising in environmental forecasting. The model is taken from [1]. Its basis is the diffusion equation in the flowing system: let the concentration of the pollutant $\Phi = \Phi(x, y, t)$ in the area of $\Omega \subset R^2$ at the moment of time $t \in (0, T)$ satisfies the following initial-boundary value problem:

$$\frac{\partial \Phi}{\partial t} + u\frac{\partial \Phi}{\partial x} + v\frac{\partial \Phi}{\partial y} - \sigma\left(\frac{\partial^2 \Phi}{\partial x^2} + \frac{\partial^2 \Phi}{\partial y^2}\right) + \tau\Phi = P,$$

$$\Phi(x, y, 0) = \Phi_0;\ \Phi(x, y, t) = \Phi_1, \qquad (1)$$

$$(x, y) \in \partial\Omega,$$

where function $P = P(x, y, t)$

$$P(x, y, t) = \sum_{s=1}^{k} p_s(t)\delta_s(x, y; x_s, y_s)$$

characterizes the capacity of pollution sources.

4 Computational experiment

Suppose that $p_s(t)$ is the capacity of i sources of pollution distributed in the vicinity of (x_s, y_s) with the density $\delta_s(x, y; x_s, y_s)$; u, v are the components of the velocity vector; σ is a turbulence diffusivity coefficient; $\tau < 0$ is the parameter determining the absorption intensity of pollutant due to its entrainment, deposition, and other chemical reactions.

Let us perform a replacement

$$\Phi(x, y, t) = \exp(Ax + By + Ct)U(x, y, t),$$

$$P(x, y, t) = \exp(Ax + By + Ct)Q(x, y, t),$$

where $A = u / 2\sigma,\quad B = v / 2\sigma,\quad C = -\tau - (u^2 + v^2) / 4\sigma,$ then the differential equation can be converted into the heat equation

$$\frac{\partial U}{\partial t} - \sigma\left(\frac{\partial^2 U}{\partial x^2} + \frac{\partial^2 U}{\partial y^2}\right) = Q. \qquad (2)$$

with initial-boundary conditions on the function U

$$U(x, y, 0) = U_0;\ U(x, y, t) = U_1, \quad (x, y) \in \partial\Omega.$$

Let us consider this transformed problem assuming that instead of pollution sources the capacity $Q(x, y, t)$, referred in the direct problem formulation, the results of experimental observations are known $U(x_j, y_j, t_j) = \varphi_j,\ j = 1, \ldots, M,$ derived from a set of sensors. The problem of identification is to restore function Q, in so doing there exists the approximate solution of the problem in the form of function U.

Let us define the input data and the affinity function of the applied AIS $\{(x_j, y_j, t_j)\}_{j=1}^{M}$ as the sampling points from the area Ω, in which the values of the function are known $U(x_j, y_j, t_j) = \varphi_j,\ j = 1, \ldots, M$, The role of affinity will be performed by the following functional, showing how well the found function satisfies the initial-boundary value problem and the known observation results [12-15]:

$$J = \sum_{j=1}^{M}\left|\frac{\partial U}{\partial t} - \sigma\left(\frac{\partial^2 U}{\partial x^2} + \frac{\partial^2 U}{\partial x^2}\right) - Q\right|^2 (x_j, y_j, t_j) +$$

$$+\lambda_0\sum_{j=1}^{M}\left|U - U_0\right|^2(x_j, y_j, 0) + \lambda_1\sum_{j=1}^{M_1}\left|U - U_1\right|^2(x_j, y_j, t_j) + \qquad (3)$$

$$\lambda_2\sum_{j=1}^{M}\left|U(x_j, y_j, t_j) - \varphi_j\right|^2.$$

This function is given piecewise: in the vicinity of the source, left and right, the lymphocyte will contain information about the respective three functions.

As a function that needs to be restored, we use the above function with the following parameters: $C = 1$, $c = 2,\ x_0 = 3,\ \delta = 1,\ \sigma = 2.$

Here $\{(x_j, y_j, t_j)\}_{j=1}^{M_1}$ are trial points in Γ; λ_0, λ_1, λ_2 are penalty parameters.

Since this problem is an approximate view of the two functions, the lymphocyte will consist of two expression trees that represent functions Q, U.

As an example the stationary solution in a one-dimensional case was used

$$U = \begin{cases} cx,\ x \in (0; x_0 - \delta], \\ \dfrac{C - c}{4\delta}(x - x_0)^2 + \dfrac{C + c}{2}x - \dfrac{C - c}{2}\left(x_0 - \dfrac{\delta}{2}\right), \\ \qquad x \in (x_0 - \delta; x_0 + \delta], \\ Cx - (C - c)x_0,\ x \in (x_0 + \delta; \omega). \end{cases} \qquad (4)$$

As a function that needs to be restored, we use the above function with parameters $C = 1,\ c = 2,\ x_0 = 3,$ $\delta = 1,\ \sigma = 2.$ Then the first part is equal to

$$Q = \begin{cases} 0,\ x \in (0; x_0 - \delta], \\ -\sigma\dfrac{C - c}{2\delta},\ x \in (x_0 - \delta;\ x_0 + \delta], \\ 0,\ x \in (x_0 + \delta; \omega). \end{cases}$$

Since the function is given piecewise: in the vicinity of the source, left and right, the lymphocyte will contain information about the respective three functions.

Then

$$U = \begin{cases} 2x,\ x \in (0; 2], \\ -\dfrac{1}{4}(x - 3)^2 + \dfrac{3}{2}x + \dfrac{5}{4},\ x \in (2; 4], \\ x + 3,\ x \in (4; 10). \end{cases}$$

$$Q = \begin{cases} 0,\ x \in (0; 2], \\ 1,\ x \in (2;\ 4], \\ 0,\ x \in (4; 10). \end{cases}$$

The form of the function U after 200 iterations of artificial immune system is shown in Fig. 2

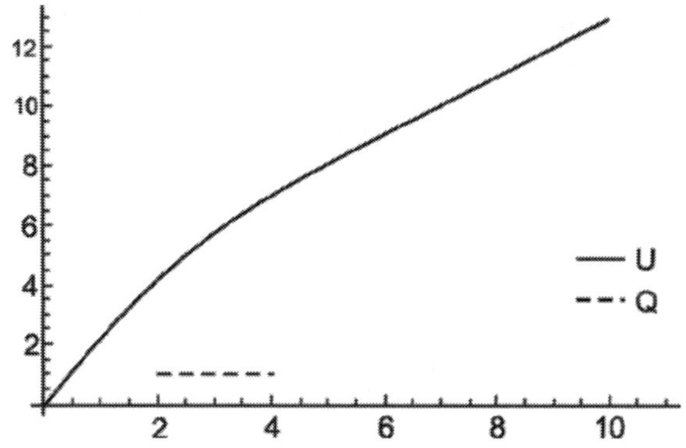

Fig. 2 Graph of the recovered functions *U* and *Q*

5 Realization

The main program consists of the following parts [16]:

1) Subsystem solution of the symbolic regression problem.

1.1 The module of a simple artificial system contains the classes of lymphocytes along with the methods to simplify expressions, to compute affinity and the algorithm of sequential implementation of the artificial immune system.

1.2 The module for the organization of the interaction between artificial immune systems in the network.

1.3 The module of implementation of a distributed artificial immune system.

1.4 The module of testing of the different functions of the subsystem, including the methods of computational experiment.

For the realization the Python language (version 3.3) has been selected [16]. This language has been chosen for the following reasons [16]:

• Cross-platform reason: during the creation of an application, running in a heterogeneous computing environment, the cross-platform factor is very important. Python allows you to create a program that does not require modification of the source code and recompiling even to work on a variety of platforms.

• The speed of development reason: Python is oriented on the increase of developer productivity and code readability. It also supports a variety of programming paradigms: structured, object-oriented, and functional [16].

• The standard library reason: includes a large number of useful functions and classes for various tasks such as networking, multithreading, process management.

During the creation process a system of version control Git has been utilized, with the source code stored on Github.

6 Conclusion

1. A unified model for the solution of identification problems is developed, where symbolic regression based on artificial immune system is using the fact that one of the main tasks of the immune system is to recognize harmful organisms and molecules.
2. A single algorithm implementing artificial immune system to solve identification problems in the symbolic regression is developed, along with its improved distribution version, using the property of immune system decentralization.
3. A software package, which includes the above-described algorithm and computational experiments, conducted for this proposed task, are developed. Further on it might be possible to use immune systems to build programs for searching antivirus and organization of the software protection against unauthorized access. Artificial immune system might happened to be just a simple device in the organization of information security process.

References

1. A.N. Vasilev, D.A. Tarhov, Parametric neural network models of building of regularization of identify problems solution in ecology. *Contemporary information technologies and IT-education* **1** (2014).
2. D. Dasgupta, *Artificial Immune Systems and their Application:* collection of papers. M.: FIZMATLIT (2006).
3. D. Dasgupta, S. Yua, F. Nino, Recent Advances in Artificial Immune Systems: Models and Applications. *Appl. Soft Computing* **11** (2011).
4. F. Freschi, M. Repetto, Multiobjective optimisation by a modified artificial immune system, *Artificial Immune Systems* (2005), 248–261.
5. N.X. Hoai, R.I. McKay, D. Essam, Solving the symbolic regression problem with tree-adjunct grammar guided genetic programming: the comparative results. Evolutionary Computation, CEC '02. *Proceedings of the 2002 Congress* **2** (200) 1326-1331.
6. I. E. Hunt, D. E. Cooke. Learning using an artificial immune system. *J. Network Comp. Appl.* **19** (1996) 189-212.
7. A.A. Ishiguro, Y. Watanabe, T. Kondo, Robot with a decentralized consensus-making mechanism based on the immune system. *Proceedings of ISADS* (1997) 231-237.
8. C.G. Johnson, Artificial immune systems programming for symbolic regression. *Genetic Programming: 6th European Conference* (2003) 345–353.
9. J.O. Kephart, A biologically inspired immune system for computers. *Proceedings of Artificial Life IV: The Fourth International Workshop on the Synthesis and Simulation of Living Systems* (1994) 130–139.
10. R.A. Schollmeier, Definition of peer-to-peer networking for the classification of peer-to-peer architectures and applications. *Proceedings of the First*

DOI: 10.1051/itmconf/20170902003

International Conference on Peer-to-Peer Computing, IEEE (2001) 101-102.
11. E. Hart, J. Timmis, Application areas of AIS: The past, the present and the future. *Appl. Soft Computing* **8** (2008) 191–201.
12. M. Bardeen, Survey of methods to prevent premature convergence in evolutionary algorithms. Workshop of Natural Computing, J. Chilenas de Computation (2013) 13–15.
13. D. Barkai, Peer-to-Peer Computing. Santa Clara: Intel Press (2002).
14 K. Bennett, M.C. Ferris, Y.E. Ioannidis, A genetic algorithm for database query optimization. *Proceedings of the fourth International Conference on Genetic Algorithms* (1991) 400-407.
15. H. Bersini, The endogenous double plasticity of the immune network and the inspiration to be drawn for engineering artifacts. *Artificial Immune Systems and Their Applications* (1999) 22-44.
16. I.F. Astakhova, S.A. Ushakov, Ju.V. Hitskova, Model and algorithm of an artificial immune systems for the recognition of single symbols and their comparison with existing methods. *WSEAS Trans. on Information Science and Applications* **13** (2016) 38-45.
17. J.R. Koza, Genetic Programming. *Cambridge: MIT Press* (1998). — 220 p.
18. W.B. Langdon, R. Poli, Foundations of Genetic Programming, *Heidelberg: Springer-Verlag* (2002).

A model of adding two edges in levels of a complete binary tree

Kiyoshi Sawada[1,*]

[1]Department of Economic Information, University of Marketing and Distribution Sciences, Kobe, Japan

Abstract. This study proposes a model of adding two edges between nodes of the same level of a complete binary tree. Firstly we add one edge between the optimal two nodes with the optimal depth N* maximizing a total shortening distance. Secondly we add another edge between nodes with the same depth M (M = 1, 2, ..., H). The total shortening distance to obtain the optimal two nodes with the optimal depth M* maximizing a total shortening distance is formulated.

1 Introduction

This paper proposes a new model of adding relations in an organization structure [1, 2]. Relations between members in an organization are added to become effective in communication of information.

This study considers a pyramid organization structure which is a complete binary tree of height H (H = 2, 3, ...). A complete binary tree is a rooted tree in which all leaves have the same depth and all internal nodes have two children [3]. Nodes and edges in the complete binary tree correspond to units and relations between units in the organization. Moreover the path between a pair of nodes in the complete binary tree is equivalent to the route of communication of information between a pair of units in the organization, and adding edges is equivalent to forming additional relations other than that between each superior and his direct subordinates [4-7].

We have proposed a model [4-7] of adding one edge between two nodes with the same depth N (N = 1, 2, ..., H) of a complete binary tree. In this model we have obtained the optimal two nodes with the optimal depth N^* maximizing a total shortening distance. The total shortening distance is the sum of shortened lengths of shortest paths between every pair of all nodes by adding edge between two nodes [4-7]. This means that the communication of information between every member in the organization becomes the most efficient.

This paper proposes a model of adding two edges between nodes of the same level of a complete binary tree. In this model firstly we add one edge between the optimal two nodes with the optimal depth N^* maximizing a total shortening distance [4-7]. Secondly we add another edge between nodes with the same depth M (M = 1, 2, ..., H). We formulate the total shortening distance to obtain the optimal two nodes with the optimal depth M^* maximizing a total shortening distance.

2 A model of adding one edge [4-7]

This section formulates the total shortening distance when an edge between nodes with the same depth N (N = 1, 2, ..., H) in a complete binary tree of height H (H = 2, 3, ...) and obtains the optimal two nodes with the optimal depth N^* maximizing a total shortening distance [4-7].

A new edge can be added between two nodes with the same depth N in a complete binary tree in N ways that lead to non-isomorphic graphs. Let $R_H(N, D)$ denote the total shortening distance by adding the new edge, where D (D = 0, 1, 2, ..., N − 1) is the depth of the deepest common ancestor of the two nodes on which the new edge is incident. When the total shortening distance for the case of D = 0 is denoted by $S_H(N)$, we obtain

$$R_H(N,D) = S_{H-D}(N-D). \tag{1}$$

We formulate $S_H(N)$ in the following.

Let v_0^X and v_0^Y denote the two nodes on which the adding edge is incident and assume that D = 0. Let v_k^X and v_k^Y denote ancestors of v_0^X and v_0^Y, respectively, with depth $N - k$ for k = 1, 2, ..., N − 1. The sets of descendants of v_0^X and v_0^Y are denoted by V_0^X and V_0^Y respectively. Let V_k^X denote the set obtained by removing the descendants of v_{k-1}^X from the set of descendants of v_k^X and let V_k^Y denote the set obtained by removing the descendants of v_{k-1}^Y from the set of descendants of v_k^Y, where k = 1, 2, ..., N − 1.

The total shortening distance can be formulated by adding up the following three sums of shortening distances: (i) the sum of shortening distances between every pair of nodes in V_0^X and nodes in V_0^Y, (ii) the sum

* Corresponding author: Kiyoshi_Sawada@red.umds.ac.jp

of shortening distances between every pair of nodes in V_0^X and nodes in V_k^Y (k = 1, 2, ..., N – 1) and between every pair of nodes in V_0^Y and nodes in V_k^X (k = 1, 2, ..., N – 1) and (iii) the sum of shortening distances between every pair of nodes in V_k^X (k = 1, 2, ..., N – 1) and nodes in V_k^Y (k = 1, 2, ..., N – 1).

The sum of shortening distances between every pair of nodes in V_0^X and nodes in V_0^Y is given by

$$A_H(N) = [W(H-N)]^2(2N-1), \tag{2}$$

where $W(h)$ denotes the number of nodes of a complete binary tree of height h (h = 0, 1, 2, ...). The sum of shortening distances between every pair of nodes in V_0^X and nodes in V_k^Y (k = 1, 2, ..., N – 1) and between every pair of nodes in V_0^Y and nodes in V_k^X (k = 1, 2, ..., N – 1) is given by

$$B_H(N) = 2W(H-N)\sum_{i=1}^{N-1}[W(H-i-1)+1](2i-1), \tag{3}$$

and the sum of shortening distances between every pair of nodes in V_k^X (k = 1, 2, ..., N – 1) and nodes in V_k^Y (k = 1, 2, ..., N – 1) is given by

$$C_H(N) = \sum_{i=1}^{N-2}[W(H-i-2)+1] \times \sum_{j=1}^{i}[W(H-N+j-1)+1](2i-2j+1), \tag{4}$$

where we define $\sum_{i=1}^{0} \bullet = 0$ and $\sum_{i=1}^{-1} \bullet = 0$.

From Equations (2), (3) and (4), the total shortening distance $S_H(N)$ for the case of D = 0 is formulated by

$$S_H(N) = A_H(N) + B_H(N) + C_H(N). \tag{5}$$

From Equations (1) and (5) we have $R_H(N, D)$. We have Theorem 1 since

$$R_H(N, D+1) - R_H(N, D) < 0. \tag{6}$$

Theorem 1. $D^* = 0$ *maximizes* $R_H(N, D)$ *for each* N [4-7].

Theorem 1 shows that the most efficient way of forming relations to two members in each level is that to two members which doesn't have common superiors except the top.

We have Theorem 2 since
(i) When H = 2, $S_H(2) - S_H(1) < 0$, and
(ii) When H = 3, 4, ..., $S_H(2) - S_H(1) > 0$ and $S_H(N+1) - S_H(N) < 0$ for N = 2, 3, ..., H – 1.

Theorem 2. [4-7]
(i) If H = 2, then N^* = 1.
(ii) If H = 3, 4, ..., then N^* = 2.

3 A model of adding two edges

This section proposes a new model of adding two edges between nodes of the same level of a complete binary tree. Firstly we add one edge between the optimal two nodes with the optimal depth N^* in Section 2 of this paper. Secondly we add another edge between nodes with the same depth M (M = 1, 2, ..., H). The total shortening distance to obtain the optimal two nodes with the optimal depth M^* maximizing a total shortening distance is formulated in this section.

Let n^X and n^Y denote the two nodes on which the optimal first adding edge with the optimal depth N^* is incident. Let m^X and m^Y denote the two nodes on which the second adding edge with the depth M is incident. Let DDCA(u, v) denote the depth of the deepest common ancestor of node u and node v.

The total shortening distance is formulated in the following cases.

(I) When H = 2, N^* = 1 and M^* = 2 trivially.

(a) When DDCA(m^X, m^Y) = 0, the total shortening distance is 2.

(b) When DDCA(m^X, m^Y) = 1, the total shortening distance is 1.

(II) When H = 3, 4, ..., N^* = 2.

(a) When DDCA(m^X, m^Y) = 0 and

(i) When DDCA(n^X, m^X) = 1 and DDCA(n^Y, m^Y) = 1, let $T_{1,H}(M)$ denote the total shortening distance for M = 1, 2, ..., H.

(ii) When DDCA(n^X, m^X) = 1 and DDCA(n^Y, m^Y) = 2 or DDCA(n^X, m^X) = 2 and DDCA(n^Y, m^Y) = 1, let $T_{2,H}(M)$ denote the total shortening distance for M = 2, 3, ..., H.

(iii) When DDCA(n^X, m^X) = 2 and DDCA(n^Y, m^Y) = 2, let $T_{3,H}(M)$ denote the total shortening distance for M = 3, 4, ..., H.

(b) When DDCA(m^X, m^Y) > 0, DDCA(m^X, m^Y) = 1 maximizes the total shortening distance for each M from Theorem 1 of Section 2. When DDCA(m^X, m^Y) = 1, let $T_{4,H}(M)$ denote the total shortening distance for M = 2, 3, ..., H.

$T_{1,H}(M)$ is given by

$$T_{1,H}(M) = [W(H-2)+1]^2, \tag{7}$$

for M = 1, and

$$\begin{aligned} T_{1,H}(M) = {} & [W(H-M)]^2(2M-1) \\ & + 2W(H-M)\sum_{i=1}^{M-2}[W(H-i-2)+1](2i+1) \\ & + 2W(H-M) + \sum_{i=1}^{M-2}[W(H-i-2)+1] \\ & \times \sum_{j=1}^{i}[W(H-M+j-1)+1](2i-2j+1), \end{aligned} \tag{8}$$

for M = 2, 3, ..., H.

$T_{2,H}(M)$ is given by

$$T_{2,H}(M) = [W(H-M)]^2(2M-2) + 2W(H-M)\sum_{i=1}^{M-2}[W(H-i-2)+1]2i + W(H-M)[W(H-2)+1] + \sum_{i=1}^{M-3}[W(H-i-3)+1] \times \sum_{j=1}^{i}[W(H-M+j-1)+1](2i-2j+2), \quad (9)$$

for $M = 2, 3, \ldots, H$.

$T_{3,H}(M)$ is given by

$$T_{3,H}(M) = [W(H-M)]^2(2M-4) + 2W(H-M)\sum_{i=1}^{M-3}[W(H-i-3)+1]2i + \sum_{i=1}^{M-4}[W(H-i-4)+1] \times \sum_{j=1}^{i}[W(H-M+j-1)+1](2i-2j+2), \quad (10)$$

for $M = 3, 4, \ldots, H$.

$T_{4,H}(M)$ is given by

$$T_{4,H}(M) = [W(H-M)]^2(2M-3) + 2W(H-M)\sum_{i=1}^{M-2}[W(H-i-2)+1](2i-1) + \sum_{i=1}^{M-3}[W(H-i-3)+1] \times \sum_{j=1}^{i}[W(H-M+j-1)+1](2i-2j+1), \quad (11)$$

for $M = 2, 3, \ldots, H$.

References

1. S.P. Robbins, *Essentials of Organizational Behavior, 7th Edition*, Prentice Hall, Upper Saddle River, N.J. (2003)
2. Y. Takahara, M. Mesarovic, *Organization Structure: Cybernetic Systems Foundation*, Kluwer Academic / Plenum Publishers, New York (2003)
3. T.H. Cormen, C.E. Leiserson, R.L. Rivest, C. Stein, *Introduction to Algorithms, 2nd Edition*, MIT Press, Cambridge, Mass. (2001)
4. K. Sawada, H. Uno, Transactions of the Japan Society for Industrial and Applied Mathematics, **10**, 335–346 (2000)
5. K. Sawada, *Proceedings of 18th International Conference on Production Research (ICPR)*, Salerno, Italy (2005)
6. K. Sawada, R. Wilson, European Journal of Operational Research, **174**, 1491–1500 (2006)
7. K. Sawada, *Proceedings of the 8th WSEAS International Conference on System Science and Simulation in Engineering*, Genova, Italy, 215–218 (2009)

ITM Web of Conferences 9, 03001 (2017) DOI: 10.1051/itmconf/20170903001
AMCSE 2016

A novel physics-inspired method for image region segmentation by imitating the carrier immigration in semiconductor materials

Xiaodong Zhuang[1,2], *Nikos* Mastorakis[2*], and *Mingyue* Ding[1]

[1]Qingdao University, Electronics Engineering College, Qingdao, 266071 China
[2]Technical University of Sofia, Industrial Engineering Department, Sofia, 1000 Bulgaria

Abstract. A novel method for image region segmentation is proposed, which is inspired by the carrier immigration mechanism in semiconductor materials. The carrier diffusing and drifting are simulated in the proposed model, and the sign distribution of net carrier at the model's balance state is exploited for region segmentation. The experiments have been done for test images and real world images, which prove the effectiveness of the proposed method.

1 Introduction

Self balancing is a mechanism existing in many natural systems. For example, the formation of the P-N junction in semiconductor materials is the result of balancing of the diffusing and drifting process of carriers. The physical P-N junction, the charge carriers in P-type and N-type semiconductor are holes and electrons respectively [1,2]. When the materials of the two are put together with compact contact, diffusion of carries will happen at the interface of contact due to carrier density difference (i.e. carrier moving from high-density side to low density side). Meanwhile, a space charge region is established. It in turn causes the drifting of carriers which is at the opposite direction of diffusing. The above process will reach a balance state [1,2]. In the self balancing mechanism shown in Figure. 1, the system's state at new balancing point may depend on the external influence (such as the external voltage applied onto the P-N junction). This mechanism may be the inspiration of novel methods for problem solving, if the self balancing mechanism suits the nature of the problem well.

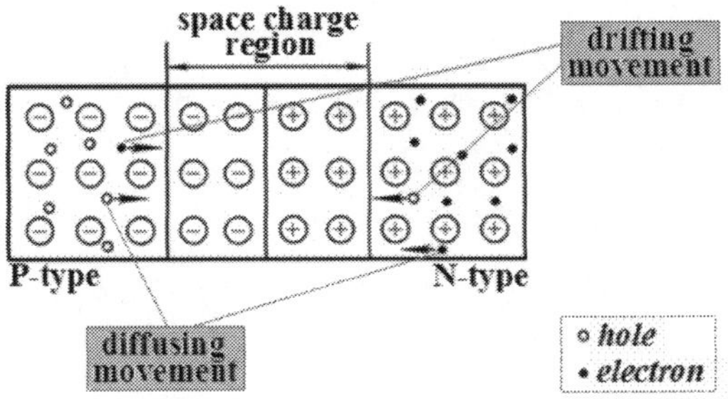

Fig. 1. The physical P-N junction.

Image segmentation is a fundamental problem in image processing, which has significant value in both theoretic and practical research [3-5]. The basis of differentiation of two adjacent regions is their difference of image characteristics. It is still an on-going and open research topic to segment image regions for various practical purposes [3-7]. In recent years, nature-inspired methods have attracted more and more research attention, in which the mechanisms in nature are imitated and adjusted in novel algorithms for image segmentation, and promising results have been obtained in such preliminary works [8,9]. In this paper, inspired by the physical P-N junction, the self-balancing mechanism of carrier diffusing and drifting is adopted in a novel segmentation framework, in which the region structure of image is formed by dynamic carrier diffusing and drifting. In the algorithm, the virtual electric field is defined artificially according to greyscale difference between adjacent image pixels, which introduces the factor of greyscale difference for segmentation.

2 The model of virtual carrier immigration in digital images

The proposed model for image segmentation is as follows. Two categories of virtual carriers are defined: positive and negative, which imitates the physical electron and hole. Each pixel is modelled as a container of virtual carriers. Each pixel has four adjacent pixels (except those on image borders), and correspondingly each carrier container has four adjacent containers. There is an interface between each pair of adjacent containers.

There are two features of the interface mentioned above. Firstly, the interface has permeability, which means the carriers at both sides of the surface can diffuse through it due to density difference. Secondly, there is a virtual electric field imposed on it, whose direction and intensity are determined by the greyscale difference between the corresponding two pixels connected by that interface. The virtual electric field is defined as:

* Corresponding author: mastor@tu-sofia.bg

DOI: 10.1051/itmconf/20170903001

$$e = K \cdot (g - g_a) \quad (1)$$

where e is the intensity of virtual electric field at the interface, K is a predefined positive coefficient, g is the greyscale of the pixel of interest and g_a is that of its adjacent pixel. The effect of each virtual electric field is limited to its corresponding interface only, and does not influence other interfaces. In such a way, the model is established consisting of virtual carrier containers and their interfaces, the virtual electric field, and also the virtual carriers.

The evolution of the system is analyzed as follows. Initially, suppose the positive and negative carriers are of the same quantity, and each container has the same amount of carriers. Also suppose all the containers have the same volume, so that the density of carrier in a container is proportional to the amount of carrier in it. Therefore, there is no density difference of carriers between adjacent containers at that time. In another word, there is no carrier diffusion at the beginning. However, due to the virtual electric field at each interface, the positive and negative carriers drift across the interfaces due to the virtual force applied by the electric field. The drifting then causes carrier density difference between two sides of the interface, which in turn makes the carriers to diffuse due to that density difference. Obviously, the diffusion has the opposite effect of drifting. For each interface and each container, such dynamic process evolves until a balance between drifting and diffusion is reached. Figure. 2 shows the details of carrier immigration between two adjacent containers, while Figure. 3 shows the overall structure of the model on digital image. The balance state is worth of study for possible use in image segmentation, and difference of the proposed algorithm and the physical process can also be clearly seen as above.

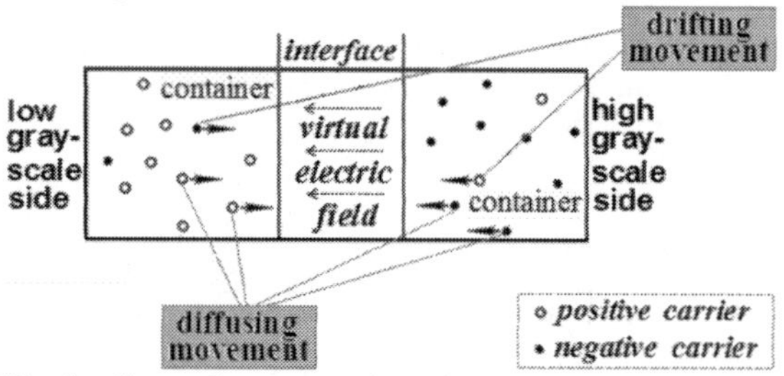

Fig. 2. Two adjacent containers in the proposed model of carrier immigration in digital images.

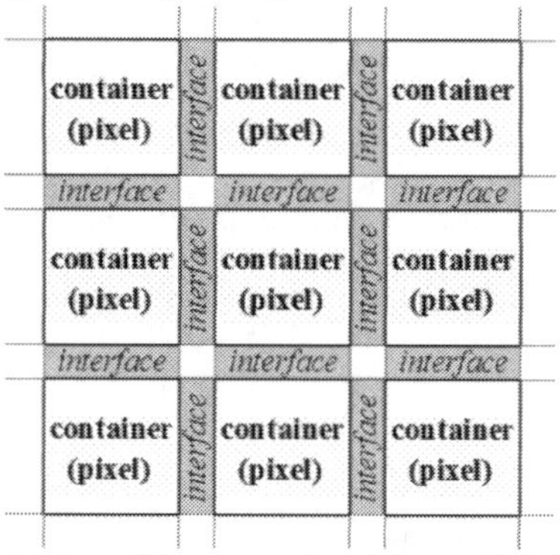

Fig. 3. The structure of the proposed model upon digital image.

3 Image segmentation based on virtual carrier immigration

3.1 Model implementation by computer simulation

In the simulation of the model on computer, the simulation must be implemented in discrete steps (i.e. iteration by iteration). In one simulation step, the drifting speed of carrier (i.e. the amount of carrier immigrating from one container to the other in an iteration of simulation, or one simulation step) is defined directly proportional to the intensity of virtual electric field:

$$\Delta c_{drifting} = K_1^{'} \cdot e \quad (2)$$

where $\Delta c_{drifting}$ is the amount of carrier drifting from one container into the other, $K_1^{'}$ is a predefined positive coefficient, e is the intensity of virtual electric field at the interface. According to Equation (1), $\Delta c_{drifting}$ is also proportional to the greyscale difference between the adjacent pixels:

$$\Delta c_{drifting} = K_1 \cdot (g - g_a) \quad (3)$$

where $\Delta c_{drifting}$ is the amount of carrier drifting from one container into the other, K_1 is a predefined positive coefficient, g is the greyscale of the pixel of interest and g_a is that of its adjacent pixel.

On the other hand, in one simulation step, the speed of diffusion has proportionality relationship with carrier density difference between the adjacent pixels. Suppose each container has the same size (or volume). Then the carrier density is proportional to the carrier amount in each container. Therefore, in the implementation the carrier density is substituted by carrier amount for diffusing:

$$\Delta c_{diffusing} = K_2 \cdot (c - c_a) \quad (4)$$

where $\Delta c_{diffusing}$ is the amount of carrier diffusing from one container into the other, K_2 is a predefined positive coefficient, c is the net carrier amount in the container of interest and c_a is that in its adjacent container. There are two types of carriers in the model: positive and negative. Each container has both types in it. In the evolving process, the two types of carrier immigrate by drifting and diffusing respectively. For each container, the net carrier is the combination of the two types after the offset between them. For simplicity in implementation, the immigration of carriers is measured by the amount of net carrier. In another word, the carrier density and the flow of carrier between containers are measured by net carrier amount.

The process of implementing the model is as follows. At the beginning, the amounts of positive and negative carriers are equal in each container. Also suppose the amount is sufficient for arbitrary amount of carrier immigration in the simulation. Then the virtual electric field is calculated at each interface between adjacent containers, which is proportional to the greyscale difference between corresponding adjacent pixels. The detailed simulation step is as follows:

Step1 For each of the four interfaces of every virtual container (or pixel), do the following: calculate the

 DOI: 10.1051/itmconf/20170903001

drifting amount of carrier due to virtual electric field; calculate the diffusing amount of carrier due to carrier density difference; sum the above two for all the 4 interfaces of a container to get its total change of net carrier amount; update the net carrier amount in that container;

***Step*2** After all the containers update their net carrier amount, calculate the average change of net carrier for all the containers. If the average change of net carrier is smaller than a predefined threshold, it is close enough to the balance state, and the simulation stops; otherwise, return to ***Step*1** to begin a new iteration of simulation.

3.2 Simulation results for simple test images

An example of the experimental results for test images is shown in Figure. 4. The simulation is implemented by programming in *C*. Figure. 4(a) shows a test image with a rectangle region. In order to demonstrate the process of carrier immigration step by step, the intermediate results at several specific simulation step numbers are recorded together with the final result. The intermediate results are shown from Figure. 4(b) to Figure. 4(f), where the white and black points represent the positive and negative sign of net carrier respectively, and the gray points represent that the net carrier is zero (i.e. not a definite sign yet). From Figure. 4(b) to Figure. 4(f), it is clear that the positive sign part expands inward, while the negative sign part expands outward from the rectangle border with increasing simulation time. The final result of net carrier distribution in Figure. 4(g) can provide a definite segmentation of Test image 1.

The above interesting results for test images can be analyzed as follows. At region borders with large greyscale variation, the dominant factor of carrier immigration is the drifting caused by the strong virtual electric field. As the overall effect, for a pair of containers, the positive carriers tend to gather into one container, while the negative carriers gather into the other. Such effect will increase the difference of net carrier amount between the adjacent containers, and the sign of the net carriers in them also tend to become opposite.

On the other hand, for a local area within a region, the greyscale difference is small. Correspondingly, the virtual electric field is also small. Here the dominant factor for carrier immigration is the density difference of carriers between adjacent containers. As the result, the amount and sign of net carrier for the containers inside a region tend to become homogeneous. Therefore, the global distribution of the sign of net carriers at balance state can provide effective basis for image segmentation.

3.3 Image segmentation based on the proposed model for real world images

In the above experimental results for the test images, it is shown that the sign of net carrier are opposite in two adjacent regions, which can provide the basis of region division in images. In order to obtain the segmentation result from the sign distribution of the net carrier, a region grouping approach is proposed as following:

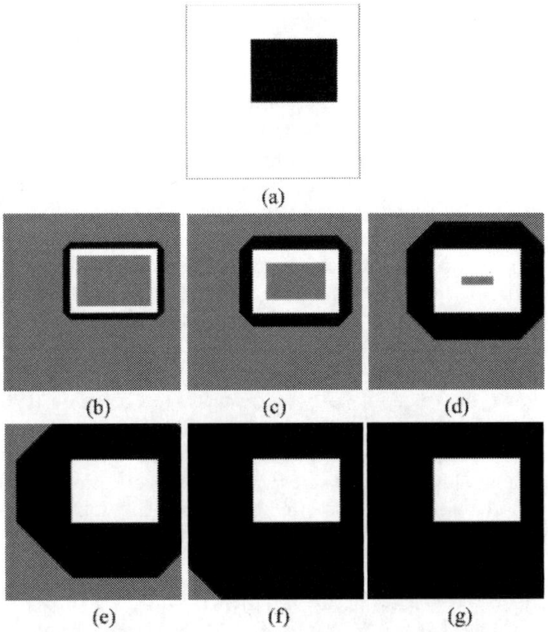

(a) the test image
(b) sign distribution of net carrier after 5 simulation iterations
(c) sign distribution of net carrier after 10 simulation iterations
(d) sign distribution of net carrier after 20 simulation iterations
(e) sign distribution of net carrier after 40 simulation iterations
(f) sign distribution of net carrier after 80 simulation iterations
(g) sign distribution of net carrier after 120 simulation iterations

Fig. 4. The simulation results for one test image.

***Step*1**: Implement the simulation of the virtual carrier immigration as proposed in section 4.1;
***Step*2**: Obtain the sign distribution of the net carrier;
***Step*3**: Group the adjacent containers (i.e. image points) with the same sign of net carrier as connected points in same region. In the region grouping process, the adjacent pixels of the 4-connection (i.e. the upper, lower, left and right pixels) for an image point *p* is investigated.

However, real world images are more complex. To investigate the effect of the proposed method, experiments are carried out for a series of real world images. For demonstration, an example of the results is shown in Figure. 5, which is for the medical heart image. The experimental results indicate that the proposed method can obtain large amount of regions (more than a hundred). There are 533 for the medical heart image. To obtain practically useful segmentation result, a region merging method is proposed based on the gray-scale similarity of adjacent regions. Given an expected number of remaining regions after merging, the following steps are carried out to merge regions:

***Step*1**: Calculate its average gray-scale value for each region.
***Step*2**: Find the pair of neighboring regions with the least difference of the average gray-scale, and merge them into one region.
***Step*3**: If current region number is larger than the expected number, return to ***Step*1**; otherwise, stop the merging.

In Figure 5, the following results show the original image, the sign distribution of net carrier at the balance state, the region segmentation results by grouping, and also the result of region merging. In the sign distribution of net carrier, the white points represent positive net carrier, and black points represent negative net carrier. In the region segmentation results and region merging results, different regions are represented by different greyscale values. For the medical image of the heart, the remained region number after merging is 50 in Figure. 5(d). Figure. 5(d) shows the heart structure clearly. Moreover, the average of the net carrier change for all the points is calculated and recorded as a measurement of the convergence degree to the balance state. Figure. 6 shows the relationship between that average value and the simulation time, which indicates that the process of carrier immigration approaches the balance state with the simulation going on. The experimental results prove that the proposed method is effective in segmentation of real world images.

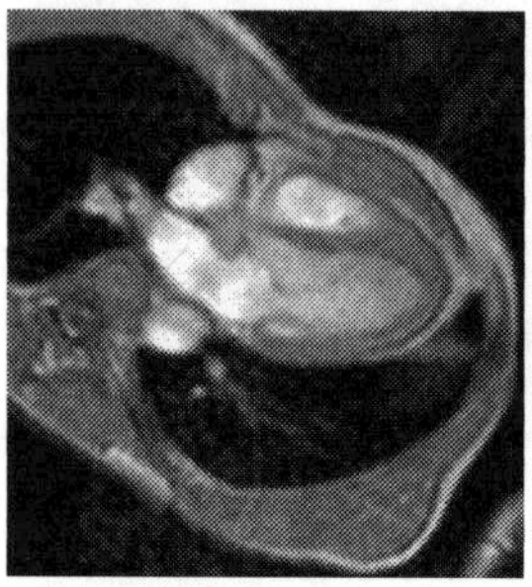

(a) the medical image

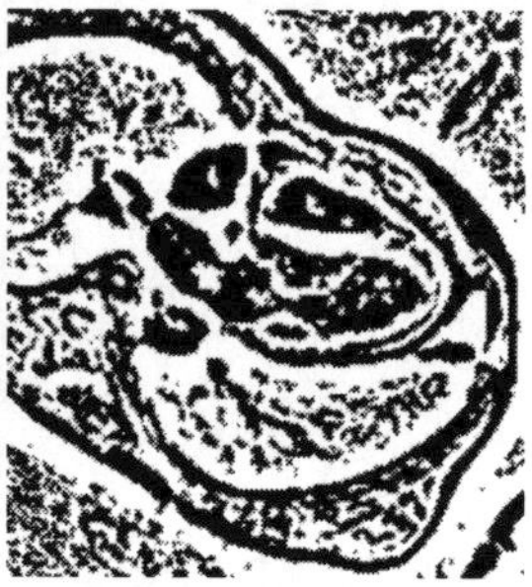

(b) the sign distribution of z coordinate at balance state

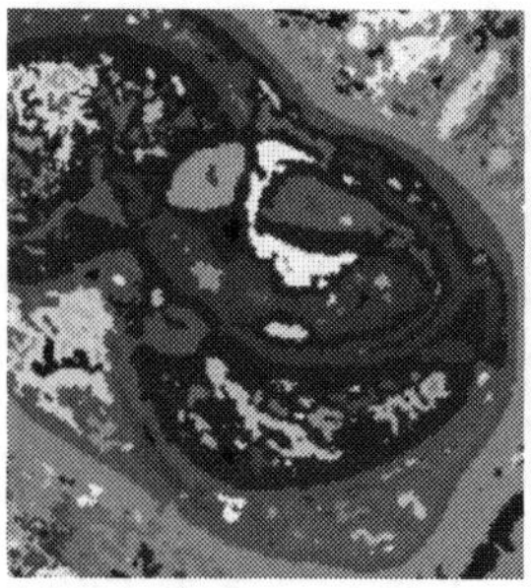

(c) the region segmentation

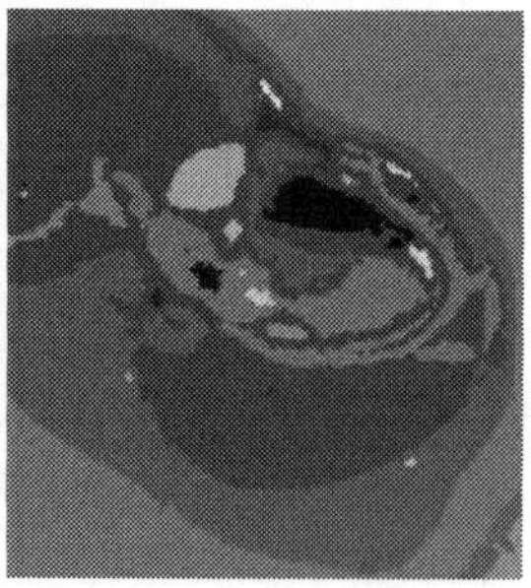

(d) the merging result for (c) for (b)

Fig. 5. The experimental results for the medical heart image.

4 Conclusion

A novel model of virtual carrier immigration is presented by imitating the diffusing and drifting of carriers in physical P-N junction. The virtual electric field between adjacent pixels is defined according to their greyscale difference, which is the major difference between the proposed model and real P-N junction. The direct local interaction and indirect global interaction of the above two carrier movements can lead to a balance state of carrier distribution, which provides clues for region segmentation. Image segmentation is implemented based on the sign distribution of net carrier at balance state, and a merging step is applied to get more comprehensible and useful segmentation results. The experimental results for test images and real world images prove the effectiveness of the proposed method. For future improvement of the segmentation results, color and texture feature will also be introduced into segmentation for possible improvement.

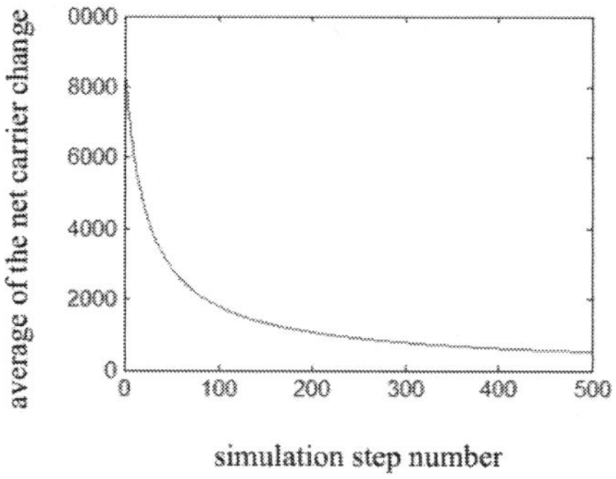

Fig. 6. The relationship between the average change of net carrier and the simulation time (for the medical heart image).

References

1. S. H. Simon, *The Oxford Solid State Basics*, (Oxford University Press, 2013)
2. W. A. Harrison, *Solid State Theory*, (Dover Publications, 2011)
3. Yujin Zhang, *Advances in Image And Video Segmentation*, (IRM Press, 2006)
4. I. V. Gribkov, P. P. Koltsov, N. V. Kotovich, A. A. Kravchenko, A. S. Koutsaev, A. S. Osipov, A. V. Zakharov, Testing of image segmentation methods, *WSEAS Transactions on Signal Processing*, **4**, 8, pp. 494-503 (2008)
5. N. Senthilkumaran, R. Rajesh, Image Segmentation – A Survey of Soft Computing Approaches, *Advances in Recent Technologies in Communication and Computing*, (ARTCom '09), pp. 844-846 (2009)
6. A. Gacsádi, L. Tepelea, I. Gavrilut, O. Straciuc, Energy based medical imaging segmentation methods by using Cellular Neural Networks, *Proceedings of the 15th WSEAS International Conference on Systems*, pp. 190-195 (2011)
7. I. A. Yusoff, N. A. Mat Isa, Two-dimensional clustering algorithms for image segmentation, *WSEAS Transactions on Computers*, **10**, 10, pp. 332-342 (2011)
8. M. S. Nixon, X. U. Liu, C. Direkoglu, D. J. Hurley, On using physical analogies for feature and shape extraction in computer vision, *Computer Journal*, **54**, 1, pp. 11-25 (2011)
9. X. D. Zhuang, N. E. Mastorakis, The relative potential field as a novel physics-inspired method for image analysis, *WSEAS Transactions on Computers*, **9**, 10, pp. 1086-1097 (2010)

Static and Dynamic Pedestrian Detection Algorithm for Visual Based Driver Assistive System

Idoko John Bush[1,*], *and Kamil Dimililer*[2]

[1]Computer Engineering Department, Near East University, Nicosia, Cyprus.
[2]Electrical & Electronic Engineering Department, Near East University, Nicosia, Cyprus.

Abstract. This paper presents a new pedestrian detection algorithm used in Advanced Driver-Assistance System with only one camera aiming to improving traffic safety. The new pedestrian detection algorithm differs from traditional pedestrian detection algorithm, which only focuses on pedestrian detection rate or pedestrian detection accuracy. Conversely, the proposed algorithm focuses on both the accuracy and the rate. Some new features are proposed to improve pedestrian detection rate of the system. Also color difference was used to decrease the false detecting rate. The experimental results show that the pedestrian detection rate can be around 90% and the false detecting rate is 3%.

Keywords. Pedestrian detection, detection rate, accuracy and false detection.

1. Introduction

Increasing concern for pedestrian safety in the last years has resulted in the flourishing of pedestrian detection algorithms. These are essential in Advanced Driving Assistance Systems for preventing accidents involving pedestrians. Car companies are considering incorporating such systems into their models. A number of research work have been done in car automatic protection systems, which is referred to as advanced driver assistance system (ADAS), to reduce the probability of the accidents.

Even though the problem was analyzed and tackled by many researchers it remains largely unsolved due to several difficulties:

i. The appearance of pedestrians in the image is always with high variability as the pose, the clothing and the articulations of the body parts can be various.

ii. Accidents always occur in urban environments where the background is complicated.

iii. When pedestrians are far from the moving vehicle, they may look similar to many background objects such as trees, telegraph pole and so on.

iv. As the speed of the vehicle is high, the pedestrian detection and the forewarning should be real time.

Several methods have been developed to improve the pedestrian detection rate and reduce the processing time. F. Garcia et al. [1] used a laser scanner and a far infrared camera to construct an ADAS. Alberto Broggi et al. [2] used four laser scanners and a camera to detect pedestrians. Junfeng Ge et al. [3] proposed a vision system for pedestrian detection and tracking during nighttime at the cost of a near-infrared camera. However, all the algorithms proposed are expensive from manufacturers' and economical angle. So this paper proposed a new algorithm to be used for ADASs with high pedestrian detection rate and less processing time. Also the new algorithm only needs one cheap camera.

2. Review of existing works

For the purpose of this paper, detail of few related works will be presented. These are papers which are strongly correlated with the approach implored in this paper and their description is needed for comparison. For a comprehensive overview of pedestrian detection algorithms the reader should consult the technical literature surveys [4], [5], [6], [7]. Despite the fact that there exist a multitude of approaches there is a tendency towards a general system architecture that is employed by most of them. This architecture is used to present and to emphasize different parts of the existing methods and the approach suggested in this paper. This paper mainly follows the description in [7] and state that the general pedestrian detection system has the following modules: pre-processing, feature extraction, region of interest selection (or foreground segmentation), object classification, post-processing (verification and refinement) and tracking.

To start off, traditional approach is described based on the method developed by Dalal [8] and extended by several other researchers [9], [10]. It is based on a fixed-size sliding window detection algorithm. To enable detection of pedestrians of different heights, the algorithm needs to resize the image and to recalculate the features for each scale. This is necessary because of two reasons: the detection window is of fixed size, and the features are not scale invariant. To obtain good results resizing must be done 4–8 times per octave and typically on 4–5 octaves. This leads to 16–40 feature recalculation steps. This is considered the weak point in similar approaches and the aim of this paper is to circumvent this situation.

* Corresponding author: john.bush@neu.edu.tr

Using two innovative ideas a recent publication by Benenson et al. [11] claim to achieve pedestrian detection at more than 100 frames per second. One of the ideas is to resize the features not images. The other is to use depth information for a successful region of interest selection. In this paper, a new pedestrian detection algorithm is proposed based on Adaboost, which is not only simple, but also with high pedestrian detection rate and high real time adjustment. Moreover, it is very cheap as only one camera is used making the system economical.

3. Proposed algorithm

The algorithm proposed in this paper contains two parts; the first one is to use feature based on Adaboost as a base algorithm, and the second one is to use color information to refining detection results.

3.1. Adaboost based algorithm

Paul Viola et al. have proposed a wavelet-based Adaboost method to integrate appearance information with motion information to detect pedestrian with great speed. The detector of this pedestrian detection system is based on the rectangle filters presented by Viola and Jones [12]. Even the wavelet-based Adaboost system contains both the static features and the motion features, the features are somewhat limited without considering the difference between the legs or arms of walking pedestrians and the background. The success rate is only 60%. So in this paper new feature shown in Fig.1 is added to express the difference between the pedestrian and background in one frame, to improve the detection rate to meet the demand of ADAS.

The sum of pixels of the white rectangles, which is multiplied by 2, is subtracted from the sum of pixels of the dark rectangles.
The information about the difference between the pedestrian and the background can be extracted as follows:

$$S2 = 3 \times S_{white} - S_{black} \quad (1)$$

$$S3 = 9 \times S_{white} - S_{black} \quad (2)$$

Where S2 is the value of the 2×2 rectangle feature as the left part in Fig.1 and S3 is the value of 3×3 rectangle feature as the right part in Fig.1. S_{white} is the sum of white pix in the rectangle and S_{black} is the sum of black pix in the rectangle.

Adaboost was used to select a set of features to build the best classifier as in [12]. The layer of the Adaboost is 15 as Markus Enzweiler has pointed out that the performance of the system runs in saturation when the number of layers of Adaboost is more than 15 [13]. For each layer the detection rate is 99.5% and the false positive is 50%.

Fig. 1. The new features stand for the difference between the legs or arms of a walking pedestrian and the background.

3.2. Refining the results of pedestrian detection maintaining the integrity of the specifications

Markus Enzweiler et al. [13] have pointed out all the pedestrian detection systems make rather similar mistakes that typical false detections between pedestrian and the things which are dominated by strong vertical structure occur in local regions. For example, there are always false positive between the pedestrian and the trees. Fig.2 (a) shows some false positives and the statistical result of different H values of the trees and Fig.2 (b) shows some true positives and the statistical result of different H values of the pedestrians. Several pedestrian detection algorithms contain one step to refining the results of pedestrian detection [14] [15] [16]. Most refinement methods use the multi-frame information to reduce the false positive, which means the target must be tracked before refining. However, these methods, which are at cost of time, can't meet the real time request of the system. So, a new method was proposed using the color information to refine the detecting results in order to reduce the false positives between the trees and the pedestrians.

Fig. 2 (a). Some false positives and the statistical result of different H values of the trees

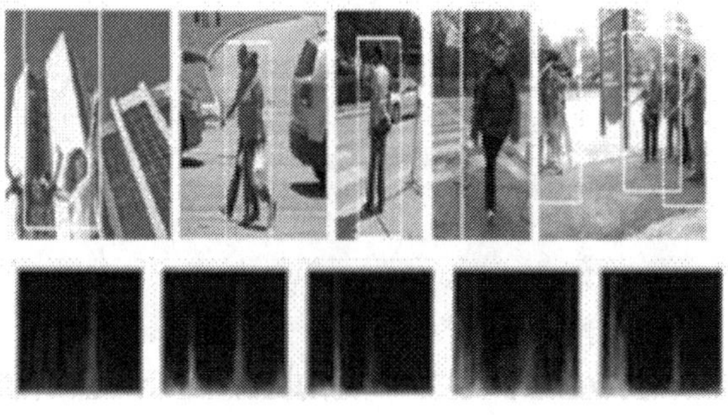

Fig. 2 (b). Some true positives and the statistical result of different H values of the pedestrians

The x axis stands for the h value from 0° to 360°, while the y axis represents for the total number for the corresponding H value. The left two figures are typical results of all the experiments.

Considering the fact that the color of the trees and the pedestrian are usually different and the H value of the objects doesn't change with the light, H value of the trees and the pedestrian was used to distinguish them.

The first step of the refining process is to calculate the H value for each pixel in the rectangle which is detected by classifier.

$$H = \begin{cases} 0, \text{ if } max = min \\ 60^0 \times \frac{G-B}{max-min} + 0^0, \text{ if } max = R,\ G > B \\ 60^0 \times \frac{G-B}{max-min} + 360^0, \text{ if } max = R,\ G < B \\ 60^0 \times \frac{B-R}{max-min} + 120^0, \text{ if } max = G \\ 60^0 \times \frac{R-G}{max-min} + 240^0, \text{ if } max = B \end{cases} \quad (3)$$

Where R,G,B is the original color information of an image, max is the maximum of the R value, G value and B value while the min is the minimum of the R value, G value and B value.

The second step is to make a count on the amount of each H value. We have done experiments on 500 pictures which contain 280 pedestrian images and 220 non pedestrian images. Fig.2 (a) and (b) shows some results of the H value of the pedestrians and the trees.

Then the last step is to remove the false positives of the trees according to the percentage of the H value within the range from 90° to 200°:

$$P_H = n_{range}/N_{all} \quad (4)$$

Where P_H is the percentage of which the H value is in the range 900 to 2000, n_{range} is the total number of the pixels, and the N_{all} is the sum of all the number with different H value. If P_H is more than a threshold, there should be a false positive.

4. Experimental results

In this section, some experimental results were presented to evaluate the pedestrian detection algorithm. The tests contain two parts, the first one is to use images captured in our university campus (Near East University, Cyprus) to evaluate detection rate and false positive rate. The second one is to test the robustness of the algorithm in real environment in the campus with a camera on a vehicle.

4.1. Evaluation of the new features

In order to test the usefulness of the new features, we give a comprehensive comparison between the different pedestrian detection system using different features. In the implementation, the number of positive examples is 280 while the number of negative examples is 220. All the positive examples are normalized to 32×80. Adaboost layer 15 was selected, the detection rate 99.5% and the false positive 50%. Fig.3 shows the comparing results. As expected, the detection rate of the pedestrian system with the new features is much higher than that without the new features at the same false positive.

4.2. Evaluation of; refining the detecting Results module

Even though the new features increase the detection rate when the false positive is small, there are still the false positives between the trees and the pedestrians. So the aim of this test is to evaluate detecting result refining module. The test lasted for 22minutes in our university campus where there are trees at both sides of the road. Then statistics was carried out on the set of 500 images. See Fig.4 for the result. We can conclude that, refining the detecting results module reduce the false positives at the same detection rate.

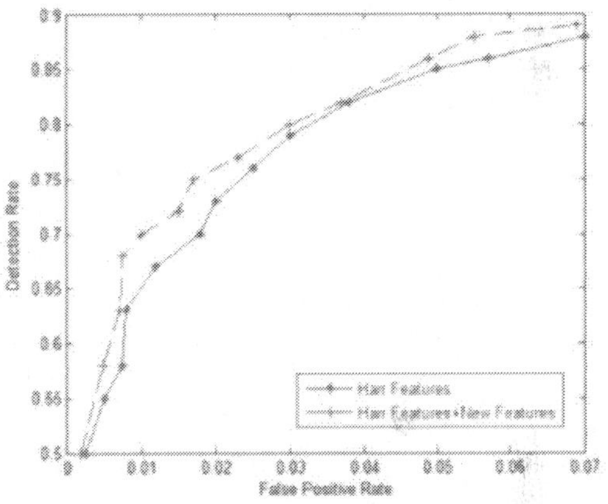

Fig. 3. Performance comparison of the pedestrian detection system using different features.

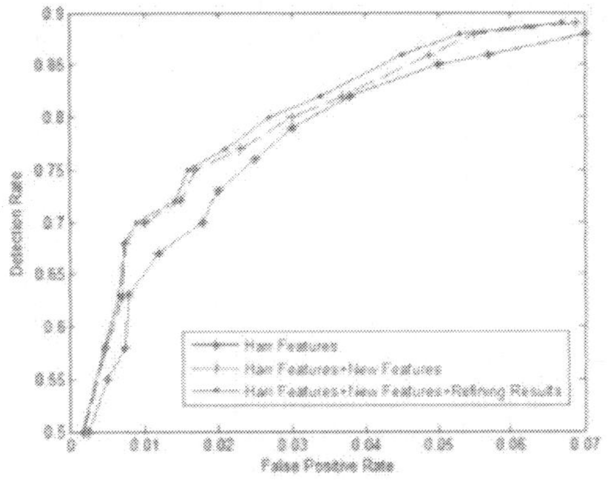

Fig. 4. Performance comparison between the system with refining detecting results module and that without refining detecting results module.

4.3. Evaluation of the robustness

Also the pedestrian detection algorithm was tested on a real car with different vehicle speed in Near East University campus. The vehicle speed of the first experiment is around 30km/h, and the second one is 35km/h. The system was built on Intel Pentium 2 processor with 2.0GHz and 4GByte Memory with Window10 system and the program was written in C++.

Table1 shows the system performance under different vehicle speeds. As expected, the detection rate of the system is around 90% in both cases, and the false alarm is under 2.

Table 1. System Performance

Vehicle Speeds	System Performance		
	Running Speeds(ms)	Detection Rate	False Alarm
30km/h	42	94.3%	1
35km/h	44	89.8%	2

5. Conclusion

This paper presents a new pedestrian detection algorithm which not only improves pedestrian detection rate but also decreases false positive rate. 500 images have been tested to evaluate the performance of the algorithm. Also we have tested the algorithm in real environment to evaluate whether the algorithm is available in real ADASs. The results indicate that the pedestrian detection rate can be improved to 90% and the false positive rate can be decreased to 3%. The real experiments in Near East University campus proved the robustness of algorithm with different speeds of cars, which showed the algorithm can meet the requirement of ADASs.

Future work in the field and environment (Near East University) will include integration of camera calibration to pedestrian detection in order to get the real distance between pedestrians and vehicles so that we can give an advanced alarm to avoid some traffic accidents. Also, comparison between the future work and other related works will be done to validate the authenticity of the pedestrian detector.

References

[1] F. Garcia, D. Olmeda, J. M. Armingol and A. de la Escalera, "Hybrid fusion scheme for pedestrian detection based on laser scanner and far infrared camera," Intelligent Vehicles Symposium, pp.1134-1139, Jun. 2010.

[2] Alberto Broggi, PietroCerri, Stefano Ghidoni, Paolo Grisleri and Ho Gi Jung, "A new approach to urban pedestrian detection for automatic braking," IEEE Trans. Intell. Transp. Syst., vol. 10, no. 4, pp.594-605, Dec. 2009.

[3] D. Gavrila, J. Giebel and S. Munder, "Vision-Based Pedestrian Detection: The PROTECTOR System," Proc. IEEE Intelligent Vehicles Symp.,pp: 13-18, 2004.

[4] P. Dollar, C. Wojek, B. Schiele, and P. Perona, "Pedestrian detection: A benchmark," in CVPR, 2009, pp. 304–311.

[5] P. Doll´ar, C. Wojek, B. Schiele, and P. Perona, "Pedestrian detection: An evaluation of the state of the art," IEEE Trans. Pattern Anal. Mach. Intell, vol. 34, no. 4, pp. 743–761, 2012.

[6] M. Enzweiler and D. M. Gavrila, "Monocular pedestrian detection: Survey and experiments," IEEE Trans. Pattern Analysis and Machine Intelligence, vol. 31, no. 12, pp. 2179–2195, Dec. 2009.

[7] D. Ger´onimo, A. M. L´opez, A. D. Sappa, and T. Graf, "Survey of pedestrian detection for advanced driver assistance systems," IEEE Trans. Pattern Anal. Mach. Intell, vol. 32, no. 7, pp. 1239–1258, 2010.

[8] N. Dalal and B. Triggs, "Histograms of oriented gradients for human detection," in CVPR, 2005, pp. I: 886–893.

[9] Q. A. Zhu, M. C. Yeh, K. T. Cheng, and S. Avidan, "Fast human detection using a cascade of histograms of oriented gradients," in CVPR, 2006, pp. II: 1491–1498.

[10] T. Watanabe, S. Ito, and K. Yokoi, "Co-occurrence histograms of oriented gradients for pedestrian detection," in PSIVT, 2009, pp. 37–47.

[11] R. Benenson, M. Mathias, R. Timofte, and L. J. V. Gool, "Pedestrian detection at 100 frames per second," in CVPR. IEEE, 2012, pp. 2903– 2910.

[12] P. Viola and M. Jones, "Rapid object detection using a boosted cascade of simple features," IEEE Conference on Computer Vision and Pattern Recognition, 2001.

[13] Markus Enzweiler, "Monocular Pedestrian Detection: Survey and Experiments," IEEE Transactions on pattern analysis and machine intelligence, vol. 31, no. 12, pp: 2179-2195, Dec. 2009.

[14] D.M. Gavrila and S. Munder, "Multi-Cue Pedestrian Detection and Tracking from a Moving Vehicle," Int'l J. Computer Vision, vol. 73, no. 1, pp: 41-59, 2007.

[15] Roland Schweiger, Heiko Neumann and Werner Ritter, "Multiple-cue data fusion with particle filters for vehicle detection in night view automotive applications," IEEE Intelligent Vehicle Symposium, pp:753758, 2005.

[16] D. Gavrila, J. Giebel and S. Munder, "Vision-Based Pedestrian Detection: The PROTECTOR System," Proc. IEEE Intelligent Vehicles Symp.,pp: 13-18, 2004.

 DOI: 10.1051/itmconf/20170903003

One-Quarter-Car Active SuspensionModel Verification

Katerina Hyniova

Faculty of Information Technology, CTU in Prague, Thakurova 9, Prague 6, 16000 Czech Republic

Abstract Suspension system influences both the comfort and safety of the passengers. In the paper, energy recuperation and management in automotive suspension systems with linear electric motors that are controlled by a designed H∞ controller to generate a variable mechanical force for a car damper is presented. Vehicle shock absorbers in which forces are generated in response to feedback signals by active elements obviously offer increased design flexibility compared to the conventional suspensions with passive elements (springs and dampers). The main advantage of the proposed solution that uses a linear AC motor is the possibility to generate desired forces acting between the unsprung (wheel) and sprung (one-quarter of the car body mass) masses of the car, providing good insulation of the car sprung mass from the road surface roughness and load disturbances. As shown in the paper, under certain circumstances linear motors as actuators enable to transform mechanical energy of the vertical car vibrations to electrical energy, accumulate it, and use it when needed. Energy flow control enables to reduce or even eliminate the demands on the external power source. In particular, the paper is focused on experiments with active shock absorber that has been taken on the designed test bed and the way we developed an appropriate input signal for the test bed that as real road disturbance acts upon the vibration absorber and the obtained results are evaluated at the end. Another important point the active suspension design should satisfy is energy supply control that is made via standard controller modification, and which allows changing amount of energy required by the system. Functionality of the designed controller modification was verified taking various experiments on the experiment stand as mentioned in the paper. .

1 Introduction

At the Czech Technical University in Prague various alternative strategies and innovations [2], [5] to classical passive suspension systems improving ride comfort of the passengers, providing steering stability, maximizing safety and improving handling properties of vehicles has been researched. In order to improve handling and comfort performance instead of a conventional static spring and damper system, semi-active and active suspension systems has been developed. Certainly there are numerous variations and different configurations of vibration suspension as well as control strategies [3], [4], [6]. In known experimental active systems hydraulic or pneumatic actuators usually provide the force input. As an alternative approach to active suspension system design, the research group has studied electromechanical actuators. Such actuators would provide a direct interface between electronic control and the suspension system.

In most active suspension systems, the biggest disadvantage consists in energy demands. Regarding linear electric motors, this drawback can be minimized or even eliminated because under certain circumstances there is a possibility to recuperate energy, accumulate it and use it later for the shock absorber when necessary. This way, it is possible to reduce the posted claims on the external power source as much as possible. In the next paragraphs also the proposed strategy how to control the energy distribution will be described. In order to regenerate electric power from the vibrations excited by road unevenness a new energy-regenerative active suspension for vehicles has been designed. The active system has been modeled and simulated to show the performance improvement and the performance experiments of the actuator prototype-testing stand have been carried out.

All suspension systems are designed to meet variable specific requirements. In suspension systems, mainly two most important points are supposed to be improved - disturbance absorbing (videlicet passenger comfort) and attenuation of the disturbance transfer to the road (videlicet car handling). The first requirement could be understood as an attenuation of the sprung mass acceleration or as a peak minimization of the sprung mass vertical displacement. The second one is characterized as an attenuation of the force acting on the road or - in the simple car model - as an attenuation of the unsprung mass acceleration. The goal is to satisfy both the above given contradictory requirements. Satisfactory results can be achieved when an active suspension systems generating variable mechanical force acting between the sprung and unsprung masses is used. Such an actuator can be a linear electric motor. In comparison with traditional actuators that use revolving electro-motors and a lead screw or toothed belt, the direct drive linear motor enables contactless transfer of electrical power according to the laws of magnetic induction. The gained electromagnetic force is applied directly without the intervention of a mechanical transmission then. Linear electric motors are easily controllable and for features like low friction, high accuracy, high acceleration and velocity, high values of generated force, high reliability and long lifetime their usage as shock absorbers seems to be ideal.

2 One-Quarter-Car Suspension Model and Shock Absorber Test Stand

A traditional one-quarter-car model has been used to design a suspension controller design and to simulate the system behavior. The basic configuration of the model is plotted in Fig.1. The model involves unsprung (wheel) and sprung (taken as one ideal quarter of the car body mass) masses, conventional passive suspension (a spring and a damper), stiffness of the tire, and linear electric motor as actuator placed in parallel to the traditional passive suspension.

In Fig. 1:

F_a control input (active suspension force) [N]
m_w unsprung mass (wheel) [kg]
m_b sprung mass supported by each wheel and taken as equal to a quarter of the total body mass [kg]
k_2 stiffness of the tire [N/m]
$z_r(t)$ road displacement (road disturbance) [m]
$z_b(t)$ displacement of the sprung mass [m]
$z_w(t)$ displacement of the unsprung mass [m]
k_1 stiffness of the passive suspension [N/m]
b_1 damping quotient of the passive suspension [Ns/m]

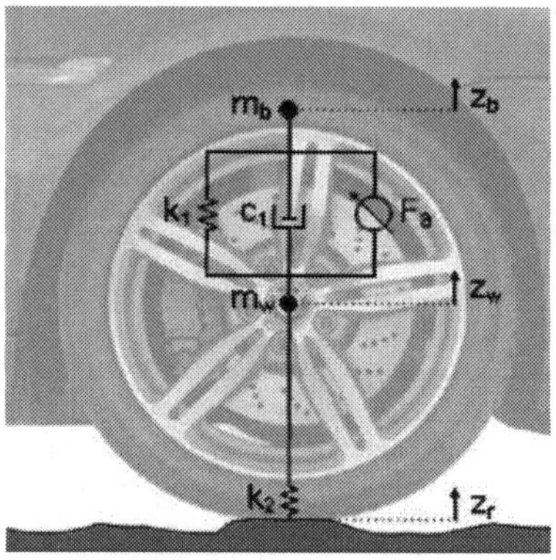

Fig.1 One-quarter-car model

The same configuration has been used for real experiments. Mechanical configuration of the test stand is obvious from Fig.2. Under the tyre there is placed another linear electric motor that uses an input experimental signal described in next paragraphs to generate road displacement (road deviations) under the running wheel.

As will be mentioned later the controller has been developed via Matlab implemented into dSpace a connected to the test stand system.

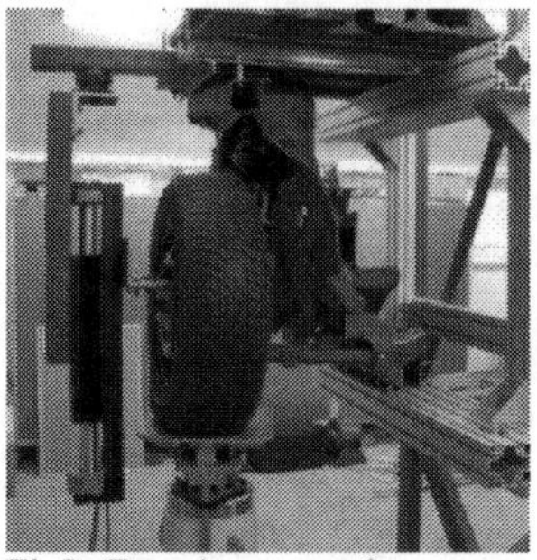
Fig2 Experiment stand

Use a two-column format, and set the spacing between the columns at 8 mm. Do not add any page numbers.

3 Linear Electric Motor

Fig.3 describes the basic principle and structure of the linear electric motor that is used as an actuator in the active suspension system. The appreciable feature of linear motors is that they directly translate electrical energy into usable linear mechanical force and motion, and vice versa. They are linear-shaped.

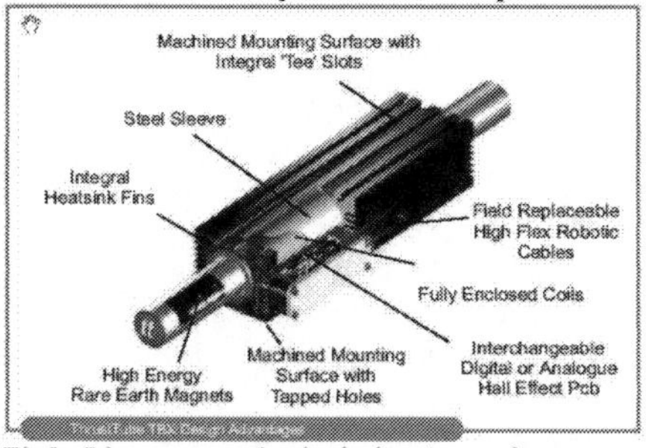

Fig3 Linear motor basic design (manufacturer spreadsheet)

Linear motor translator movements reach high velocities (up to approximately 200m/min), accelerations (up to g multiples), and forces (up to kN). The electromagnetic force can be applied directly to the payload without the intervention of a mechanical transmission. In practice, the most often used type is the synchronous three-phase linear motor.

For the suspension system design it is necessary to deal with one important question - whether it is more advantageous to include the model of the linear electric motor into the model for active suspension synthesis or if it should be used only for simulations.

Comparing advantages and disadvantages of the model inclusion, it can be said that the closed-loop provides more information so that better control results can be achieved. Unfortunately, there are also some significant disadvantages in such a situation. The first one insists in the rank of the system (and consequently the rank of the controller that increases up to 5 or more)

DOI: 10.1051/itmconf/20170903003

and the second one is that the D matrix in the state space description of the motor model does not have full rank and that is why implementation functions are limited or too complicated. For these reasons the linear electric motor model has not been integrated in the model for active suspension synthesis.

There is another important question - whether the entire linear motor model could be omitted and a linear approximation of the desired force could be taken into account. As the mechanical and the electrical constant of the motor are inconsiderable – just about 1ms each, the linear approximation is adequate and satisfactory. Moreover the linear approximation is supported by robustness of the H∞ control design what has been verified by various simulation results and experiments.

4 Energy balance

As said above, linear electric motors can recuperate energy. When the generated force is of the same direction as the suspension velocity, the energy has to be supplied into the system. Otherwise, it can be recuperated and accumulated for some future usage.

In fact, there are some non-linearity's in the recuperation process and that is why the energy management (control) is a bit difficult. Just for imagination, the 3-D plot (shown in Fig. 4) represents the force-velocity profile of the recuperated energy. It shows how much recuperated (and only recuperated) energy can be obtained under the given forces and velocities. In the plot, when the recuperated energy is equal to zero or bigger it is necessary to supply the energy into the system.

This characteristic surface gives important information regarding one of the requirements on the control system. Optimization objectives are equal to maximization of the recuperated energy (with necessary trade-offs).

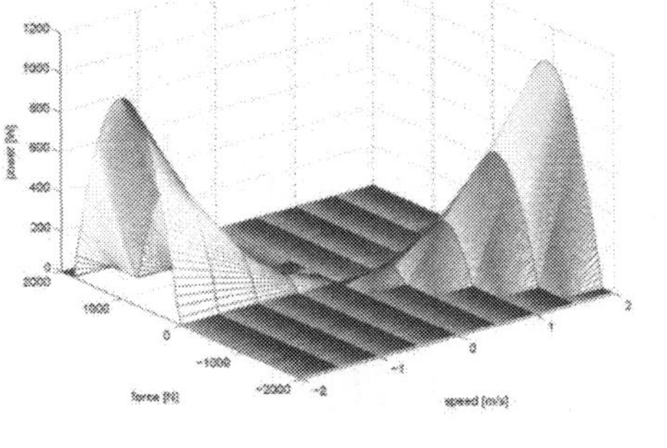

Fig.4 Recuperation aria of the linear electric motorl

5 Experimental Signal Used to Simulate Road Displacement

In order to bring about simulation and practical experiments at the test stand it is necessary to find a proper experimental signal that represents the road profile and excites the active suspension). Although the simplified suspension model seems to be linear the truth is that there are many nonlinear parts in the system. Now the question is what signal to generate for experimental testing to reach a true model of the uneven road under the wheel. We can define two types of input signals regarding objectives:

- to prove results of simulations and pre-calculations
- to test real behavior on the road

Let's start with the first objective – to verify simulation results. The best signal might be considered probably white noise because full frequency spectrum could be analyzed then. But it should be noted that the system is nonlinear and even white noise is not satisfactory. Moreover it is not possible to generate easily white noise by the test stand.

For these reasons a bump has been chosen as a signal, that often occurs on the road profiles and that can be generated by the test bed easily. This signal allows observing both directions – bump up and bump-down. Since it is not possible to generate infinite slope the following signal approximation has been used (Eq.1)(widely known approximation). Different magnitudes of the signal have been tested because the system is nonlinear. Magnitudes have been chosen according to mechanical dimensions of the suspension system:

$$\dot{z}_r(t) = 0.5\pi \sin 20\pi(t - 0.1) \qquad (1)$$

Thus this signal is used to verify the quality of the simulation model, it does not confirm the usability of the controller on a real road surface.

More significant results can be obtained from the input signal, which is alike the road profile. A deterministic random signal is used to approximate it. The input signal for simulation is described by the following equation (2):

$$z_r = \sum_{i=1}^{n} \sqrt{\frac{\dot{\omega}_i}{\pi \cdot v_x}} \{ Re(\frac{b_o}{-\omega_i^2 + a_1 j\omega_i + a_o}) \cdot cos(\omega_i t + \alpha) + Im(\frac{b_o}{-\omega_i^2 + a_1 j\omega_i + a_o}) \cdot sin(\omega_i t + \alpha) \}$$

$$b_o = 0.121 \cdot v_x$$
$$a_o = 2.249 \cdot v_x \qquad (2)$$
$$a_1 = 30.36 \cdot v_x$$

where v_x represents the car velocity.

Thus resulted signal is obtained as a superposition of the sinusoids with deterministic "random" angles (α in Eq.2). In this case 128 random angles have been calculated. The used pseudo-random signal is plotted in Fig.5.

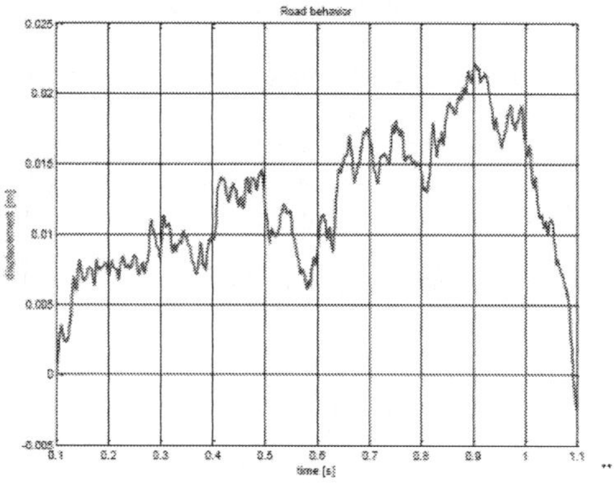

Fig5 Random signal approximation

6 On The Control loop

A robust controller is necessary to design for the suspension system, because the system parameters often vary in a wide range. Especially the body mass varies for every single drive. For this reason H∞ control theory [3] has been chosen for controller design [2] as the ideal controller design. Matlab Toolbox procedures have been used for H∞ controller computation.

Nevertheless the standard H∞ controller cannot handle energy consumption. Modification of the controller had to be done during experiments. Additional input, which controls energy demands was supposed to be connected to the controller. Master controller then can use this input to keep energy balance (Design of the master controller is not involved in the paper.). The structure of the modified controller is shown in Fig.6.

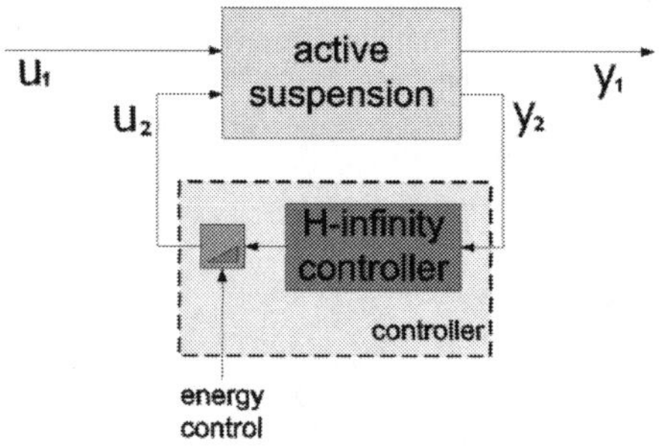

Fig6 Modified controller structure

Than energy management is supposed to be controlled by an external signal (Fig.6) depending on the car and road parameters, i.e. on the energy accumulator capacity and the road profile, respectively.

First possible way to control energy fed into the suspension system consist in the analysis of the driving conditions and cyclic re-computing of the control signal in real time. For high sampling frequency (over 1 kHz) and because of the performance of the controller can not be guaranteed for all operating conditions this approach has been rejected.

The second possibility is to control the energy consumption by controller deterioration. Then the designed controller is reliably robust and the active suspension system is relatively stable. Simply said, the controller deterioration sort of devalues the suspension performance, but enables to gain some energy to be stored. Strictly speaking, let us assume two kinds of driving conditions:

- the terrain /road surface the car is going on is very rough and uneven and there is enough energy stored in the accumulator system- then the controller works in the standard mode, the linear motor consumes energy from the accumulator (supercapacitors) and the suspension performance is preserved.
- the terrain /surface under the car wheels is relatively smooth and there is not enough energy stored in the accumulator system (supercaps) as it was consumed because of the situation described above. The external signal provides the information to the controller to deteriorate its performance and to reduce the energy consumption. The deterioration is stated by the desired force attenuation so that the suspension gets devaluated.

If the force is completely attenuated the suspension system works only via the passive suspension part while the linear electric motor works as a generator generating electrical energy to be stored in the accumulator system. Of course, the suspension performance is devaluated now (to the passive suspension level in the worst case).

That is the basic description how the control loop with the modified controller works. Now, let us show some results (percentage indicators) of the real experiments taken at the test stand that endorse presumed above.

7 Results

Deterministic random signal stated in (2) has been used for real experiments taken on the shock absorber test stand. Two important things should be followed during experiments – suspension comfort improvement and energy consumption (see Fig. 7). In Fig 7, the road profile input, energy consumption, and corresponding body displacement for standard controller setting are displayed.

As an indicator of the comfort improvement body (sprung) mass displacement can be taken.

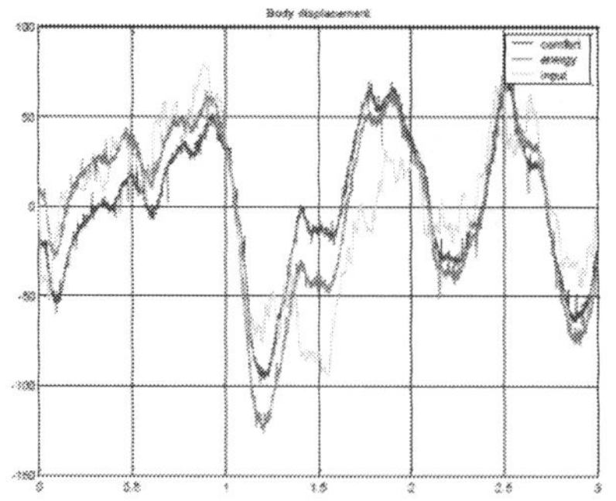

Fig7 Sprung mass displacement

Fig.8 shows two curves - energy demand for standard energy consumption setting (called "comfort setting") and energy demands for lower consumption (called "energy setting").Both curves were measured excited with the same input- the deterministic random signal (road profile). Actually, negative values of energy in the figure represent the recuperated energy.

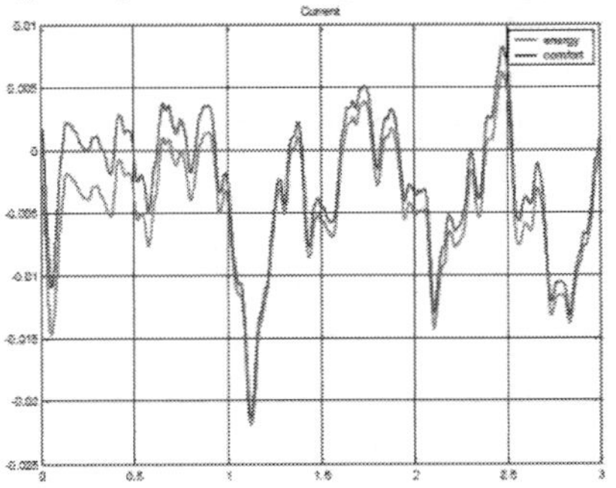

Fig8 Energy demands (demands on electrical current) [1] *K.*

Let's show the results as mean values of "comfort" and "energy" indicators stated in [1]. Tab.1 involves mean values for the body displacement as absolute values of the defined body displacement indicator and also as a percentage of its improvement.

Table 1 Displacement mean values

	Mean value	Percentage
Body displacement – comfort	38.6	100%
Body displacement – energy	47.4	123%

The first line shows that for standard controller setting the comfort indicator (38.6) is taken as 100%. In the second line, when the controller was deteriorated, the comfort is devaluated up to 47.4, i.e. to 123% .

Tab.2 involves mean values of energy indicator. Lower value corresponds to the lower energy demand. Similarly the first line shows that for standard controller setting the comfort indicator (2.689) is taken as 100%. In the second line, when the controller was deteriorated, the energy indicator is devaluated up to 1.598, i.e. to 59% then.

Briefly, controller deterioration causes comfort devaluation to 123% while energy consumption decreases to 59 %.

Table 2 Power mean values

	Mean value	Percentage
Comfort setting	2.689	100%
Energy setting	1.598	59%

8 Conclusion

In this paper the H∞ controller for active suspension with linear electric motor has been used for experiment done on the experiment stand. An experiment signal for real road profile simulation has been developed and then it has been used for experiments. The method for the direct real-time energy control with respect to reduction of the energy consumption has been used. Experiments verified results of the simulations and showed that it is possible to change energy demands according to the road profile and status of the energy storage in the car (battery or super-capacitor). The method can be extended to general plants with considerable energy demands, where the decreasing actuator signal in a given range can preserve the system stability. Thus this controller with linear motor as an actuator can be used in any suspension system.

This research has been supported by the MSMT project No. 1M6840770002 "Josef Bozek's Research Center of Combustion Engines and Automobiles II" and MSMT project INGO No. LG14005.

References

1 *K* Hyniova and J. Honcu: Active Suspension System -- Experiments, *in Proceedings of the 4th WSEAS/IASME International Conference on DYNAMICAL SYSTEMS and CONTROL*, Corfu, Greece, October 26- 28, 2008

2 A. Kruczek and A. Stribrsky, A: Full-car Model for Active Suspension – Some Practical Aspects, *in Proccedings of IEEE International Conference on Mechatronics,* Istanbul, Turkey, 2004

3 Nastac, S.: Simulation of active Vibration Isolation with Fuzzy Logic Device, *In: Proceedings of the 10th international Conference on mathematical and computational methods in Science and Engineering (MACMESE '08)*, pp. 346-351*) Brasov*,2008.

4] G.C.D. Sousa and B.K. Bose, A fuzzy set theory based control of a phase-controlled converter DC machine drive, *IEEE Trans. Industrial Applications 30* (1994)

5] Stribrsky, A. - Hyniova, K. - Honcu, J.: *Reduction of Vibrations in Mechanical Systems*, Proceedings of Workshop 2001, Part A, pp.174-175, CTU Prague, 2001

6 Zhou and J. C. Doyle*: Essentials of Robust Control*, Prentice Hall, 1998

DOI: 10.1051/itmconf/20170903004

Plasma Arc Cutting Dimensional Accuracy Optimization employing the Parameter Design approach

John Kechagias[1,*], *Markos* Petousis[2], *Nectarios* Vidakis[2], and *Nikos* Mastorakis[3]

[1]TEI Thessaly, Mechanical Engineering Department, 41110 Larissa, Grecee
[2]TEI Crete, Mechanical Engineering Department, 71004 Heraklion, Crete, Greece
[3]Technical University of Sofia, Department of Industrial Engineering, Sofia, Bulgaria

Abstract. Plasma Arc Cutting (PAC) is a thermal manufacturing process used for metal plates cutting. This work experimentally investigates the influence of process parameters onto the dimensional accuracy performance of the plasma arc cutting process. The cutting parameters studied were cutting speed (mm/min), torch standoff distance (mm), and arc voltage (volts). Linear dimensions of a rectangular workpiece were measured after PAC cutting following the full factorial design experimental approach. For each one of the three process parameters, three parameter levels were used. Analysis of means (ANOM) and analysis of variances (ANOVA) were performed in order for the effect of each parameter on the leaner dimensional accuracy to be assessed.

1 Introduction

Plasma-Arc Cutting (PAC) is a thermal manufacturing process used for processing various electrical conducting materials, such as carbon steel, stainless-steel, aluminium, cast iron and non-ferrous metals [1-3]. PAC mechanism (Fig. 1) is characterized by an electric arc established between the electrode and the workpiece. The electrode acts as the cathode, and the workpiece material acts as the anode in the process.

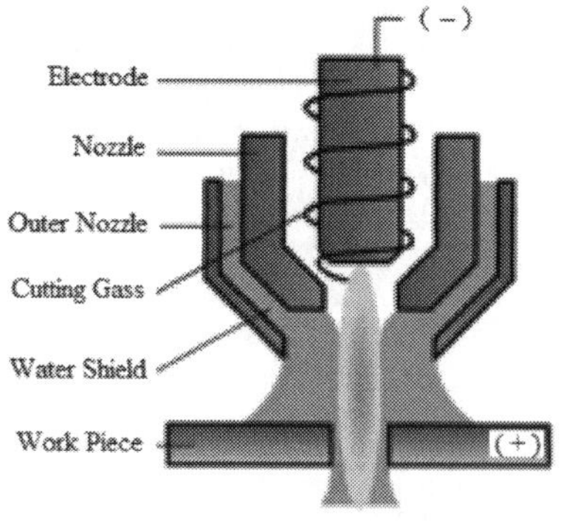

Fig. 1. Plasma-arc cutting mechanism.

Other thermal processes which are antagonistic to the PAC are the laser beam machining process (LMP) and the flame cutting. Selecting the most suitable process for industrial applications depends on several factors such as the type of the material, the layer thickness, the cutting speed, the quality indicators of each process and the cost. Plasma arc cutting can be used for the cutting of metal plates with thickness varying from 5 to 40mm (Fig. 2).

The multi-parameter optimization of the PAC process according to quality indicators, such as the kerf characteristics, the dimensional accuracy and the quality of the cut surface has been experimentally studied by several researchers in several materials and experimental regions [3-11].

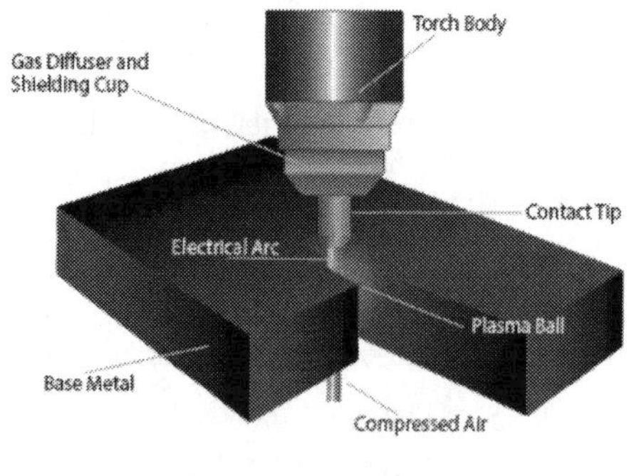

Fig. 2. Plasma Arc Cutting (PAC) process.

The current research work investigates the influence of plasma-arc cut process parameters onto the dimensional accuracy of St37 mild steel workpieces. A multi-parameter optimization was carried out using the full factorial design of experiments method.

2 Parameter Design

In this work experiments were conducted on St37 carbon steel (mild steel), which is widely used for industrial applications. The Yildirim CNC cutting machine was employed having a Burny phantom MCU and HPR 260

* Corresponding author: jkechag@teilar.gr

air gas torch (Fig. 3). All experiments were performed in 16mm thickness sheets (Fig. 4).

Fig. 3. CNC plasma arc cutting machine (Yildirim).

A three parameter full factorial design method was implemented (Table 1). Columns A, B, and C, are assigned to cutting speed (mm/min), torch standoff distance (mm) and arc voltage (Volt), respectively.

Fig. 4. Workpieces (rectangle shape: 30mmX40mmX16mm).

A 30mmx40mm rectangular cut was performed for all 27 combinations as indicated in the orthogonal matrix of the experiment (Table 1). Clockwise (CW) cut direction was followed in all experiments.

The leaner X dimension of the rectangular workpiece was measured on its top surface (Fig. 5). For each workpiece the average of five measurements as well as the deviation (maximum minus minimum value of the five measurements) were calculated during this work.

The measurements were taken along the X direction using a 1 micron accuracy digital calliper.

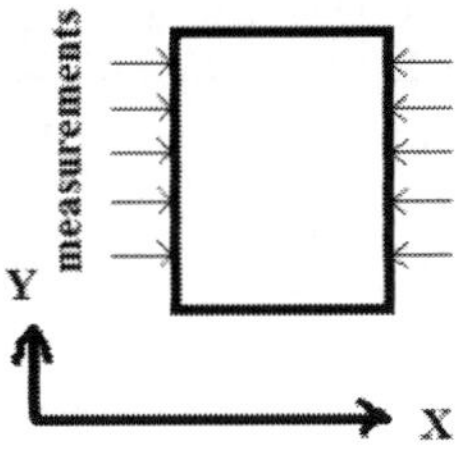

Fig. 5. Direction of measurements

3 Data Analysis

3.1 Analysis of Means - ANOM

After all the experiments were conducted and measured, an analysis of means (ANOM) was performed onto the experimental results (Table 2 and 3) and relevant diagrams were exported (Fig. 6 and 7) in order to optimize the process according to the Leaner Dimension and the minimum deviation of the leaner dimensions criteria.

Table 1. Experimental cases studied.

	A.	B.	C.		
No	**Cutting Speed (mm/min)**	**Torch Standoff Distance (mm)**	**Arc Voltage (Volts)**	**Average X mm**	**Dev (max-min) mm**
1	2000	3	145	39.14	0.26
2	2000	3	150	39.114	0.35
3	2000	3	155	38.938	0.25
4	2000	3.6	145	38.926	0.81
5	2000	3.6	150	38.816	0.22
6	2000	3.6	155	38.962	0.33
7	2000	4.2	145	39.356	0.49
8	2000	4.2	150	39.136	0.37
9	2000	4.2	155	39.078	0.61
10	2600	3	145	39.29	0.25
11	2600	3	150	39.654	0.11
12	2600	3	155	39.5	0.45
13	2600	3.6	145	39.714	0.4
14	2600	3.6	150	39.432	0.43
15	2600	3.6	155	39.306	0.66
16	2600	4.2	145	39.392	0.5
17	2600	4.2	150	39.408	0.46
18	2600	4.2	155	39.228	0.41
19	3200	3	145	39.612	0.52
20	3200	3	150	39.498	0.41
21	3200	3	155	39.64	0.21
22	3200	3.6	145	39.634	0.77
23	3200	3.6	150	39.668	0.29
24	3200	3.6	155	39.604	0.72
25	3200	4.2	145	39.714	0.65
26	3200	4.2	150	39.646	0.45
27	3200	4.2	155	39.51	0.37
Mean				39.367	0.435

Table 2. Mean values for each process parameter level.

Leaner dimension: X – Direction			
Means	Level 1	Level 2	Level 3
m_A	39.05	39.44	39.61
m_B	39.38	39.34	39.39
m_C	39.42	39.37	39.31

Table 3. Mean values for each process parameter level.

Deviation: (Maximum – Minimum)			
Means	Level 1	Level 2	Level 3
m_A	0.41	0.41	0.49
m_B	0.31	0.51	0.48
m_C	0.52	0.34	0.45

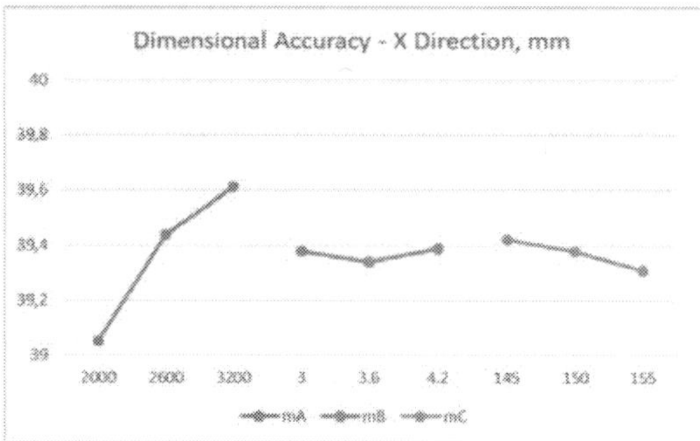

Fig. 6. ANOM Diagram (leaner dimensional accuracy).

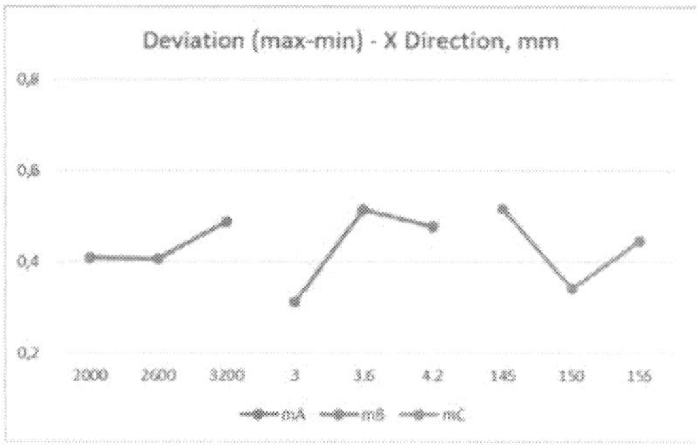

Fig. 7. ANOM Diagram (deviation of measurements).

According to the ANOM analysis and figure 6, the combination for optimizing the leaner X dimension is (best target value [13]); Cutting Speed: 3200 mm/min, Torch standoff distance: 3 mm, and Arc Voltage: 145Volt. Minimization of the Deviation ([13]) is achieved when; Cutting Speed: 2000 mm/min, Torch standoff distance: 3 mm, and Arc Voltage: 140Volt.

3.2 Analysis of Variances - ANOVA

Another approach to assess the different parameters impact onto the performance indicators is by using the Analysis of Variances (ANOVA). ANOVA is also needed for estimating the error variances for the parameters effects and the error prediction variances [13].

F is the variance ratio, i.e. the mean square ratio due to a parameter and the mean square of the error (calculated in Tables 4 and 5). A parameter with an F ratio greater than 4 is considered as very important. If F ratio is smaller than 1, it is considered as unimportant.

ANOVA analysis for the leaner dimension in the X direction shows that the variance due to all the considered parameters is about 80 percent of the total error. In this case, cutting speed is the key factor affecting about 76% the leaner dimension.

ANOVA analysis for the calculated results deviation shows that the overall variance due to all the considered parameters is about 46%. This indicates that the process noise parameters cause larger variance, than the error caused due to the selected process parameters. Torch standoff distance is the most important factor affecting Deviation about 25%.

Table 4. ANOVA Analysis – Leaner dimension

Leaner dimension: X – Direction					
	DoF	SoS	MS	F	%
A	2	1.486	0.743	38.202	76.5%
B	2	0.010	0.005	0.263	0.5%
C	2	0.058	0.029	1.482	3.0%
Total	6	1.554			80.0%
Error	20	0.389	0.019		
Tot Error (Σei)	26	1.943	0.075		

Table 5. ANOVA Analysis - Deviation

Deviation: (Maximum – Minimum)					
	DoF	SoS	MS	F	%
A	2	0.037	0.019	0.840	4.5%
B	2	0.210	0.105	4.714	25.3%
C	2	0.137	0.068	3.070	16.5%
Total	6	0.384			46.3%
Error	20	0.445	0.022		
Tot Error (Σei)	26	0.829	0.032		

4 Conclusions

In this work the plasma-arc cut process parameters

DOI: 10.1051/itmconf/20170903004

impact onto the X-direction dimensional accuracy as well as onto the deviation (maximum minus minimum of five measurements) was experimentally investigated.

The optimum parameter values for the leaner dimensions on the X axis were; Cutting Speed: 3200 mm/min, Torch standoff distance: 3 mm, and Arc Voltage: 145Volt. The optimum parameter values for the deviation were; Cutting Speed: 2000 mm/min, Torch standoff distance: 3 mm, and Arc Voltage: 140Volt.

Finally, the ANOVA analysis shows that the leaner dimension in the X axis is affected the most by the cutting speed (about 76%), and the deviation is affected the most by the torch standoff distance (25%).

References

1. S. Kalpakjian, Manufacturing Engineering and Technology, Addison-Wesley, 1995, pp.844-845.
2. S.M. Ilii , M. Coteata, A. Munteanu, "Experimental results concerning the variation of surface roughness parameter (Ra) at plasma arc cutting of a stainless steel workpiece", International Journal of Modern Manufacturing Technologies, II/1 (2010) 31-36.
3. K. Salonitis, S. Vatousianos, "Experimental Investigation of the Plasma Arc Cutting Process", Procedia CIRP, 3 (2012) 287 – 292
4. J. Kechagias, M. Billis, S. Maropoulos, "A parameter design of CNC plasma-arc cutting process using robust design", Int. J. Experimental Design and Process Optimisation, 2010, 1(4):315-326.
5. A. Tsiolikas, J. Kechagias, K. Salonitis, N. Mastorakis, "Optimization of cut surface quality during CNC Plasma Arc Cutting process", Int. J. of Systems Applications, Engineering & Development, 2016, Vol 10, pp.305-308.
6. J. Kechagias, P. Stavropoulos, S. Maropoulos, K. Salonitis, "On the multi – paramenter optimization of CNC plasma-arc cutting process quality indicators using Taguchi Design of Experiments", Recent Advances in Electrical Engineering, 2014, pp.128-133.
7. B. Asiabanpour, D.T. Vejandla, J. Jimenez, C. Novoa, "Optimising the automated plasma cutting process by design of experiments", Int. J. Rapid Manufacturing, 2009, Vol. 1, No. 1, pp.19-40.
8. R. Bini, B.M. Colosimo, A.E. Kutlu, M. Monno, "Experimental study of the features of the kerf generated by a 200A high tolerance plasma arc cutting system", Journal of Materials Processing Technology, 2008, Vol.196, pp.345–355.
9. J.C. Chen, Y. Li, and R.I. Cox, "Taguchi-based Six Sigma approach to optimize plasma cutting process: an industrial case study", International Journal of Advanced Manufacturing Technology, 2009, Vol.41, pp.760–769.
10. E. Gariboldi, and B. Previtali, "High tolerance plasma arc cutting of commercially pure titanium", Journal of Materials Processing Technology, 2005, Vol.160, pp.77–89.
11. A. Gullu, and U. Atici, "Investigation of the effects of plasma arc parameters on the structure variation of AISI 304 and St 52 steels", Materials & Design, 2006 Vol.27, pp.1157–1162.
12. Q. Zhou, H. Li, F. Liu, S. Guo, W. Guo, and P. Xu, "Effects of nozzle length and process parameters on highly constricted oxygen plasma cutting arc", Plasma Chemistry & Plasma Processing, 2008, Vol.28, pp.729–747.
13. M.S. Phadke, Quality Engineering using Robust Design, Prentice-Hall, Englewood Cliffs, NJ, 1989.

Low-velocity impact response of a pre-stressed isotropic Uflyand-Mindlin plate

Yury Rossikhin[1,*], *Marina* Shitikova[1,**], and *Phan Thanh* Trung[1,2]

[1] *Voronezh State Technical University, Research Center on Dynamics of Solids and Structures, Voronezh 394006, Russia*
[2] *HCMC University of Technology and Education, Vietnam*

Abstract. The low-velocity impact response of a precompressed circular isotropic elastic plate is investigated in the case when the dynamic behavior of the plate is described by equations taking the rotary inertia and transverse shear deformations into account. Contact interaction between the rigid impactor and the target is modeled by a generalized Hertz contact force, since it is assumed that the viscoelastic features of the plate represent themselves only in the place of contact and are governed by the standard linear solid model with fractional derivatives due to the fact that during the impact process decrosslinking occurs within the domain of the contact of the plate with the sphere, resulting in more free displacements of molecules with respect to each other, and finally in the decrease of the plate material viscosity in the contact zone.

1 Introduction

An impact response analysis requires a good estimate of contact force throughout the impact duration. Low velocity impact problems, which also took the local indentation into account, have been solved by many authors. Reference to the state-of-the-art reveiws [1,4] shows that in most studies it was assumed that the impacted structure was free of any initial stresses. But this does not adequately reflect the real multidirectional complex loading states that the materials experience during their service life. The detailed reveiw of papers wherein prestressing of targets was taken into account during the solution of dynamic problems dealing with impact interaction could be found in [5]. But there is practically no investigations on viscoelasticity influence during the impact interaction.

In the present paper, we generalize the approaches suggested in [5] for study of the impact response of a pre-stressed elastic circular plate and in [6] for investigating the impact response of a plate, viscoelastic features of which are induced within the contact domain. But distinct to [6] , in the problem under consideration the generalized Hertz contact law is utilized by considering time-dependent operators describing rigidity and Poisson's ratio of plate's material.

2 Problem formulation and methods of solution

Let us consider the problem of the impact of rigid sphere upon a pre-stressed circular isotropic Uflyand-Mindlin plate which is presumed to be of infinite extent in order to ignore the waves reflected from its edges. The plate out of the contact zone is considered to be elastic, while within the contact domain its microstructure changes and it gains viscoelastic properties.

2.1 Dynamic response of a circular elastic plate

The equations of motion of a pre-stressed circular isotropic elastic Uflyand-Mindlin plate could be written in the polar coordinate system with the origin at the center of the plate in the following form [5]:

$$D\left(\frac{\partial^2\varphi}{\partial r^2}+\frac{1}{r}\frac{\partial\varphi}{\partial r}\right)-D\frac{\varphi}{r^2}+K\mu h\left(\frac{\partial w}{\partial r}-\varphi\right)=\rho\frac{h^3}{12}\ddot{\varphi}, \quad (1)$$

$$K\mu h\left(\frac{\partial^2 w}{\partial r^2}-\frac{\partial\varphi}{\partial r}+\frac{1}{r}\frac{\partial w}{\partial r}-\frac{\varphi}{r}\right)=\rho h\ddot{w}+N\left(\frac{\partial^2 w}{\partial r^2}+\frac{1}{r}\frac{\partial w}{\partial r}\right), \quad (2)$$

where w is the plate deflection, φ is the angle of inclination of the normal to plate's middle surface, r is the polar radius, h is the plate thickness, ρ is its density, $K = 5/6$ is the shear coefficient, N is the constant compression force acting in the radial direction, $D = Eh^3/12(1-\nu^2)$ is the bending rigidity, E, μ, and ν are the longitudinal modulus, shear modulus, and Poisson's ratio, respectively, and an overdot denotes a derivative with respect to time t.

It is assumed that as a result of impact upon the plate, two transient waves (the surfaces of strong discontinuity) arise in it at the point of impact, which further propagate with velocities G_α ($\alpha = 1, 2$) along the plate in the form of diverging cylindrical surfaces-strips. Let us interpret the surface of strong discontinuity as a layer of small thickness δ, within which the desired value Z changes monotonically and continuously from the magnitude Z^+ to the magnitude

*e-mail: yar@vgasu.vrn.ru
**e-mail: shitikova@vmail.ru

 DOI: 10.1051/itmconf/20170903005

Z^-. Suppose that the ahead and back fronts of the shock layer arrive at a certain point with the fixed radius at the time instants t and $t+\Delta t$, respectively, where Δt is small. Inside the shock layer the following relationship

$$\partial Z/\partial r \approx -G^{-1}\partial Z/\partial t \tag{3}$$

is fulfilled.

A certain desired function Z behind the fronts of wave surfaces $\Sigma^{(1)}$ and $\Sigma^{(2)}$ could be represented in terms of the ray series [7]

$$Z(r,t) = \sum_{\alpha=1}^{2}\sum_{k=0}^{\infty}\frac{1}{k!}[Z_{,(k)}]^{(\alpha)}\Big|_{t=\frac{r}{G_\alpha}}\left(t-\frac{r}{G_\alpha}\right)^k \times H\left(t-\frac{r}{G_\alpha}\right), \tag{4}$$

where $[Z_{,(k)}]^{(\alpha)} = \left[\partial^k Z/\partial t^k\right]^{(\alpha)} = Z_{,(k)}^{+(\alpha)} - Z_{,(k)}^{-(\alpha)}$ are discontinuities in the kth-order time-derivatives of the desired function Z on the waves surfaces Σ_α, i.e., at $t = r/G_\alpha$, the upper indices " + " and " − " denote that the values are calculated immediately ahead of and behind the wave fronts, respectively, and $H(t - r/G_\alpha)$ is the unit Heaviside function.

Since the impact process is of short duration, then it is possible to restrict only by zero-order terms, i.e.,

$$Z(r,t) = \sum_{\alpha=1}^{2}[Z]^{(\alpha)}\Big|_{t=\frac{r}{G_\alpha}} H\left(t-\frac{r}{G_\alpha}\right). \tag{5}$$

Writing Eqs. (1) and (2) within the shock layer with due account for (3), then integrating the obtained relationships over t from t to $t+\Delta t$ and tending $\Delta t \to 0$ yield

$$\rho G_1^2 = \frac{E}{1-\nu^2}, \tag{6}$$

$$\rho G_2^2 = K\mu - N/h. \tag{7}$$

For further treatment we need to determine the transverse force Q_r. For this purpose we would split Eq. (2) in two equations

$$\frac{\partial Q_r}{\partial r} + \frac{1}{r}Q_r = \rho h\ddot{w} + N\left(\frac{\partial^2 w}{\partial r^2} + \frac{1}{r}\frac{\partial w}{\partial r}\right), \tag{8}$$

$$Q_r = K\mu h\left(\frac{\partial w}{\partial r} - \varphi\right). \tag{9}$$

Differentiating (8) and (9) one time with respect to time, applying the condition of compatibility (3) with due account for (7), we obtain

$$Q_r = -K\mu h G_2^{-1} W, \tag{10}$$

where $W = \dot{w}$.

2.2 Equations of motion of the contact domain and the impactor

At $t > 0$ the displacement of the sphere's center could be represented in the form

$$z = w + \alpha, \tag{11}$$

where α is the impactor's indentation due to the local bearing of target material within the contact domain.

Then the equation of motion of the part of the plate being in contact with the sphere and the equation of the sphere have the form

$$2\pi a(t)Q_r + P(t) = \rho h\pi a^2(t)\ddot{w} + 2\pi a(t)N\frac{\partial w}{\partial r}, \tag{12}$$

$$m\ddot{z} = -P(t), \tag{13}$$

where $a(t)$ is the radius of the contact domain, and $P(t)$ is the contact force.

Equations (12) and (13) could be solved with due account for formula (10), as well as considering the following initial conditions:

$$z\Big|_{t=0} = 0, \quad \dot{z}\Big|_{t=0} = V_0, \quad w\Big|_{t=0} = 0, \quad \dot{w}\Big|_{t=0} = 0. \tag{14}$$

It is assumed that the viscoelastic features of the plate represent themselves only in the place of contact and are governed by the standard linear solid model with fractional derivatives. The matter is fact that during the impact process, decrosslinking occurs within the domain of the contact of the plate with the sphere, resulting in more free displacements of molecules with respect to each other, and finally in the decrease of the plate material viscosity in the contact zone. This circumstance allows one to describe the behaviour of the plate material within the contact domain by the standard linear solid model involving fractional derivatives, since variation in the fractional parameter (the order of the fractional derivative) enables one to control the viscosity of the plate material.

In this case, the generalized contact Hertz theory could be used to define the contact force

$$P(t) = \frac{4\sqrt{R}}{3}\tilde{k}\alpha^{3/2}, \tag{15}$$

and the time-dependent radius of the contact domain is defined as

$$a(t) = \sqrt{\alpha(t)R}, \tag{16}$$

where R is the radius of the impactor, $\tilde{k}$ is the time-dependent operator

$$\tilde{k} = \frac{\tilde{E}}{1-\tilde{\nu}^2} \tag{17}$$

dependent on the viscoelastic operators $\tilde{E}$ and $\tilde{\nu}$ which are valid within the contact domain and based on the fractional derivative standard linear solid model.

Thus, the operator corresponding to the Young's modulus has the form [8]

$$\tilde{E}_1 = E_\infty\left[1 - \nu_\varepsilon \,\vartheta_\gamma^*(\tau_\varepsilon^\gamma)\right] \quad (0 \le \gamma \le 1), \tag{18}$$

where E_∞ and E_0 are the non-relaxed (instantaneous modulus of elasticity, or the glassy modulus) and relaxed elastic (prolonged modulus of elasticity, or the rubbery modulus) moduli which are connected with the relaxation time τ_ε and retardation time τ_σ by the following relationship:

$$\left(\frac{\tau_\varepsilon}{\tau_\sigma}\right)^\gamma = \frac{E_0}{E_\infty}, \tag{19}$$

$$\nu_\varepsilon = \frac{E_\infty - E_0}{E_\infty} = \frac{\Delta E}{E_\infty}, \tag{20}$$

$$\ni_\gamma^* (\tau_i^\gamma) Z(t) = \int_0^t \ni_\gamma \left(\frac{t-t'}{\tau_i}\right) Z(t')dt' \quad (i = \varepsilon, \sigma), \tag{21}$$

$$\ni_\gamma \left(\frac{t}{\tau_i}\right) = \frac{t^{\gamma-1}}{\tau_i^\gamma} \sum_{n=0}^{\infty} \frac{(-1)^n (t/\tau_i)^{\gamma n}}{\Gamma[\gamma(n+1)]}, \tag{22}$$

$\Gamma[\gamma(n+1)]$ is the Gamma-function, $\ni_\gamma (t/\tau_i)$ is Rabotnov's fractional exponential function [9] which at $\gamma = 1$ goes over into the ordinary exponent, and operator $\ni_\gamma (\tau_i)$ transforms into operator $\ni_1^* (\tau_i)$. When $\gamma \to 0$, the function $\ni_\gamma (t/\tau_i)$ tends to the Dirac delta-function $\delta(t)$.

As numerous experiments with volume stresses and strains show, for the majority of materials the operator of volume extension-compression $\widetilde{K}$ is a constant value, that is

$$\widetilde{K} = K_\infty, \tag{23}$$

where K_∞ is a certain constant.

Now we could calculate the Poisson's operator $\widetilde{\nu}$ according to formula (23), which could be rewritten in the form

$$\frac{\widetilde{E}}{1-2\widetilde{\nu}} = \frac{E_\infty}{1-2\nu_\infty} = 3K_\infty, \tag{24}$$

where ν_∞ is the non-relaxed magnitude of the Poisson's ratio.

Considering (18), from formula (24) we have

$$\widetilde{\nu} = \nu_\infty + \frac{1}{2}(1-2\nu_\infty)\nu_\varepsilon \ni_\gamma^* (\tau_\varepsilon^\gamma). \tag{25}$$

For further treatment we should know the following operators:

$$\frac{1}{1+\widetilde{\nu}} = \frac{1}{1+\nu_\infty + \frac{1}{2}(1-2\nu_\infty)\nu_\varepsilon \ni_\gamma^* \left(\tau_\varepsilon^\gamma\right)}, \tag{26}$$

$$\frac{1}{1-\widetilde{\nu}} = \frac{1}{1-\nu_\infty - \frac{1}{2}(1-2\nu_\infty)\nu_\varepsilon \ni_\gamma^* \left(\tau_\varepsilon^\gamma\right)}. \tag{27}$$

In order to calculate the operators in the right-hand side of (26) and (27), we assume that they have the following form:

$$\frac{1}{1+\widetilde{\nu}} = \frac{1}{1+\nu_\infty}\left[1 - B \ni_\gamma^* \left(t_1^\gamma\right)\right], \tag{28}$$

$$\frac{1}{1-\widetilde{\nu}} = \frac{1}{1-\nu_\infty}\left[1 + D \ni_\gamma^* \left(t_2^\gamma\right)\right], \tag{29}$$

where B, t_1 and D, t_2 are yet unknown constants.

Equating the right sides of relationships (26), (28) and (27), (29), reducing the obtained expressions to the common denominator with due account for formula [8]

$$\ni_\gamma^* \left(\tau_\varepsilon^\gamma\right) \ni_\gamma^* \left(\tau_\sigma^\gamma\right) = \frac{\tau_\varepsilon^\gamma \ni_\gamma^* \left(\tau_\varepsilon^\gamma\right) - \tau_\sigma^\gamma \ni_\gamma^* \left(\tau_\sigma^\gamma\right)}{\tau_\varepsilon^\gamma - \tau_\sigma^\gamma} \tag{30}$$

yield

$$\frac{1}{2}\frac{(1-2\nu_\infty)\nu_\varepsilon}{1+\nu_\infty}\left[1 - B\frac{\tau_\varepsilon^\gamma}{\tau_\varepsilon^\gamma - t_1^\gamma}\right] \ni_\gamma^* \left(\tau_\varepsilon^\gamma\right) - B\left[1 - \frac{1}{2}\frac{(1-2\nu_\infty)\nu_\varepsilon}{(1+\nu_\infty)} \times \frac{t_1^\gamma}{(\tau_\varepsilon^\gamma - t_1^\gamma)}\right] \ni_\gamma^* \left(t_1^\gamma\right) = 0, \tag{31}$$

$$-\frac{1}{2}\frac{(1-2\nu_\infty)\nu_\varepsilon}{1-\nu_\infty}\left[1 + D\frac{\tau_\varepsilon^\gamma}{\tau_\varepsilon^\gamma - t_2^\gamma}\right] \ni_\gamma^* \left(\tau_\varepsilon^\gamma\right) + D\left[1 + \frac{1}{2}\frac{(1-2\nu_\infty)\nu_\varepsilon}{(1-\nu_\infty)} \times \frac{t_2^\gamma}{(\tau_\varepsilon^\gamma - t_2^\gamma)}\right] \ni_\gamma^* \left(t_2^\gamma\right) = 0. \tag{32}$$

Vanishing to zero the expressions in square brackets in (31) and (32), we determine unknown constants

$$\begin{aligned} B &= \frac{(1-2\nu_\infty)\nu_\varepsilon}{2(1+\nu_\infty) + (1-2\nu_\infty)\nu_\varepsilon} = \frac{\tau_\varepsilon^\gamma - t_1^\gamma}{\tau_\varepsilon^\gamma}, \\ t_1^{-\gamma} &= \tau_\varepsilon^{-\gamma}\left[1 + \frac{(1-2\nu_\infty)\nu_\varepsilon}{2(1+\nu_\infty)}\right], \quad t_1^\gamma = \frac{\tau_\varepsilon^\gamma}{A}, \\ D &= \frac{(1-2\nu_\infty)\nu_\varepsilon}{2(1-\nu_\infty) - (1-2\sigma_\infty)\nu_\varepsilon} = \frac{\tau_\varepsilon^\gamma - t_2^\gamma}{\tau_\varepsilon^\gamma}, \\ t_2^{-\gamma} &= \tau_\varepsilon^{-\gamma}\left[1 - \frac{(1-2\nu_\infty)\nu_\varepsilon}{2(1-\nu_\infty)}\right], \quad t_2^\gamma = \frac{\tau_\varepsilon^\gamma}{C}, \\ A &= \frac{2(1+\nu_\infty) + (1-2\nu_\infty)\nu_\varepsilon}{2(1+\nu_\infty)}, \\ C &= \frac{2(1-\nu_\infty) - (1-2\nu_\infty)\nu_\varepsilon}{2(1-\nu_\infty)}. \end{aligned} \tag{33}$$

Now we could calculate the operator

$$\widetilde{k} = \frac{\widetilde{E}}{1-\widetilde{\nu}^2} = \frac{\widetilde{E}}{2}\left(\frac{1}{1+\widetilde{\nu}} + \frac{1}{1-\widetilde{\nu}}\right). \tag{34}$$

For this purpose, we substitute (18), (28) and (29) in (34) with due account for formula (30), as a result we obtain

$$\widetilde{k} = \frac{E_\infty}{1-\nu_\infty^2}\left[1 - \sum_{j=1}^{2} m_j \ni_\gamma^* \left(t_j^\gamma\right)\right], \tag{35}$$

where

$$m_1 = \frac{3}{2}\frac{B(1-\nu_\infty)}{(1-2\nu_\infty)}, \quad m_2 = \frac{1}{2}\frac{D(1+\nu_\infty)}{(1-2\nu_\infty)}.$$

Considering Eqs. (35) and (21), the contact force is defined as

$$P(t) = k_\infty\left[\alpha^{3/2}(t) - \sum_{j=1}^{2} m_j \int_0^t \ni_\gamma\left(-\frac{t-t''}{t_j}\right)\alpha^{3/2}(t'')dt''\right], \tag{36}$$

where

$$k_\infty = \frac{4\sqrt{R}}{3}\frac{E_\infty}{1-\nu_\infty^2}.$$

Now integrating Eq. (13) yields

$$z = -\frac{1}{m}\int_0^t P(t')(t-t')dt' + V_0 t. \tag{37}$$

Utilizing (36), it is possible to rewrite (37) in the form

$$z(t) = V_0 t - \frac{k_\infty}{m}\int_0^t \left[\alpha^{3/2}(t') - \sum_{j=1}^{2} m_j \int_0^{t'} \ni_\gamma\left(-\frac{t'-t''}{t_j}\right)\alpha^{3/2}(t'')dt''\right](t-t')dt'. \tag{38}$$

3 Solution of governing equations

Now considering formulas (10) and (16), as well as relationship

$$\frac{\partial w}{\partial z} = -G_2{}^{-1} W, \tag{39}$$

which is obtained from (3) if we substitute there the function Z by the function w, Eq. (12) could be rewritten in the form

$$M\alpha \dot{W} + g\alpha^{1/2} W = P(t), \tag{40}$$

where contact force $P(t)$ is defined by (36), $M = \rho\pi hR$, and $g = 2MG_2R^{-1/2}$.

Substituting (11) in (13) yields

$$m\dot{W} + m\ddot{\alpha} = -P(t). \tag{41}$$

Note that since the impact process is of short duration, then in the integrals entering in Eqs. (36) and (38) could be represented as [10]

$$\ni_\gamma\left(-\frac{t}{t_j}\right) \approx \frac{t^{\gamma-1}}{t_j^\gamma \Gamma(\gamma)} \quad (j = 1, 2). \tag{42}$$

The set of governing equations (40) and (41) with due account for (42) is reduced to the following:

$$M\alpha\dot{W} + g\alpha^{1/2}W = k_\infty\left[\alpha^{3/2} - \Delta_\gamma \int_0^t (t-t')^{\gamma-1}\alpha^{3/2}(t')dt'\right], \tag{43}$$

$$m\dot{W} + m\ddot{\alpha} = -k_\infty\left[\alpha^{3/2} - \Delta_\gamma \int_0^t (t-t')^{\gamma-1}\alpha^{3/2}(t')dt'\right], \tag{44}$$

where

$$\Delta_\gamma = \frac{1}{\Gamma(\gamma)}\left(\frac{m_1}{t_1^\gamma} + \frac{m_2}{t_2^\gamma}\right).$$

3.1 Analysis of the system's critical state

The most interesting is the case of $N \to N_{\text{crit}} = K\mu h$, i.e., $G_2 \to 0$, and the plate occurs in the critical state, since all energy of shock interaction is concentrated in the contact region, what may result in damage of the structure within the contact zone.

Let the compression force N in the plate attain its critical magnitude. Then as a result of impact of a rigid body upon the plate, only one wave is generated in the plate which further propagates with the velocity G_1, but the second wave turns out to be locked within the contact region. In this case, coefficient $g = 0$ and Eq. (43) is reduced to

$$M\alpha\dot{W} = k_\infty\left[\alpha^{3/2} - \Delta_\gamma \int_0^t (t-t')^{\gamma-1}\alpha^{3/2}(t')dt'\right]. \tag{45}$$

If we consider

$$\alpha \approx V_0 t \tag{46}$$

as a first approximation, then Eq. (45) with due account for [11]

$$\int_0^t (t-t')^{\gamma-1}t'^{3/2}(t')dt' = \frac{3}{\gamma}\left(\frac{1}{3} - \frac{1}{5}\gamma\right)t^{3/2+\gamma} \tag{47}$$

takes the form

$$\dot{W} = \frac{k_\infty V_0^{1/2}}{M}\left[t^{1/2} - \Delta_\gamma \frac{3}{\gamma}\left(\frac{1}{3} - \frac{1}{5}\gamma\right)t^{1/2+\gamma}\right]. \tag{48}$$

Integrating (48) yields

$$W = \frac{k_\infty V_0^{1/2}}{M}\left[\frac{2}{3}t^{3/2} - \Delta_\gamma \frac{3}{\gamma}\left(\frac{1}{3} - \frac{1}{5}\gamma\right)\frac{1}{3/2+\gamma}t^{3/2+\gamma}\right]. \tag{49}$$

3.1.1 The case $\gamma = 0$

In the pure elastic case, i.e. at $\gamma = 0$, $\Delta_0 = \Delta_\gamma|_{\gamma=0} = 0$, and thus Eq. (45) takes the form

$$\dot{W} = \frac{k_\infty}{M}\alpha^{1/2}. \tag{50}$$

The solution of (50) with due account for (46) has the form

$$W = \frac{2k_\infty V_0^{1/2}}{3M}t^{3/2} \tag{51}$$

whence it follows that the velocity of deflection increases as time goes on.

3.1.2 The case $\gamma = 1$

In the case of conventional viscosity, i.e. at $\gamma = 1$, Eq. (45) takes the form

$$\dot{W} = \frac{k_\infty}{M}\left[\alpha^{1/2} - \frac{2}{5}\Delta_1\alpha^{3/2}\right], \tag{52}$$

the solution of which is

$$W = \frac{2k_\infty V_0^{1/2}}{M}\left[\frac{1}{3} - \frac{2}{25}\Delta_1 V_0 t\right]t^{3/2}, \tag{53}$$

where

$$\Delta_1 = \Delta_\gamma|_{\gamma=1} = \frac{m_1}{t_1} + \frac{m_2}{t_2}.$$

Vanishing relationships (52) and (53) to zero, we could estimate the time at which velocity W attains its maximal value and the contact duration, respectively,

$$t_{\max} = \frac{5}{2\Delta_1 V_0}, \tag{54}$$

$$t_{\text{cont}} = \frac{25}{6\Delta_1 V_0} = \frac{5}{3}t_{\max}. \tag{55}$$

Substituting (54) in (53) provides the maximal magnitude of the velocity

$$W_{\max}|_{t=t_{\max}} = \frac{4}{15}\frac{k_\infty V_0^{1/2}}{M}t_{\max}^{3/2} \tag{56}$$

whence it follows that viscosity softens the impact response of the plate as compared with the elastic case.

DOI: 10.1051/itmconf/20170903005

4 Conclusion

The low-velocity impact response of a precompressed circular isotropic elastic plate is investigated. The dynamic behavior of the plate is described by equations taking the rotary inertia and transverse shear deformations into account. Longitudinal compressing forces are uniformly distributed along the plates median plane.

Contact interaction between the rigid impactor and the target is modeled by a generalized Hertz contact force, since it is assumed that the viscoelastic features of the plate represent themselves only in the place of contact and are governed by the standard linear solid model with fractional derivatives. This is explained by the fact that during the impact process, decrosslinking occurs within the domain of the contact of the plate with the sphere, resulting in more free displacements of molecules with respect to each other, and finally in the decrease of the plate material viscosity in the contact zone. This circumstance allows one to describe the behaviour of the plate material within the contact domain by the standard linear solid model involving fractional derivatives, since variation in the fractional parameter (the order of the fractional derivative) enables one to control the viscosity of the plate material.

From the results obtained the following conclusions may be drawn:

1. If a circular plate is subjected to the action of a constant compression force uniformly distributed in its middle plane along the boundary circumference, then during impact upon such a pre-stressed plate the nonstationary wave of a transverse shear (surface of strong discontinuity) is generated and then propagates with the velocity dependent on the compression force.

2. At certain critical magnitude of the compression force, the velocity of the transient wave of transverse shear vanishes to zero, resulting in 'locking' of this wave within the contact domain.

3. 'Locking' of the wave, in its turn, leads to the fact that energy during impact does not dissipate (as it takes place in the case of the generation and propagation of the transverse shear wave) but remains inside the contact zone, what could result in damage of the contact domain.

4. It is shown that for an elastic plate the critical compressional force leads to the increase in the velocity of the contact spot with time, resulting in cut off of the rigid washer (the contact zone) with further its knocking out of the plate.

5. If viscosity of the plate material is induced within the contact domain, then it softens the impact, and in this case the velocity of the contact spot continuously increases from zero to a certain maximal magnitude and then decreases to zero.

6. At fractional order viscosity, the maximal velocity of the contact spot could be controlled by the choice of the value of the fractional parameter.

Acknowledgement

The research described in this publication has been supported by the Ministry of Education and Science of the Russian Federation.

References

[1] S. Abrate, *Impact on laminated composite materials*, 2nd edition (Cambridge University Press, 2005)

[2] Yu.A. Rossikhin, M.V. Shitikova, The Shock and Vibration Digest **39**, 273–309 (2007)

[3] A.C.J. Luo, Y. Guo, *Vibro-impact dynamics* (Wiley, 2013)

[4] R.A. Ibrahim, *Vibro-impact dynamics: Modeling, mapping and applications* (Springer, 2009)

[5] Yu.A. Rossikhin, M.V. Shitikova, Shock Vibr. **13**, 197–214 (2006)

[6] Yu.A. Rossikhin, M.V. Shitikova, J. Sound Vibr. **330**, 1985–2003 (2011)

[7] Yu.A. Rossikhin, M.V. Shitikova, Acta Mech. **102**, 103–121 (1994)

[8] Yu.A. Rossikhin, M.V. Shitikova, Comp. Math. Appl. **66**, 755–773 (2013)

[9] Yu.N. Rabotnov, Prikladnaya Matematika i Mekhanika (in Russian) **12**, 53–62 (1948) (English translation of this paper could be found in Fract. Calc. Appl. Anal. **17**, 684–696 (2014), doi:10.2478/s13540-014-0193-1)

[10] Yu.A. Rossikhin, M.V. Shitikova, M.G. Estrada Meza, SpringerPlus **5**:206 (2016), DOI 10.1186/s40064-016-1751-2

[11] Yu.A. Rossikhin, M.V. Shitikova, D.T. Manh, WSEAS Trans. Appl. Theor. Mech. **66**, 125–128 (2016)

 DOI: 10.1051/itmconf/20170903006

On Scalable and Efficient Security Risk Modelling of Cloud Computing Infrastructure based on Markov processes

Dimitrios A.Karras

Sterea Hellas Institute of Technology, Automation Dept, Psachna, Evoia, 34400, Greece, dakarras@teiste.gr, dimitrios.karras@gmail.com, dimitrios.karras@ieee.org

Abstract. While cloud computing infrastructures proliferates in nowadays computing and communications technology there are few reports investigating models for their security. In this paper, new efficient models are developed and evaluated for analyzing the security-related behavior of cloud computing architectures and networks comprising complex interconnected communication systems adapted towards a generalized analysis. These cloud related models, based on Markov processes, allow calculation of critical security factors for the cloud infrastructure, related to intrusion detection, of such interconnected and distributed systems components and the evaluation of the associated security mechanisms. Although, at this step an architecture of at least three interconnected systems is analyzed, the systematic model introduced allows for a generalized model of N interconnected systems in a cloud architecture under reasonable assumptions. We herein show the principles of such an analysis. Security parameters calculation and Security mechanisms evaluation may support the risk analysis and the decision making process in resolving the trade-offs between security and quality of service characteristics corresponding to the complex interconnected computing and communication systems.

Keywords. Cloud infrastructures, Security Risk Analysis, Interconnected Systems, Markov Processes, Intrusion Detection

1. Introduction

The increasing role of communication services makes crucial the issue of ensuring the security attributes of the underlying computing and communication infrastructures in terms of secrecy, integrity and availability. The security attacks in computer and communication systems may result in [1]: information disclosure, unauthorized modification of files, messages and transactions, masquerading or successful break-in, decreasing communication services availability, repudiation in sending and receiving messages of electronic orders or in creating and modifying files, and the possibility of traffic analysis and the creation of user/consumer profiles. These attacks may emanate from legitimate users, unauthorized users and processes, such as malicious software.

Security is often cited as one of the greatest barriers to communications services, including Internet commerce. Of course, security is important to communication services in many ways, but it is really part of the way that business is enabled by the technology. Indeed, the security of communication systems, for instance for electronic commerce, is a business problem, not merely a technology one. Technologies such as public key encryption provide critical components of an overall solution, but they are not enough. Such technologies can be applied both to systems designed from scratch as well as to systems built around off-the-shelf products for Internet commerce. The important issue is to properly design the whole interconnected communication system so that security technologies could be applied. To this end a significant help could be provided by attempting to model the system computing and communication infrastructure. This is precisely the goal of this paper, namely, to model such interconnected infrastructures in terms of security.

Security violations leave abnormal patterns of system usage and accounting [2,3]. To cope with intrusions or attempted break-ins, system monitoring techniques or intrusion-detection mechanisms and audit trails are used, that rely on the collection of audit data and their comparison with the usage and accounting profiles maintained by the system [4]. The conditional probability of detecting an intrusion given that the intrusion has occurred is called intrusion coverage and used as a measure of the effectiveness of the intrusion-detection mechanism. The number of normal and abnormal usage and accounting types (patterns) is extremely high and they can be differentiated only partially so that it is very difficult to have an intrusion coverage close to 1. An alarm is triggered if certain thresholds are reached. The detection sensitivity level and the false alarm rate depend on the thresholds set [5]. Increasing the detection sensitivity level leads to higher false alarm rates, i.e., better intrusion coverage appears to be in trade-off with false alarms.

Audit trails, i.e., data that allow tracing from users and transactions of related processes aim at detecting or deterring system intrusion and helping assessing the damage caused by intrusions in the case of successful ones. Issues regarded in research efforts in the context of audit trails include the analysis and specification of auditable events and the quality improvement of the mechanisms related to efficiency, protection and the prevention of denial of service. They, also, include the association and analysis of related events and the automation of intrusion detection and damage assessment functions [4].

Intrusion detection mechanisms can be used in stand-alone or networked systems. They are based on the development of user and system or network resources usage profiles and knowledge-oriented or statistically oriented methods. They have limitations, since the absence of rules for all possible intrusion scenarios or inaccurate statistical distributions do not lead to detection of intrusions or attempted break-ins. On the other hand, they may lead to false alarms, if unexpected user actions or resource usage patterns occur, which are not foreseen by the rules or the distributions used.

To study the behavior of security attacks or intrusion processes, models have to be developed and used, since it is quite impossible to directly analyze real computer systems and networks or information infrastructures to this respect.

In section 2, the model is described and the mathematical notations and the system equations are discussed. In section 3, we apply the model and discuss the various results obtained for a set of parameter values. Finally, section 4 summarizes this paper with conclusions and future directions.

2. Cloud Security Models Description and Analysis

In this research we develop and use Markov models by considering the states of each system component of the interconnected information infrastructure, which reflect system functioning with respect to the above stated possible attacks. These states are explicitly associated with the security attributes of secrecy, integrity and availability. On the other hand, the existing dependencies between the component systems comprising the cloud infrastructure are taken into account in the proposed models. While single system security models exist in the literature [4,6], the suggested models for analyzing security parameters in infrastructures is one of the first research efforts for investigating the effects of multiple dependent systems operation in the interconnected communication and information infrastructure security planning.

We assume constant arrival rates of attacks and constant state transition rates, which allow the use of exponential or geometrical distributions, since there are no exact analytical solution methods for non-Markovian models. (Approximation techniques could be used in the case of non-constant rates.)

Model A- the cloud as a single system being in attack

Figure 1 shows the model, which relates to a single system and consists of 7 states. The system is in state 0 when there are no security violations or attempted attacks. All security attributes are well maintained. With the first attempted attack, the system enters in state 1. The system remains in this state as long as it is under attack, the attacks are not detected and the system has not been penetrated. From this state, transition back to state 0 takes place if the attacks are detected or to state 2, if the attacker obtains authentication information and penetrates the system. The attacker remains in state 2 as long as he obtains (disclosures) confidential information and may move to state 3 if he starts to modify files, programs and messages or to state 4 if he chooses to hinder the access of authorized users to programs, hardware and data. When the attacker is detected, the system enters in the state 5, where it is reconfigured and transition back to state 0 occurs. Transition from state 0 to state 6 may take place if a false alarm is triggered. After the reconfiguration the inverse transition occurs. Transitions between states 2, 3 and 4 take place according to the actions of the attacker, which lead to unauthorized information disclosure, modification and access to system or network resources, respectively.

Notation and system of equations

In this research we use the following notation, which is common in textbooks on stochastic processes, queueing theory and Markovian chains in particular [7].

λ_{ij}, is the transition rate from state i to state j, τ_{ij}, is the transition probability from state i to state j and P_i, is the probability of the system or network or infrastructure to be in state i (steady state).

From the state-transition-rate diagram shown in Fig. 1, it is obvious that the Markov chain is irreducible and we accept the limit that $P_k = lim\ P_k(t)$ as $t \to \infty$. In the equilibrium case we are interested in that the flow must be conserved in the sense that the input flow must equal the output flow for any given state. By inspection we can establish the following equilibrium (steady-state) equations for the cloud model A.

$$(\lambda_{01}\tau_{01} + \lambda_{06}\tau_{06})P_0 = \lambda_{10}\tau_{10}P_1 + \lambda_{50}\tau_{50}P_5 + \lambda_{60}\tau_{60}P_6 \quad (1)$$

$$\lambda_{10}\tau_{10}P_1 + \lambda_{12}\tau_{12}P_1 = \lambda_{01}\tau_{01}P_0 \quad (2)$$

$$(\lambda_{23}\tau_{23} + \lambda_{24}\tau_{24} + \lambda_{25}\tau_{25})P_2 = \lambda_{12}\tau_{12}P_1 + \lambda_{32}\tau_{32}P_3 + \lambda_{42}\tau_{42}P_4 \quad (3)$$

$$(\lambda_{32}\tau_{32} + \lambda_{34}\tau_{34} + \lambda_{35}\tau_{35})P_3 = \lambda_{23}\tau_{23}P_2 + \lambda_{43}\tau_{43}P_4 \quad (4)$$

$$(\lambda_{42}\tau_{42} + \lambda_{43}\tau_{43} + \lambda_{45}\tau_{45})P_4 = \lambda_{24}\tau_{24}P_2 + \lambda_{34}\tau_{34}P_3 \quad (5)$$

$$\lambda_{50}\tau_{50}P_5 = \lambda_{25}\tau_{25}P_2 + \lambda_{35}\tau_{35}P_3 + \lambda_{45}\tau_{45}P_4 \quad (6)$$

$$\lambda_{60}\tau_{60}P_6 = \lambda_{06}\tau_{06}P_0 \quad (7)$$

By means of this model we may analyze the systems comprising an interconnected information infrastructure separately. The security-related dependence between these systems can be taken into account if we adapt the probability transitions from state 1 to state 2 of the controlled system by adding to its initial value the equilibrium probability of the controlling system being in state 2.

We assume that successful attacks in the various systems are independent. However, if the controlling system is penetrated, the controlled system may be penetrated immediately or with higher probability than when it is attacked directly and not through the controlling system.

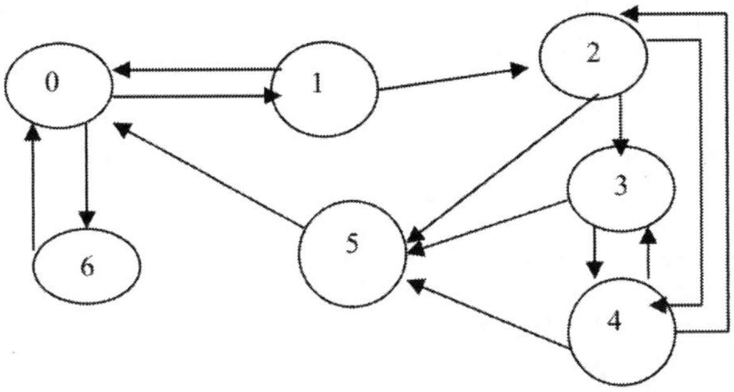

Fig. 1. State-transition-rate for the diagram of model A for the cloud modelled as a single system

However, the cloud is an interconnected system of let's say N components. In order to find out the related probabilities for every component we could assume that all components are independent, each corresponding to a probability Pc(state-k), with probabilities Pc(state-k) being equal for all components c, and for every state k of the above defined system of equations. In order to estimate Pc(state-k) from the relevant P(state-k) of the cloud system, after solving the previously mentioned equations, we have to model the events involved for c=1..N and k=0..6. Under these assumptions we could have, involving the theory of total probability for independent and mutually disjoint events, since each cloud component state could be considered as such compared to the rest of cloud components,

P(state-k)= P(all possible combinations of events for c=1..N components being in state k) =>

$$P(\text{state-k}) = C(N,1) * Pc(\text{state-k})\ (1-Pc(\text{state-k}))^{(N-1)} + C(N,2) * Pc(\text{state-k})^2\ (1-Pc(\text{state-k}))^{(N-2)} + C(N,3) * Pc(\text{state-k})^3\ (1-Pc(\text{state-k}))^{(N-3)} + \ldots C(N,r) * Pc(\text{state-k})^r\ (1-Pc(\text{state-k}))^{(N-r)} + \ldots\ C(N,N) * Pc(\text{state-k})^N\ (1-Pc(\text{state-k}))^{(N-N)} \quad \textbf{(8)}$$

where it is known that,

$C(n,r)=n!(r!(n-r)!)$

If P(state-k) is known by solving the previously mentioned Markov process based system of Model A, then every Pc(state-k) can be calculated solving equation (8).

Initial Ad-Hoc Model B for cloud in intrusion

The interconnected communication and information infrastructure is modeled by a Markovian chain again for two non local systems under the same cloud. In this case an Ad Hoc analysis and model is presented, where some states are omitted. In the general form, the model relates to n systems and m states of each system, which may lead to mxn states of the Markovian chain if transitions from all states to all others are possible. We assume Markov chains which are irreducible and for which exists the limit Pk =lim Pk(t) as t->∞ for all states k.

Figure 2 shows the initial model B, which relates to two systems or networks comprising an information infrastructure and consists of 12 states. The systems are in state (0,0) when there are no security violations or attempted attacks. With the first attempted attack, the attacked systems enter in state (1,0) or (0,1) if it is the first or the second system attacked. From this, state transition to state (1,1) may occur if both systems are under attack. Transition to state (2,0), (2,1) or (0,2), (1,2) takes place if the attempted intrusion leads to successful penetration of the first or the second system, respectively. If one of the systems is occupied then the second system is penetrated as well, (2,2). From this, state transition to state (3,3) occurs when the penetration is detected. After the reconfiguration of the systems, state (0,0) is entered. From state (0,0) transition may occur to state (4,0) or (0,4) if a false alarm of the first or the second system is flagged.

After the false alarm is resolved current state becomes the (0,0). From Fig. 2 we obtain the following equilibrium equations by simplifying the numbering of the states in an ad hoc way as follows: (0,0) – 0, (1,0) – 1, (0,1) – 2, (1,1) – 3, (2,0) – 4, (0,2) – 7, (2,1) – 5, (1,2) – 6, (2,2) – 8, (3,3) — 9, (4,0) – 10, (0,4) – 11.

If p is the matrix of the transition probabilities and P the vector of the steady state probabilities then, the following equation holds, as it is known: pP=P

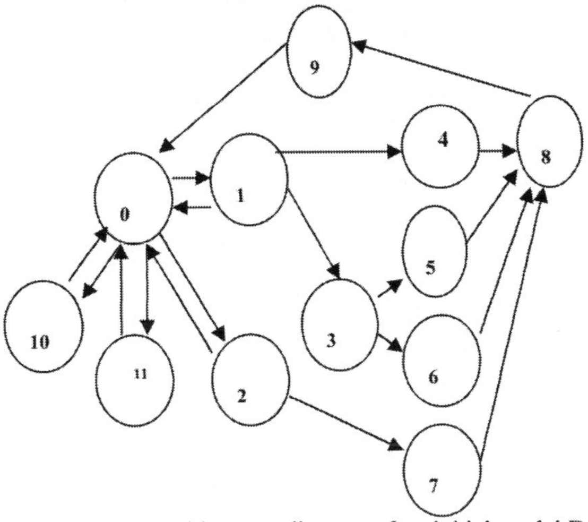

Fig. 2. State-transition-rate diagram of an initial model B for two interconnected systems or networks of the same cloud infrastructure.

$$\left(\lambda_{01}\tau_{01}+\lambda_{02}\tau_{02}+\lambda_{10,0}\tau_{10,0}+\lambda_{11,0}\cdot\tau_{11,0}\right)P_0$$
$$=\lambda_{10}\tau_{10}P_1+\lambda_{20}\tau_{20}P_2+\lambda_{30}\tau_{30}P_3+\lambda_{10,0}\tau_{10,0}P_{10}+\lambda_{11,0}\tau_{11,0}P_{11} \quad (9)$$
$$(\lambda_{13}\tau_{13}+\lambda_{14}\tau_{14})P_1=\lambda_{01}\tau_{01}P_0 \quad (10)$$
$$(\lambda_{23}\tau_{23}+\lambda_{27}\tau_{27})P_2=\lambda_{02}\tau_{02}P_0 \quad (11)$$
$$(\lambda_{35}\tau_{35}+\lambda_{36}\tau_{36})P_3=\lambda_{13}\tau_{13}P_1+\lambda_{23}\tau_{23}P_2 \quad (12)$$
$$\lambda_{48}\tau_{48}P_4=\lambda_{14}\tau_{14}P_1 \quad (13)$$
$$\lambda_{58}\tau_{58}P_5=\lambda_{35}\tau_{35}P_3 \quad (14)$$
$$\lambda_{68}\tau_{68}P_6=\lambda_{36}\tau_{36}P_3 \quad (15)$$
$$\lambda_{78}\tau_{78}P_7=\lambda_{27}\tau_{27}P_2 \quad (16)$$
$$\lambda_{89}\tau_{89}P_8=\lambda_{48}\tau_{48}P_4+\lambda_{58}\tau_{58}P_5+\lambda_{68}\tau_{68}P_6+\lambda_{78}\tau_{78}P_7 \quad (17)$$
$$\lambda_{90}\tau_{90}P_9=\lambda_{89}\tau_{89}P_8 \quad (18)$$
$$\lambda_{10,0}\tau_{10,0}P_{10}=\lambda_{0,10}\tau_{0,10}P_0 \quad (19)$$
$$\lambda_{11,0}\tau_{11,0}P_{11}=\lambda_{0,11}\tau_{0,11}P_0 \quad (20)$$

We solve the above equations for steady-state probabilities. From these we may calculate the probabilities for each system of the underlying interconnected cloud communication and information infrastructure.

However, again, this model B based cloud infrastructure is an interconnected system of let's say N components. In order to find out the related probabilities for every such component we could assume that all components are independent, as in model A, each corresponding to a probability P_{Bc}(state-k), with probabilities P_{Bc}(state-k) being equal for all components c, and for every state k of the above defined system of equations. In order to estimate P_{Bc}(state-k) from the relevant P_B(state-k) of the cloud system, after solving the previously mentioned equations, we have to model the events involved for c=1..N and k=0..12. Under these assumptions we could have, involving the theory of total probability for independent and mutually disjoint events, since each cloud component state could be considered as such compared to the rest of cloud components,

P_B(state-k)= P(all possible combinations of events for c=1..N components being in state k) =>

P_B(state-k) = C(N,1)* P_{Bc}(state-k) (1-P_{Bc}(state-k))$^{(N-1)}$ + C(N,2) * P_{Bc}(state-k)^2 (1-P_{Bc}(state-k))$^{(N-2)}$ + C(N,3) * P_{Bc}(state-k)^3 (1-P_{Bc}(state-k))$^{(N-3)}$ + …C(N,r) * P_{Bc}(state-k)^r (1-P_{Bc}(state-k))$^{(N-r)}$ + ….. C(N,N) * P_{Bc}(state-k)^N (1-P_{Bc}(state-k))$^{(N-N)}$ **(21)**

where it is known that,
C(n,r)=n!(r!(n-r)!)

If P_B(state-k) is known by solving the previously mentioned Markov process based system of Model A, then every P_{Bc}(state-k) can be calculated solving equation (21).

A systematic Model B for cloud in intrusion-Towards a Scalable Analysis for interconnected cloud subsystems

In this interconnected cloud model, again, the communication and information cloud infrastructure is considered as a Markovian chain moxdel. In the general form, the model relates to n systems and m states of each system, which may lead to mxn states of the Markovian chain if transitions from all states to all others are possible. We herein employ, however, a scalable model B, which leads to more unknown variables than the previous initial model B but it leads to a better, scalable and more systematic model B of two interconnected system than before. We assume again Markov chains which are irreducible and for which exists the limit Pk =lim Pk(t) as t->∞ for all states k.

Figure 3 shows the model, which relates to two systems or networks comprising an information infrastructure and consists of 14 states. Figure 3 can be obtained from figure 1 and it is its generalization for two interconnected systems. It bares similarities with figure 2 architecture, which is ad hoc. Such a systematic view could lead to other possible meaningful generalizations. Taking into account that mn states of the Markovian chain if transitions from all states to all others are possible, this means that in our case 72 = 49 states would exist. However, the proposed meaningful generalization of model A, in the case of two interconnected systems, leads, as we will see in m x n = 14 states only. The systems are in state (0,0) when there are no security violations or attempted attacks. With the first attempted attack, the attacked systems enter in state (1,0) or (0,1) if it is the first or the second system attacked. From this, state transition to state (1,1) may occur if both systems are under attack. Transition to state (2,0), (2,1) or (0,2), (1,2) takes place if the attempted intrusion leads to successful penetration of the first or the second system, respectively. If one of the systems is occupied then the second system is penetrated as well, (2,2). The attacker remains in state (2,2) as long as he obtains (disclosures) confidential information and may move to state (3,3) if he starts to modify files, programs and messages or to state (4,4) if he chooses to hinder the access of authorized users to programs, hardware and data. When the attacker is detected, the system enters in the state (5,5), where it is reconfigured and transition back to state (0,0) occurs. After the reconfiguration the inverse transition occurs. Transition from state (0,0) to state (6,0) or (0,6) may take place if a false alarm of the first or the second system is flagged.

After the false alarm is resolved current state becomes the (0,0). From Fig. 3 we obtain the following equilibrium equations by simplifying but in a systematic way easily shown below, the numbering of the states:

(0,0) – 0, (1,0) – 1, (0,1) – 2, (1,1) – 3, (2,0) – 4, (2,1) – 5, (0,2) – 6, (1,2) – 7, (2,2) – 8, (3,3) — 9, (4,4)-10, (5,5)-11, (6,0) – 12, (0,6) – 13.

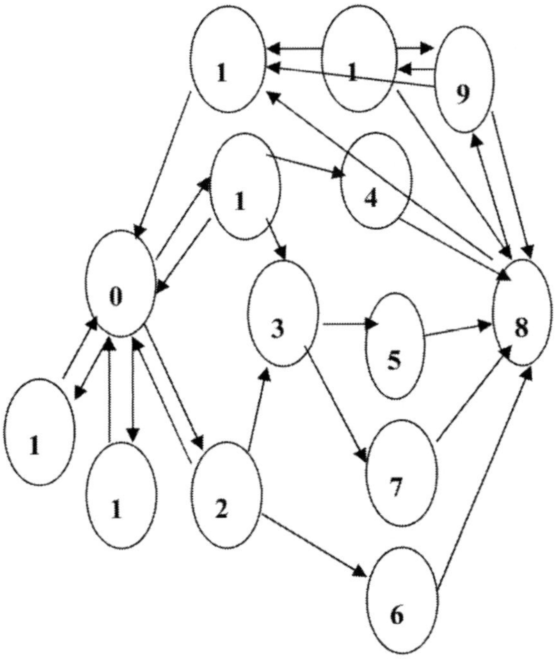

Fig. 3. State-transition-rate diagram of model B for two interconnected subsystems of the cloud infrastructure.

$$(\lambda_{01}\tau_{01} + \lambda_{02}\tau_{02} + \lambda_{0,12}\tau_{0,12} + \lambda_{0,13}\tau_{0,13})P_0$$
$$= \lambda_{10}\tau_{10}P_1 + \lambda_{20}\tau_{20}P_2 + \lambda_{11,0}\tau_{11,0}P_{11} + \lambda_{12,0}\tau_{12,0}P_{12} + \lambda_{13,0}\tau_{13,0}P_{13} \quad (22)$$

$$(\lambda_{13}\tau_{13} + \lambda_{14}\tau_{14})P_1 = \lambda_{01}\tau_{01}P_0 \quad (23)$$

$$(\lambda_{23}\tau_{23} + \lambda_{26}\tau_{26})P_2 = \lambda_{02}\tau_{02}P_0 \quad (24)$$

$$(\lambda_{35}\tau_{35} + \lambda_{37}\tau_{37})P_3 = \lambda_{13}\tau_{13}P_1 + \lambda_{23}\tau_{23}P_2 \quad (25)$$

$$\lambda_{48}\tau_{48}P_4 = \lambda_{14}\tau_{14}P_1 \quad (26)$$

$$\lambda_{58}\tau_{58}P_5 = \lambda_{35}\tau_{35}P_3 \quad (27)$$

$$\lambda_{68}\tau_{68}P_6 = \lambda_{26}\tau_{26}P_2 \quad (28)$$

$$\lambda_{78}\tau_{78}P_7 = \lambda_{37}\tau_{37}P_3 \quad (29)$$

$$\lambda_{89}\tau_{89}P_8 + \lambda_{8,11}\tau_{8,11}P_8 = \lambda_{48}\tau_{48}P_4 + \lambda_{58}\tau_{58}P_5 + \lambda_{68}\tau_{68}P_6 + \lambda_{78}\tau_{78}P_7 \quad (30)$$

$$\lambda_{98}\tau_{98}P_9 + \lambda_{9,10}\tau_{9,10}P_9 + \lambda_{9,11}\tau_{9,11}P_9 = \lambda_{89}\tau_{89}P_8 + \lambda_{10,9}\tau_{10,9}P_{10} \quad (31)$$

$$\lambda_{10,9}\tau_{10,9}P_{10} + \lambda_{10,11}\tau_{10,11}P_{10} + \lambda_{10,8}\tau_{10,8}P_{10} = \lambda_{9,10}\tau_{9,10}P_9 \quad (32)$$

$$\lambda_{11,0}\tau_{11,0}P_{11} = \lambda_{10,11}\tau_{10,11}P_{10} + \lambda_{9,11}\tau_{9,11}P_9 + \lambda_{8,11}\tau_{8,11}P_8 \quad (33)$$

$$\lambda_{12,0}\tau_{12,0}P_{12} = \lambda_{0,12}\tau_{0,12}P_0 \quad (34)$$

$$\lambda_{13,0}\tau_{13,0}P_{13} = \lambda_{0,13}\tau_{0,13}P_0 \quad (35)$$

We solve again the above equations for steady-state probabilities. From these we may calculate the probabilities for each system of the underlying interconnected communication and information cloud infrastructure.

As in the previous initial model B, if we define P_B(state-k) the estimated steady state probabilities acquired by solving the system of equations 22-35 above, then every P_{Bc}(state-k), which is the relevant probability of state k=0..13 of each cloud infrastructure component c=1..N can be calculated solving equation (21) again.

2. Preliminary Numerical Examples using Excel

The selection of the parameter values is based on the tests and results of [4,5]. For model A, we assume transition rates equal to 1 per day from states 0 and 1, transition rates equal to 25 from states 2, 3, 4, 5, and 8 to all others and transition probabilities, τ01 = 1-τ06, τ10 = 1-τ = 0.1, τ23= τ24= τ32= τ34= τ42= τ43=(1-τ)/2, τ12=τ25=τ35=τ45=τ, τ50=τ60=1, τ=0.2,……,1.0 (intrusion coverage). In the same way, for model B we assume transition rates per day λ01= λ13= λ14= λ02= λ27=λ23= λ89= λ0,10= λ10,0= λ0,11= λ11,0=1, λ10= λ20=12, λ48= λ35= λ36= λ58= λ68=25, λ78= λ90=3 and transitions probabilities, τ01=(1- τ0,10)/2 , τ13= τ14= τ27= τ23=0.1, τ02=1- τ0,10 , τ10= τ20=0.9, τ48= τ68= τ58= τ78= τ89= τ90= τ10,0= τ11,0=1, τ35= τ36=0.08, τ0,11= τ0,10= τ (false alarm rate) and τ = 0.0,…,0.08

With these assumptions we have obtained preliminary numerical results, involving Excel, shown in the next two diagrams, which validate our interconnected communication and information cloud infrastructure modelling approach, in terms of results compatible with that of literature for single systems.

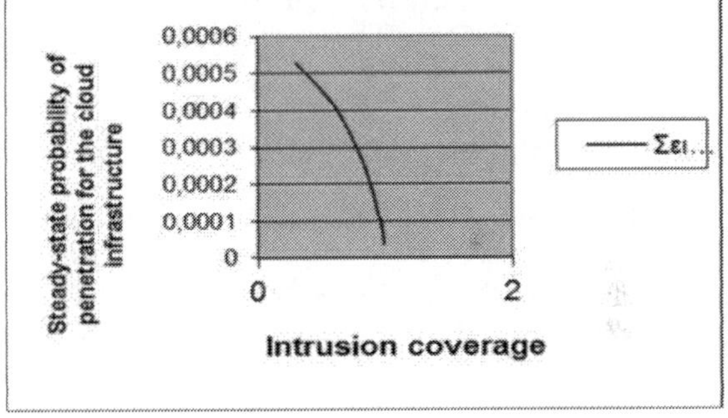

Fig. 4. Steady state probability of intrusion for model A as a function of intrusion coverage

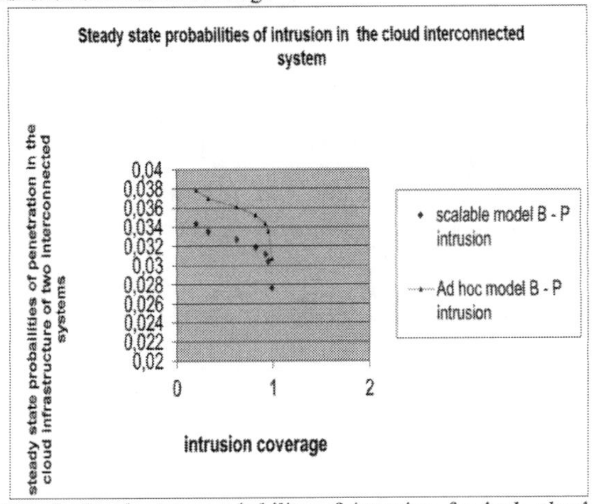

Fig. 5. Steady state probability of intrusion for both cloud models B as a function of intrusion coverage

DOI: 10.1051/itmconf/20170903006

Discussion and Prospects

In this research we presented three models for the analysis of cloud security-related attack processes by means of Markovian chains. The first model is proposed for use in the analysis of the cloud considered as a single system or network, while the second in the analysis of the cloud considered involving two interconnected systems or networks. The second model is an ad hoc initial model aimed at minimizing analysis costs, while the third one is a more detailed model defined towards a generalized model of security analysis for cloud involving interconnected systems. The models allow for the calculation of the expected probabilities of the systems to be in various states such as safe-state, under attack, in intrusion state and in false-alarm-state. For each such state and for each model we have estimated cloud components relevant probabilities. Future work will aim at generalizing, especially the third model, for N cloud interconnected subsystems as well as at expanding the models with respect to the probability distributions used. Also, future work will aim at the development of simulation models for the analysis of the security-related behaviour of cloud information infrastructures in complex communication systems, and as a validation tool for the analytical models. Furthermore, the involvement of neural networks and computational intelligence techniques for approximating the generalized probability distributions in the analytical models, might be investigated.

References

1. P. Helman and G. Liepins, "Statistical foundations of audit trail analysis for the detection of computer misuse", IEEE Trans. On Software Engineering, SE-19, 1993, pp. 886-901.
2. D.E. Denning, 'An Intrusion-detection Model', IEEE Trans. On Software Engineering, SE-12, 1987, pp. 222-232.
3. C. Stoll, 'Stalking the Wily Hacker', Communications of the ACM, 1988, pp. 484-497.
4. B. C. Soh and T. S. Dillon, "Setting optimal intrusion-detection thresholds", Computers & Security, Vol. 14, 1995, pp. 621-631.
5. G.E. Liepins and H.S. Vaccaro, 'Intrusion Detection: Its Role and Validation', Computers & Security, Vol. 11, 1992, pp. 347-355.
6. B. C. Soh and T. S. Dillon, "System intrusion processes: a simulation model", Computers & Security, Vol. 16, 1997, pp. 71-79.
7. L. Kleinrock. "Queueing Systems, Volume I: Theory, John Wiley and Sons, New York, 1975

ITM Web of Conferences **9**, 03007 (2017) DOI: 10.1051/itmconf/20170903007
AMCSE 2016

Synchronization Error Correction for Asynchronous Channels Data Transmission

Nikolaos Bardis[1,a], *Nikolaos Doukas*[1,b] and *Oleksandr P .Markovskyi*[2,c]

[1] Department of Mathematics and Engineering Sciences, Informatics LAB, Hellenic Military Academy, Greece
[2] Department of Computer Engineering,National Technical University of Ukraine, (Polytechnic Institute of Kiev), Ukraine

Abstract. Loss of synchronization is a common source of errors in asynchronous data channels. Minute differences in the operating frequencies between the transmitter and the receiver result in data bits being lost or false bits being inserted. This paper presents an innovative technique, especially designed for detecting and correcting errors of this type. Data packets are preprocessed and areas that are susceptible to are determined. Suitable redundancy is introduced in the form of control symbols. On the receiver side, similar calculations take place and decisions are made on the occurrence and positions of the transmission errors due to loss of synchronization, which are hence corrected. The proposed method is computationally simple, since it uses a small number of simple mathematical operations, contrary to existing, general purpose techniques. The transmission overheads it imposes do not vary significantly from existing error control codes. Additionally, the number of errors that may be corrected is not subject to the same limits as existing techniques.

1 Introduction

Information transmission is an enabling technology for a variety of systems, most of which are not classified as purely ICT applications, e.g. vehicles, aircraft, ships, weapons systems, production machinery, home appliances etc. The operation of virtually all hardware equipment is heavily dependent on the existence of reliable transmission channels interconnecting subsystems and capable of achieving high transmission rates. Data transmission infrastructure in this context is commonly based on sequential channels that are embedded in the system. The user may often be oblivious to the existence of such embedded transmission channels.

Commonly, serial asynchronous data transmission is the preferred technological basis for embedded channels of the type described above. The characterizing feature of these channels is the fundamental frequency that is employed for controlling this asynchronous transmission. Cost, space and other constraints impose limitations to the number of interface lines used, as well as minimal interface line changes. Serial interfaces that are widely used in such contexts include the Universal Serial Bus (USB) and FireWire. Within the USB protocol, the Non Return to Zero Invert (NRZI) encoding scheme, which is also described by the term *bit stuffing*.

Serial data transmission has been the object of significant technological evolvements and provides ever increasing transmission rates. Typical examples of increasing rates becoming widely available include SATA interfaces, that has during recent years accelerated by a factor of five and the USB 3.0 connection that has increased the standard transmission rate of its previous version by a factor of ten, achieving transmission at 5Gb/s.

Increased transmission rates imply higher error rate values, since the physical media used for the transmission are not in general susceptible to technological evolution and / or are exposed to several types of interference. Hence, low level data transmission is not a very reliable process within modern information processing hardware. Complex physical processes are simultaneously affecting the physical medium, leading to distortions of the transmitted information. It may be stated that in general, larger transmission rates are associated with the existence of more physical processes and phenomena that distort and impede transmission

Classical error correction techniques [1, 2, 3] are not suitable for the case of transmission errors due to loss of synchronization, since they do not take advantage of the nature of the errors in order to increase efficiency. A commonly used method for the prevention of the appearance of synchronization errors is Bit stuffing [4]. This may be implemented by introducing a zero bit after each sequence of binary digits whose length reaches a critical limit, thus changing the dynamics of the

[a] Nikoloas Bardis: bardis@ieee.org
[b] Nikolaos Doukas: ndoukas@sse.gr
[c] Oleksandr P. Markovskyi: markovskyy@i.ua

 DOI: 10.1051/itmconf/20170903007

communications channel. The consequence of this mechanism is that extra bit sequences are transmitted and the non- synchronized bit sequences are interrupted. As a result, the analysis of the received block, including the detection and the correction of the interrupted bits requires a significant amount of overhead computation time.

Specialized methods for the correction of single and double synchronization errors are primarily founded on the principal and the arithmetic of weighted checksums.

More specifically for the correction of single synchronization errors, the two component control code [5, 6] algorithm is used. The first component is computed as the sum modulo 2 of all the bits of the information block. The difference in the first component bit sent by the transmitter and the one calculated by the receiver is used to characterize the type of the error, either the reception of a single extra bit or the loss of a single bit. The second component is calculated as the arithmetic sum of the sequence numbers of all the positions on the original block where unit bits appear in the original block. Using the difference between the received and the calculated second part of the control block, the location of the positions of the erroneous synchronized bits is performed.

An alternative existing method [7, 8, 9] is capable of correcting double synchronization errors and uses a two component correction code. With the bit values of the first component, the fragments of the transmitted packet that contain errors. The second component bit determines the type of each synchronization error, namely the increase or decrease of the received data by a bit with value 1.

The disadvantage of these methods is that the number of the errors that may be corrected is limited. At the same time, there exist multiple applications for which the transmitter and the receiver operate at significantly different temperature conditions, hence presenting increased probability of error appearance, with multiplicities greater than 2.

Therefore, existing methods do not provide in a large number of instances, effective solutions for the problem of detection and correction of transmission errors due to the loss of synchronization.

The purpose of this research is to develop an effective method for the correction of all errors due to loss of synchronization in the transmission rates of data between computer systems that use asynchronous data.

2 Development of the method

For the purpose of attaining this aim, a simple and efficient method for correcting all errors due to loss of synchronization that occur during the sequential transmission of an n – bit packet in an asynchronous channel.

Let $B_R = \{b_1, b_2,\ldots,b_n\}$, $\forall k \in \{1..n\}$: $b_k \in \{1, 0\}$be the data that are taken into account in the analysis of the control sequence at the transmitter and the receiver. The method enables the detection of errors due to synchronization, determines their type and executes the required corrections, independent of the number of errors in the packet.

Modern sequential data exchange protocols are synchronized via the sign bits – zero or one. The transmission of bits of the opposite sign do not incur a sign change on the line, which implies that their transition is not synchronized. For example, USB channels are not synchronized during the transmission of unit bits. The total length of the sequence of units is determined at the receiver by the duration of the constant voltage on the line. Эта длительность на приемнике и передатчике измеряется схемой таймера The timers at the transmitter and the receiver may differ. Измеренная длительность постоянного потенциала на приемнике и передатчике может отличаться. This problem becomes more pronounced when the transmitter and receiver timers operate in different temperatures. The result of this discrepancy is that the number of the subsequent orders of units at the receiver may differ from the number of units at the transmitter. The synchronization errors hence appear.

In practice, the length difference of the transmitted and received packets, that contain a sequence of unit bits, does not contain more units. The probability of error appearance increases with increasing lengths of unit bit sequences. If the length l tends to a critical threshold h there is a realistic risk of synchronization errors appearing. For this reason, it becomes more difficult to use specialized devices. For a USB port and approximately equal transmitter and receiver temperatures, the critical threshold is six units: h=6 [8].

For the transmission of an n-bit data packet, the appearance of synchronization errors is possible exclusively for packets that contain the sequence of unity bits and their length is greater than or equal to the critical length h.

For this reason, the developed approach proposes the determination of a fragments transmitted within the block of data BS that contain sequential unit bits and have a length greater than or equal to $h-1$.

The determined fragments may be assigned names E_{1S}, E_{2S}, …, E_{mS}, where m is the number of n – bit fragments within the block BS. The lengths of the fragments are correspondingly assigned to the variables: l_{1S}, l_{2S}, …, l_{mS}, $\forall i = 1, \ldots, m$; $l_{iS} \geq h-1$.

On the receiver side, it is proposed that a control sequence S is calculated that consists of two bit symbols c_{1S}, c_{2S}, …,c_{mS}, where $c_{iS} \in \{0,1,2,3\}$, $i \in \{1,..,m\}$ and each one of those symbols is calculated as the remainder of the division of the length of the ith fragment by four.

$$c_{iS} = l_{iS} \bmod 4 \qquad (1)$$

The control sequence S is transmitted along with the information block BS.

The received information block (BR) is analyzed correspondingly to the structure of the transmitted block: the fragments E_{1R}, E_{2R}, ... E_{mR} are parsed that should be populated by sequences of unit bits whose length is not shorter that h –1. The lengths of these fragments are

DOI: 10.1051/itmconf/20170903007

assigned to variables l_{1R}, l_{2R}, …, l_{mR}, $\forall i = 1, \ldots, m$; $l_{iR} \geq h-1$.

On the receiver side, using the received block BR, it is proposed that the control sequence R is calculated, consisting of the two bit symbols c_{iR}. In this case $i = 1 \ldots m$ and each c_{iR} is calculated as the remainder of the division of the length l_{iR} of the ith fragment E_{iR} by four:

$$c_{iR} = l_{iR} \bmod 4 \qquad (1)$$

The detection of the errors in the received information block, the determination of their type, as well as the correction are implemented on the receiver side. In the ith received fragment, E_{iR}, it is possible to observe the following two types of errors due to timing: the appearance of additional unity bits and the subsequent increase in the length of the received fragment compared to the transmitted fragment, as well as the loss of a unity bit which means that the received fragment is one bit shorter than the transmitted one, i.e. $l_{iR} = l_{iR} - 1$.

The solution of the exercise of the detection of synchronization errors is proposed to be performed via the comparison and analysis of two control sequences: the one received from the transmitter and the one calculated at the receiver. If all the corresponding symbols are equal, then the solution implies is that there were no transmission errors: $\forall i \in \{1,2,..,m\}$: $c_{iS} = c_{iR}$. In other occasions, if for every i symbol pairs $c_{iS} \neq c_{iR}$, the result is interpreted as the existence of synchronization errors for the ith fragment. For the determination of the types of errors, the following procedure is proposed: the hypothesis is tested that the length of the sequence of units in the ith fragment has been increased on the receiver side. For this reason a unit is added in the ith control symbol c_{iS} of the sequence of the transmitter and the calculation of the remainder of the division of the augmented symbol modulo 4. The result is compared with the corresponding symbols of the control sequence of the receiver c_{iR}. In case these are equal i.e. $(c_{iS} + 1) \bmod 4 = c_{iR}$, the above hypothesis is assumed to be true and for this reason the decision is made to correct the ith fragment of the received packet by eliminating the excess unit bit. If on the other hand, $(c_{iS} + 1) \bmod 4 \neq c_{iR}$ and $c_{iS} \neq c_{iR}$ the decision is made that a different type of synchronization transmission error occurred and the number of ones of the ith fragment was reduced on the receiver side: for the correction of this error a unit bit is added to the ith fragment of the packet on the receiver side.

Formally, the proposed method for the correction of transmission synchronization errors may be summarized as the following sequence of steps:

1. In the transmitted packet BS, the fragments E_{1S}, E_{2S}, …, E_{mS} are determined in which the number of sequential unit bits is greater than $h - 2$.
2. Determination of the length l_{iS} of the ith fragment E_{iS}, $i \in (1 \ldots m)$ for the transmitted packet.
3. The two bit sequence control symbols $c_{1S}, c_{2S}, \ldots, c_{mS}$ are calculated for the sequence S on the sender side, by obtaining the remainder of the division of the length l_{iS} by four for every fragment i.
4. The packet BS is transmitted together with the control sequence symbols of S c_{1S}, c_{2S}, …, c_{mS}.
5. In the received packet BR, correspondingly to step 1 the fragments E_{1R}, E_{2R}, …, E_{mR} are determined that consist of sequences of unit bits the length of which is larger than $h - 2$.
6. Correspondingly to step 2, the number of unit bits in each fragment is determined: E_{1R}, E_{2R}, …, E_{mR}. These are essentially equal to l_{lR}, $l_{2R}, \ldots, l_{mR}$, $\forall i=1, \ldots, m$; $l_{iR} \geq h - 1$.
7. The control sequence R of the receiver consists of the two bit symbols c_{1R}, c_{2R}, …, c_{mR}. Each symbol c_{iR} is calculated as the remainder of the division of the length l_{iR} by four for every fragment i.
8. The counter j used for processing symbols c_{lR}, $c_{2R}, \ldots, c_{mR}$ is initialized to 1.
9. If $c_{jS} = c_{jR}$ then proceed to step 12
10. If $(c_{jS} + 1) \bmod 4 = c_{jR}$, then the fragment j was transmitted with a synchronization error: at the receiver the number of unit bits of the jth fragment was increased by 1. Consequently, the error is corrected by decreasing the unit bits of the jth fragment E_{jR} by 1. Proceed to step 12.
11. The fragment E_{jR} suffered the opposite synchronization error: during transmission, the number of unit bits of the jth fragment was reduced by 1. Hence a unit bit is added to the jth fragment E_{jR}.
12. If $j < m$, increment the counter j and return to step 9.

End of procedure.

The proposed method may be illustrated via the following example:

Let the critical limit h for the existence of risk for the occurrence of synchronization errors be defined as h=6. Furthermore, assume that the transmitted packet size is 64 bits, BS = {0011 **1111 1100** 0010 0**111 1111 1110** 0100 1**011 1111** 0001 **1111 1**101 0101 0**011 1110**}.

According to the proposed method, the transmitted packet BS is processed to determine the unit bit sequences the length of which is not shorter than $h - 1 = 5$. The packet BS contains 5 (m = 5) such sequences E_{1S}, E_{2S}, …, E_{5S} that are marked with boldface. The numbers of units in each fragment are calculated as: $l_{1S} = 8$, $l_{2S} = 10$, $l_{3S} = 6$, $l_{4S} = 7$, $l_{5S} = 5$. The symbols c_{1S}, $c_{2S} \ldots$, c_{5S} of the control sequence S are calculated as the as the remainder of the division of l_{iS} by 4 for each fragment i: $c_{1S} = l_{1S} \bmod 4 = 8 \bmod 4 = 0$, $c_{2S} = l_{2S} \bmod 4 = 10 \bmod 4 = 2$, $c_{3S} = l_{3S} \bmod 4 = 6 \bmod 4 = 2$, $c_{4S} = l_{4S} \bmod 4 = 7 \bmod 4 = 3$, $c_{5S} = l_{5S} \bmod 4 = 5 \bmod 4 = 1$. Consequently, the control sequence S consists of the following five symbols: $S = \{0, 2, 2, 3, 1\}$.

Consider for example the incidence where the receiver receives the following erroneous data packet: BR = {0011 **1111 1110** 0001 0011 **1111 1110** 0100 1**011 1111 1**000 **1111 1**101 0101 0**011 1110**}. In the received packet BR, the fragments are determined as sequences of unit bits of length not less than $h - 2$. The received packet BR

DOI: 10.1051/itmconf/20170903007

contains 5 such sequences ($m = 5$) to be considered, E_{1R}, E_{2R} ..., E_{5R}, that are marked in boldface. The length of these fragments is correspondingly equal to l_{1R}=9, l_{2R}=9, l_{3R}=7, l_{4R}=6, l_{5R}=5. The control sequence R, consisting of the symbols c_{1R}, c_{2R}, ..., c_{5R}, is formulated by calculating the remainder of the division of l_{iR} by four for each fragment i: $c_{1R}=l_{1R}$ mod 4=9 mod 4=1, $c_{2R}=l_{2R}$ mod 4=9 mod 4=1, $c_{3R}=l_{3R}$ mod 4=7 mod 4=3, $c_{4R}=l_{4R}$ mod 4=6 mod 4=2, $c_{5R}=l_{5R}$ mod 4=5 mod 4=1. The obtained control sequence R is hence calculated as containing the following symbols: R = {1, 1, 3, 2, 1}.
The counter is initialized: $j = 1$.
According to Step 9, the relationship between corresponding symbols of the two control sequences S and R: in the case of c_{1S} and c_{1R}, since $c_{1S} \neq c_{1R}$ ($0 \neq 1$), the 1st fragment of the received packet BR contains a synchronization error. The determination of the type of the error is performed according to Step 10, by testing the hypothesis $(c_{1S}+1)$ mod $4=c_{1R}$. For the first pair of control symbols the hypothesis is verified since (0+1) mod 4=1 (1=1). It is hence determined that the received packet contains one extra unit bit in the fragment E_{1R}. The correction of this error is performed by removing the surplus bit from the fragment E_{1R}.
Since the value of the counter is smaller than the number of fragments m, the counter is incremented, i.e. $j = 2$.
The relationship between the second pair of symbols of the control sequences S and R, c_{2S} and c_{2R} is investigated: $c_{2S} \neq c_{2R}$ ($2 \neq 1$) and a synchronization error is determined to exist in the second fragment of the received packet BR. The condition defined in Step 10 is tested $(c_{2S} + 1)$ mod 4 = c_{2R}, that in this case is found not to be satisfied: (2 + 1) mod $4 \neq 1$ ($3 \neq 1$). According to Step 11, it is concluded that the fragment E_{2R} has been received with a synchronization error: the number of unit bits in the fragment has been reduced by one on the receiver side. The elimination of the error is achieved by inserting a unit bit in the fragment E_{2R}.
S $j < m$ the counter is incremented to $j = 3$.
According to Step 9 the relationship between the third pair of control symbols c_{3S} and c_{3R} of the control sequences S and R is investigated. Since these are not equal, $c_{3S} \neq c_{3R}$ ($2 \neq 3$), the decision is made that the received fragment E_{3R} contains a synchronization error. For the determination of the type of the error the condition $(c_{3S} + 1)$ mod $4=c_{3R}$ is investigated, that is in this occasion verified: (2 + 1) mod 4=3 (3=3). According to Step 10 the error is determined to be manifested as an extra unit bit in the received fragment E_{3R}. A correction is performed in the 3rd fragment of the received packet BR, by removing one non – synchronized bit.
The value of the counter j is less than m: $3 < 5$; hence the counter is incremented ($j = 4$).
The relationship between the fourth ($j = 4$) pairs of control symbols c_{4S} and c_{4R} of the control sequences S and R is investigated: since $c_{4S} \neq c_{4R}$ ($3 \neq 2$), the 4th fragment is detected to contain a synchronization error and requires correction. According to Step 10, the condition $(c_{4S}+1)$ mod $4=c_{4R}$ is not satisfied since (3+1) mod $4 \neq 2$ ($0 \neq 2$). This error is classified as a synchronization error whereby the number of unit bits in the 4th fragment has been reduced by one during transmission. For the correction of the detected error, a unit bit is inserted into the fragment E_{4R} of the received packet.
According to Step 12, the value of the counter is incremented: j = 5.
According to Step 9: since $c_{5S} = c_{5R}$, it is hence verified that the fragment E_{5R} was transmitted without any synchronization errors. The calculation jumps to Step 12.
As the value of the counter j is equal to the number of fragments m, the error correction process terminates.
In summary, for the presented example it was detected that the received packet BR contained four synchronization errors in five fragments:

- In fragment E_{1R}, an additional unit bit was created
- In fragment E_{2R}, a unit bit was deleted,
- In fragment E_{3R}, an additional unit bit was created
- In fragment E_{4R}, a unit bit was deleted
- In fragment E_{5R}, there were no errors appearing

The information packet at the receiver side will hence be reconstructed as: BR = {0011 1111 1100 0010 0111 1111 1110 0100 1011 1111 0001 1111 1101 0101 0011 1110}, which is an exact match of the transmitted packet BS.

3 Efficiency analysis

The principle advantage of the proposed synchronization error correction method, compared to existing techniques is the lack of limitation in the number of errors that may be corrected. All current techniques for error correction have been designed so as to correct up to two transmission errors. The proposed method does not require any assumptions on the numbers of errors that may have occurred during data transmission. This fact is particularly important for real time system and procedure control applications, components of which operate in different temperatures. More specifically, relevant applications include aircraft control, weapons systems and C4ISR. In such applications, the number of digital sensors involved may reach 250,000. Systems of this scale may be described as complex digital systems for the control of devices and technological processes that are associated with extreme temperature conditions.

The number k of control symbols that are transmitted together with the information packet, is a random variable that depends on the length n.

Assuming that the values of zero and one appear with equal probability for any bit in the information packet, it may be shown that the average number of fragments that consist of exactly j unit bits is equal to $n/2^{j+2}$. Consequently, the mean number k of the fragments of the information packet that consist of a sequence of unit bits of length that is not less than $h - 2$, is estimated as the sum:

$$k \approx \sum_{j=h-1}^{\log_2 n+5} \frac{n}{2^{j+2}} = \frac{n}{2^h} \quad (3)$$

Hence, taking into account that for each sequence at least $h - 2$ unit bits convey the digital symbol, the mean number of redundant information bits is $n / 2^{h-1}$ or the percentage of the redundancy $2^{-(h-1)}\cdot 100\%$. For example, if a common value of $h = 6$, for the controlled transmission of a 256 byte long packet ($n = 2048$), according to (3), the mean number k of the length of the sequence of units of length of no less than $h - 1 = 5$ is $k = 32$. Consequently, the proposed method requires the transmission of an additional number of $k = 32$ double bit symbols or 64 bits that correspond to an overhead of $2^{-(h-1)}\cdot 100\% = 3.125\%$.

Existing systems that are capable of only correcting double synchronization errors, use for similar circumstances $2\cdot\log_2 k = 64$ control bits and therefore impose a level of redundancy that is similar to the proposed method. Using bit stuffing, an additional bit is required for each sequence of length greater than $h - 2$ unit bits, i.e. $k/2 = 16$ extra bits. The proposed method is therefore shown to possess high applicability for the correction of transmission synchronization errors and the level of redundancy imposed as overhead is not significantly different in practice, compared to existing methods.

One of the principle advantages of the proposed method, in contrast with existing ones, is the use of simple mathematical operations. In total, the detection of synchronization errors requires on average two comparison operations, one addition plus one more logical operation. Hence the proposed error correction procedure is mathematically simple and requires significantly smaller computational complexity, compared to existing ones. Compared to bit stuffing, the proposed method offers, for medium synchronization error occurrence rates, a significant acceleration of the error control procedure, because of the use of much simpler mathematical operations.

4 Conclusions

Asynchronous serial transmission is a fundamental technology that enables the use of embedded information processing equipment in diverse environments. Transmission errors due to loss of synchronization are a special class of errors, to which serial asynchronous lines are especially susceptible and represent an important impediment to the achievement of increased transmission speeds in such transmission lines. Classical error correction techniques can only correct limited numbers of errors and impose significant computational overheads. A technique especially designed for correcting transmission errors due to loss of synchronization was presented. The technique is based on segmentation of the information packet and the insertion of suitable error control codes. This technique provides capabilities for correction of larger numbers of errors than classical techniques, is computationally more efficient and imposes less transmission overheads than bit stuffing, which is the alternative, specialized technique for correcting transmission errors due to loss of synchronization. The technique was illustrated by means of an example and was shown to be capable of correcting four such errors, of two different types within a 64 bit packet. Simulation and experimental results will be pursued so as to demonstrate the applicability of the proposed method.

References

1. Klove T. Error Detecting Codes: General Theory and Their Application in Feedback Communication Systems / T. Klove, V. Korzhik.- Norwell, MA: Kluwer, 1995. – 433 p.
2. A. J. McAuley, Weighted sum codes for error detection and their comparison with existing codes, IEEE/ACM Trans. On Networking, vol. 2, no. 1, pp. 16–22, Feb. 1994.
3. Zumbragel J. Skachek V., Flanagan M. On the Pseudocodeword Redundacy of Binary Linear Codes J. Zumbragel, V. IEEE Trans. of Information Theory.-2012.-Vol.58.- № 7.- P.4848-4861.
4. R. Prévost, M. Coulon, D. Bonacci, J. LeMaitre, J. P. Millerioux and J. Y. Tourneret, "CRC-assisted error correction in a trellis coded system with bit stuffing," 2011 IEEE Statistical Signal Processing Workshop (SSP), Nice, 2011, pp. 381-384. doi: 10.1109/SSP.2011.5967710
5. Bardis N.G, Markovskyy A.P, "Utilization of Avalanche Transformation for Increasing of Echoplex and Checksum Data Transmission Control Reliability", ISITA 2004 - INTERNATIONAL SYMPOSIUM ON INFORMATION THEORY AND ITS APPLICATIONS, IEEE / SITA, Parma, Italy, October 10-13, ISBN: 4-902087-08-1, 2004
6. Nikolaos G. Bardis, Athanasios Drigas and Oleksandr P. Markovskyi, Performance Increase of Error Control Operation on Data Transmission, 3rd International Conference on New Technologies, Mobility and Security, IEEE, IEEE COMMUNICATIONS SOCIETY, Egypt – Cairo, 20-23 December 2009.
7. Bardis N. G, Doukas, N., Markovskyi O.P, Transmission Error Correction Based on the Weighted Checksum, Proceedings of the APPLIED COMPUTING CONFERENCE 2009 (ACC '09), MATHEMATICAL METHODS AND APPLIED COMPUTING Volume II, Vouliagmeni-Athens, Greece, pp.758-763, September 28-30, 2009.
8. Nikolaos Bardis, Nikolaos Doukas, Olexander Markovskyi, "Efficient Burst Error Correction Method for Application in Low Frequency Channels and Data Storage Units", DSP/IEEE 2011 17th International Conference on Digital Signal Processing, 6-8 July 2011 Corfu, Greece.

9. Olga Fedorechko, Olexandr Markovskyi, Nikolas Doukas, Nikolas Bardis. Synchronization Error Detection of Data Transmission Errors in Asynchronous Channels. Proceedings of the 18th International Conference on Systems (part of CSCC '14) Santorini Island, Greece July 17 - 21, 2014

Program Tools for Estimation of the Green Software Reliability

Dmitry Maevsky*, *Elena* Maevskaya, *Ludmila* Shapa and *Dmitry* Stetsyuk

Odessa National Polytechnic University, 65044 Boulevard of Shevchenko, 1, Ukraine

Abstract. The paper deals with a review of the current software of estimation of the green software reliability. The paper demonstrates that any software is capable of becoming a "green" one if it is at the high reliability level, which has to be continuously measured and controlled. In order to implement these actions, the paper offers the results of reviewing the most frequently used software performing the reliability estimation. On the basis of the analysis of this software characteristics a taxonomic system of classification features describing its (software) functionality and the ease of presentation of the estimation results with the help of this software is offered. By using the offered taxonomy a system of classification of software reliability estimation tools was created. The conclusions of the current trends of research in the area of reliability estimation and green software development were made.

1 Introduction

At present any software is considered to be a green one if the electric energy is saved in using this type of software. It is really so. However as to the critical-purpose software systems the saving factor is just one (and not the most important) of the requirements of greenness. On the first place in the critical-purpose software systems there is the factor of reliability and resilience. Indeed, the defects containing in a software can destabilize the performance of many critical infrastructures such as nuclear power, transport, communications, health service. Herein the material and energy resources required to reconstruct the mentioned critical infrastructures or to eliminate the consequences of accidents occurring at them can be many times greater than the cost of electrical energy, which is saved in the process of their usage.

So, the provision of stated indicators of Green Software reliability is of vital importance for creating the qualitative and safe software systems. At present there exist about sixty various software reliability grows models (SRGM) to calculate these values. To obtain the numerical values of the reliability indicators these models uses time series of data of the identified defects. The indicators themselves are found by doing the complicated math and using the nonlinear programming unit. The manual calculation of these coefficients takes a lot of time, and this fact became an impulsive cause for creation of a large number (more than twenty) of program tools for software reliability estimation. Each of these software tools as a rule permits for estimating the reliability according to not one but a significant number of SPMs. The availability of the variety of program tools makes some difficulties for a developer in choosing the best and most suitable SRGM of the amount of implemented in it (program tool). The complexity of the task is aggravated by the two factors. Firstly, as a rule the software reliability theory is unknown to developers. Secondly, there is no single all-purpose SRGM, which accurately determines the reliability indicators in all cases. In choosing the model the indicators classification system offered for the first time in [1] and described in detail in monograph [2] can be useful for developers. However, the classification of software reliability estimation program tools is absent at the moment. That is why its development is timely and topical.

2 The review of reliability estimation program tools

The review of program tools found in the available literary sources containing the information, the amount of which allows to understand their functionality and user interface, is given below.

SRMP (Software Reliability Modeling Programs) [3] – a program tool for reliability estimation created by Bev Littlewood in 1988. It includes the following reliability models: Goel-Okumoto, Musa-Okumoto, Duane as well as Littlewood-Werall quadratic and linear models. SRMP presents the results of estimation in the form of tables and graphs. The estimation accuracy indicators are formed in a separate way. SRMP accepts only time between failure as input data. The results provided in graphic form with some statistics.

SMERF [3, 4] (Statistical Modelling and Estimation of Reliability Functions for Software) – a tool developed by William Farr in 1988. The reliability estimation according to such SRMs as Littlewood-Werall quadratic and linear models, Musa's basic and logarithmic models, geometric one, Jelinsky-Moranda, Yamada, Schneidewind, Brooks and Motley's binomial and

* Corresponding author: Dmitry.A.Maevsky@gmail.com

DOI: 10.1051/itmconf/20170903008

Poisson's models, generalized Poisson model is implemented in it.

RGA (Reliability Growth Analysis and Repairable System Analysis)– a program tool created in 1992 by ReliaSoft Corporation and Dr. Larry Crow. According to [5], RGA includes full support for traditional reliability growth analysis using the applicable models – Crow-AMSAA (NHPP), Duane, Gompertz, Modified Gompertz, Lloyd-Lipow or Logistic – for a variety of developmental data types – time-to-failure, discrete (success/failure) and reliability data. As described in [5], RGA offers all of the major reliability growth models, plus advanced analysis methods that are not available in another program tools. RGA also provides opportunities for fielded repairable system analysis, including a Design of Reliability Test utility and a method for analyzing the system's reliability behaviour over time in order to calculate optimum overhaul times and other metrics of interest. The latest version of the program tool was published in March 2016.

CASRE (Computer-Aided Software Reliability Estimation Tool) [3, 6] – a tool, which was implemented in 1993 by Allen P. Nikora. It includes the following reliability models: Brooks-Motley binominal model, Schneidewind, Jelinsky-Moranda, Littlewood-Verrall, Musa-Okumoto, basic and logarithmic Musa models. CASRE provides the developed automated support of software reliability modelling. The main advantage of CASRE is its various reliability estimations under a paradigm which linearly combines the component models. This tool features an enhanced graphical user-interface which greatly facilitates the process of software reliability estimation.

ROBUST (Reliability of Basic and Ultra-reliable Software sysTems) [7, 8] – a program tool created by Y. K. Malaiya, Jason Denton and Naixin Li in 1995. It supports four SRGMs: Muas, Musa-Okumoto, Duane and S-shaped. It gives the possibility to take into account the metrics containing the information of a program, crew of developers, etc. ROBUST provides a wide range of support for traditional software reliability growth models. It is capable of working with failure data in either a cumulative time to failure or failure interval format.

CARATS (Computer-Aided Reliability Assessment Tool for Software Based on Object-Oriented Design) [9] – a tool created by Chien-Chia Chen, Chu-Ti Lin, Hen-Hsen Huang and Shih-Wei Huang. The following SRGMs are implemented: Yamada, Musa-Okumoto, Goel-Okumoto, Duane. There is a chance to compare SRGMs with the help of maximum likelihood method and the one of least deviation. CARATS can use both traditional SRGMs and neural-network methods to assess software reliability.

Refis (A New Statistical Software Reliability Tool) [10] – a program tool developed by M. A. A. Boon, E. Brandt, Corro Ramos, A. Di Bucchianico and R. Henzen in 2006. It supports the following SRGMs: Littlewood-Verall, geometrical one, Musa, Jelinsky-Moranda, Muas-Okumoto, Goel-Okumoto, Schick and Wolverton, Schneidewind, S-shaped, Duane.

SRATS (Software Reliability Assessment Tool on Spreadsheet) [11] – a tool, which was created in 2005. SRATS functions as a Microsoft Excel add-in, that is why it cannot be used as an independent tool. The latest version came into existence in 2012.

SafeMan [12] – a program tool developed by Takaji Fujiwara and Mitsuhiro Kimura in 2014. Only two RSGMs have been implemented – Musa model and S-shaped model. This tool has function of automatic selection of the optimum SRGM.

3 Classification features formation

The presented classification features aimed at the software reliability estimation by program tools are to reflect two factors, which are important for users: the ease of a program tool usage and its functionality. The first factor is stipulated by the interface characteristics of a program tool, and the second – by mathematical apparatus implemented in it.

The authors offer to attribute the method of source data statement and the one of result presentation to the interface features. Let us consider them in detail.

The ease (and frequently the use) is identified with the help of a method of source data statement. E.g. if a user has got the cumulative series of distribution of the failures detected in the course of time, and a program tool is able only to fix the moments of their detection at the input then such a program tool cannot be used.

That is why the author offer to use a type of source data, which is appropriate for this particular program tool, as the **first classification feature**. We distinguish such types of source data presentation:

– TF (Time of Failure) – the astronomical time of the failure emergence;

– TBF (Time Between Failures) –time interval between two successive failures;

– CF (Cumulative Failure) –the cumulative amount of failures;

– FI (Failures experienced in a time Interval) – a number of failures detected at the predetermined time interval.

The second feature of Software reliability estimation tools classification is a method of results presentation of their run. The method of presentation is of great importance since it is the method itself that determines the ease for a user, who deal with a Software tool. The coefficients of applied reliability models do not provide a user with any useful information. On the basis of models coefficients, the main reliability indicators - failure rate, MTBF and hazard function - can be only identified. These indicators provide the users, who are not informed enough in the theory of Software reliability, with little data as well. The users of the Software reliability estimation tools are to be either the specialists in the program debug area or the ones in the Software engineering field. For them the results of run of Software reliability estimation tools should be presented in the familiar terms "a number of failures at a time" and "the law of changes of the number of failures over time". As the analysis carried out by the authors has shown the majority of the current Software reliability estimation tools provides the user with the two mentioned types of results – the main indicators of reliability and the ones of the number of failures (Fig. 1).

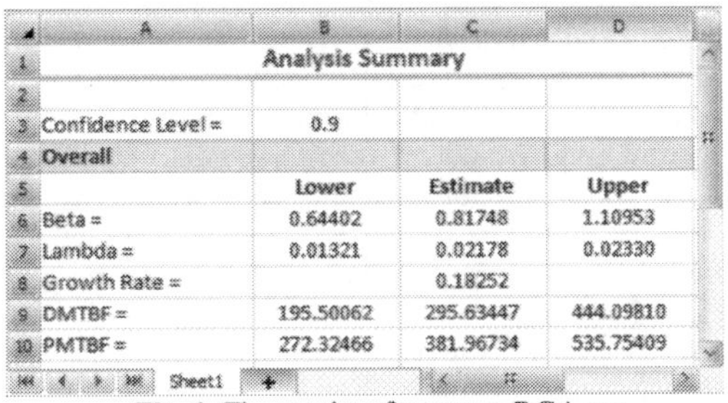

	A	B	C	D
1	Analysis Summary			
2				
3	Confidence Level =	0.9		
4	Overall			
5		Lower	Estimate	Upper
6	Beta =	0.64402	0.81748	1.10953
7	Lambda =	0.01321	0.02178	0.02330
8	Growth Rate =		0.18252	
9	DMTBF =	195.50062	295.63447	444.09810
10	PMTBF =	272.32466	381.96734	535.75409

Fig. 1. The results of program RGA run

There are two types of presentation of the indicators of the number of failures. The first type is a table in which the times and the number of failures in Software over times are given. The second type are the graphs displaying the data of the table. The graphic presentation of results is more demonstrative but the tabular form is more functional. Using the table data one can accurately identify the number of failures in software and evaluate the rate of its changes. We distinguish three features demonstrating the presentation of the results of Software tools run:

– R –the results are the indicators of reliability;

– T –the results are presented in a tabular form;

– G –the results are given in a graphical form.

In each case the simultaneous usage of several features is possible.

The third classification feature displays the functionality of Software reliability estimation tools. At present the functionality is indicated on the basis of the SRGM list implemented in this model. However, a mere listing of models cannot be informative for a user in principle. The users of Software reliability estimation tools are not specialized in the theory of Software reliability. They do not know the peculiarities of existing models and the rules of their usage. The ideal case is when a user does not generally suspect about the existence of the models. In this case the model itself is not of any importance for him (her) to obtain some result. The only thing a user need is to obtain a reliable result. However, all the considered Software reliability estimation tools are implied to be utilized by a user for the manual choice of one or several models from the list (Fig. 2).

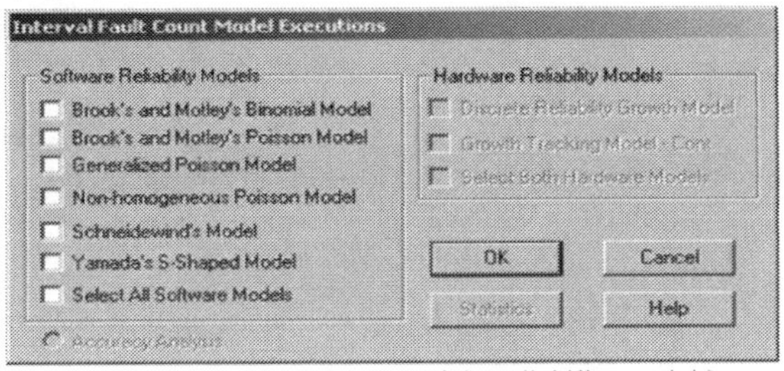

Fig. 2. The window of choice of the reliability model in program SMERFS

Taking into account the fact that the different set of models are implemented in different Software reliability estimation tools the choice of the necessary tool is the impossible task for an unexperienced user today. This task can be carried out only with the help of a specialist in this field or at random.

In the given paper the authors offer to change the approach to the choice of Software reliability estimation tool. The users are not informed in reliability models area but they know the peculiarities of their own Software, the conditions of its development and testing. However, each of the reliability models is created taking account the definite assumptions, which are generally determined by testing conditions. E.g. in the majority of models an assumption is taken that in correcting a failure the new failures are not included in Software. That is why the choice of a particular reliability model in every particular case is made on the basis of comparison of assumptions of a model and the conditions of testing. On the other hand, a set of models implemented in each Software reliability estimation tool, creates a set of assumptions, with the help of which it meets this Software tool. So it becomes possible to choose a Software tool on the basis of comparing the testing conditions and a set of assumptions.

In the fullest form the assumptions of the majority of RSGMs have been analysed and generalized in the paper [13]. In this paper the authors offer a matrix of RSGM assumptions in the lines of which all models assumptions are grouped and in the columns – the models, which meet these assumptions. All assumptions are united in the four groups:

– I –take into account the way of detecting and correcting the failures;

– II –display the models classification according to [1];

– III –take into account the way of distributing the detected failures in time;

– IV –the assumptions, which are inherent only to some individual models.

Each of the groups in its turn is divided into subgroups. The complete list of all subgroups with the detailed explanations one can find in the above mentioned paper. The matrix of matches between assumptions and reliability models is presented ibidem. Based on all said above as the third functional feature of Software reliability estimation tools the authors offer to use the assumptions of models implemented in them (tools) according to the matrix of assumptions.

The authors of article [13] propose the following sequence of steps for software reliability evaluation using assumption matrix:

1. Analysis of software requirements and software engineering process.
2. Obtaining of the information about assumptions usable to the developed software.
3. SRGM choice using assumption matrix.
4. Determination of SRGM input parameters and analysis of their complexing opportunities.
5. SRGM parameters fitting.

DOI: 10.1051/itmconf/20170903008

6. Calculation of software reliability measures and statistical analysis of obtained data.

Fig. 3 shows a fragment of the assumptions matrix (according to [13], Table 1).

Attributes / Category	Finite												Infinite				
Type	Binomial		Poisson									Bino-mial	Poisson				
Family/Class	Exponential							Weibull & Gamma						Exponential			
Group	1. De-eutrophication		2. NHPP			3. Hyper-exponential		4. S-shaped				5. Weibull		6. Geometric			7. Logarithmic
Model	Jelinski-Moranda	Shooman	Goel-Okumoto	Schneidewind	Musa	Basic	Lapri	Basic	Hyper-exponential	Ohba	Test Effort	Shick-Wolverton	Duane	1-st Moranda	2-nd Moranda	Lipow	Musa-Okumoto
	1.1	1.2	2.1	2.2	2.3	3.1	3.2	4.1	4.2	4.3	4.4	5.1	5.2	6.1	6.2	6.3	7.1
Assumptions	1	1	1	1	1	1	1	1	1	1	1	1	1	1	1	1	1
I 1.1																	
I 1.2	1	1	1	1	1	1	1	1	1	1	1	1	1	1	1	1	1
I 1.3	1	1	1	1	1	1	1	1	1	1	1	1	1	1	1	1	1
I 1.4	1	1	1	1	1	1	1	1	1	1	1	1	1	1	1	1	1
II II.1 II.1.1	1	1	1	1	1	1	1	1	1	1	1	1					
II II.1 II.1.2													1	1	1	1	1
II II.2 II.2.1 II.2.1.1			1	1	1	1	1	1	1	1	1		1	1	1	1	1
II II.2 II.2.1 II.2.1.2			1	1	1	1	1	1	1	1	1		1	1	1	1	1
II II.2 II.2.1 II.2.1.3			1	1	1	1	1	1	1	1	1		1	1	1	1	1
II II.2 II.2.2 II.2.2.1	1	1										1					
II II.2 II.2.2 II.2.2.2	1	1										1					
II II.3 II.3.1	1	1	1	1	1	1	1							1	1	1	1
II II.3 II.3.2								1	1	1	1	1	1				

Fig. 3. Fragment of Assumption Matrix

The forth classification feature, which also indicates a Software tool functionality, is to perform the prediction of reliability indicators changes in the subsequent periods of time. Broadly speaking the reliability indicators prediction is not a complicated mathematical problem. The point is that a reliability model is finally a mathematical formula determining the law of changes of the number of failures over time.

The formula contains the coefficients identified on the basis of the time series of failure detection. That is why one should indicate the model coefficients only once and then, substituting different time values into the formula, identify the predicted indicators of reliability. However, taking into account the fact that a final user is badly (or not at all) informed in the reliability models area a Software tool is to take the task of calculation of the predicted indicators. We are able to distinguish two values of the feature:

– E – a Software tool performs only the reliability estimation;

– P – a Software tool performs both the estimation and prediction of reliability.

We can present the classification features (taxons) system graphically (Fig. 4).

The group of interface features determines two aspects in the system. First, it can be called a traditional ease when using a Software tool.

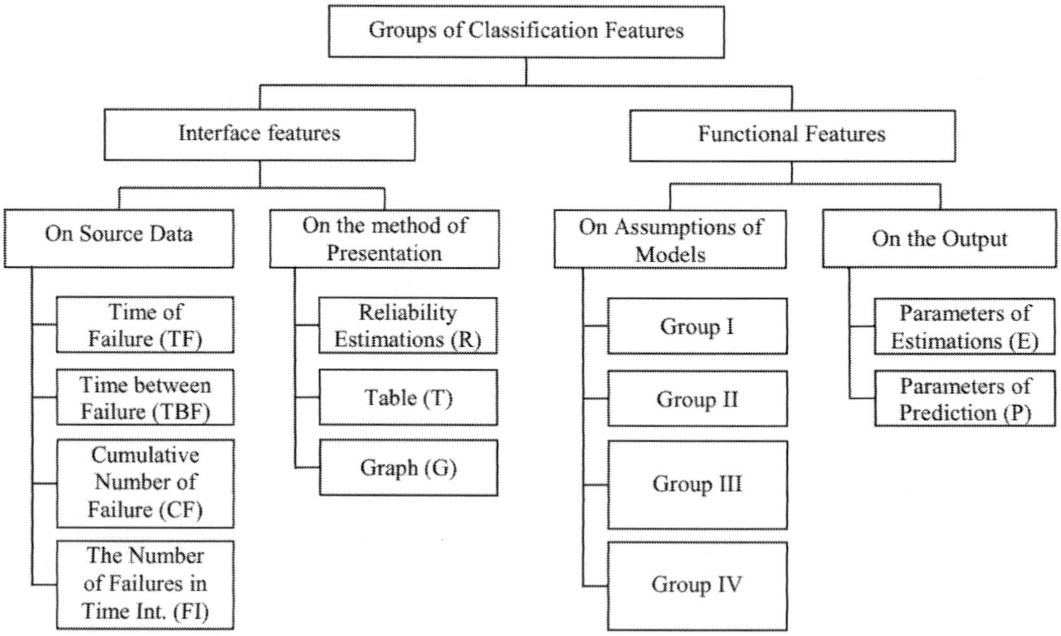

Fig. 4. Classification Features of Software Reliability Estimations Tools

However, the way of setting the source data, which can be also attributed to the source data, which can be also attributed to the interface features, determines finally a reliability model used That is why we can say that in the developed system of classification the interface features determine one more – methodological aspect of usage of Software reliability estimation tools.

The group of functional features identifies the potential of a Software tool to estimate the reliability. The usage of simple assumptions of models understandable for a user allows to significantly ease the choice of a suitable Software tool of all the diversity of the current Software tools.

On the basis of the offered classification features the classification of Software reliability estimation tools described above has been carried out. The classification is given in table 1 at the previous page. The distinguished features have been presented in the columns of the table according to the hierarchy levels shown in Fig.4.

Table 1. Classification of Software Reliability Estimation Tools

Software Tool	Interface Features						Functional Features					
	Source Data				Results		Assumptions of Models				Prediction	
	TF	TBF	C	FI	G	T	I	II	III	IV	E	P
SRMP		v			v	v	I	1, 2.1, 3	2, 5	2.1, 5.2	v	
SMERFS	v	v			v	v	I	II	1, 2, 4, 5, 6, 7	1.1, 2.1, 2.3, 4.1, 5.2, 7.1	v	v
RGA			v	v	v	v	I	1.2, 2.1, 3.2	5	5.2	v	
CASRE	v	v	v	v	v	v	I	1, 2, 3.1	1, 2, 7	1.1, 2.1, 7.1	v	v
ROBUST	v	v					I	1, 2.2, 3	2, 4, 5, 7	2.3, 4.1, 5.2, 7.1	v	v
CARATS	v	v			v		I	1, 2.2, 3	2, 4, 5, 7	2.1, 4.1, 5.2, 7.1	v	v
Refis		v	v		v	v	I	II	1, 2, 4, 5, 6, 7	1.1, 2, 4.1, 5, 6.1, 7.1	v	
SRATS			v	v	v	v	I	1, 2.1, 3	2, 4, 7	2.1, 4.1, 7.1	v	
SafeMan	v		v		v	v	I	1.1, 2.2, 3	2, 4	2.3, 4.1	v	

Conclusion

The paper offers an hierarchic group of classification features – taxons, which represent the Software reliability estimation tools and their main characteristics in the most complete and correct way. Both interface and functional (performance) characteristics of such kind of Software tools have been clearly highlighted in the system. The usage of the developed taxonometric features allows to create a classification of the current Software tools. The developed classification system is able to facilitate a user's efforts in choosing a software reliability estimation tool necessary for him (her) among the variety of tools. The advantage of the classification system offered is as follows: it can be employed by users, who are not specialists in the field of Software reliability theory. An ordinary user does not know and he (she) must not, which reliability models are realized in this or that software tool, and understand the peculiarities of these models operation. He (she) does not need to know such "terrible" specific terms as "Вейбулл's distribution", «Software Reliability Growth Models», «Maximum Likelihood Estimation» and others. Furthermore a user is even unaware of the existence of Software reliability models itself. On the basis of the system offered a user is able to make a choice of the most suitable Software tool, having only understandable information available for him (her). The information only reflects the conditions of Software testing and creating as well as a defect detection and correction method. The usage of the reliability estimation software tool classification demonstrated in the article will allow to make the reliability theory closer to everyday life and give an opportunity to specialists to develop more reliable and stable Software..

References

1. J.D. Musa, K. Okumoto, Electronic Systems Effectiveness and Life Cycle Costing, NATO ASI Series, **3**, 395, (1983)
2. M.R. Lyu *Handbook of Software Reliability Engineer-ing* (New-York: McGraw-Hill, 1996)
3. G.E. Stark *International Symposium on Software Reliability Engineering*, 90, (1991)
4. D.R. Wallace, *26th Annual NASA Software Engineering Workshop*, 147, (2001)
5. L. Crow, ReliaSoft, (http://www.reliasoft.com/rga/)
6. M.R. Lyu, A.P. Nikora, *Fifth International Workshop on Computer-Aided Software Engineering*, 264, (1992)
7. Y. Malaiya, J. Denton, N. Li, (http://citeseerx.ist.psu.edu/viewdoc/summary?doi=10.1.1.5.5397)
8. Li, N. *Sixth International Symposium on Software Reliability Engineering*, 375, (1995)
9. C.C. Chen, C.T. Lin, H.H. Huang, S.W. Huang, C.Y. Huang, *IEEE Region 10 Conference TENCON 2006*, 1, (2006)
10. M.A.A. Boon, E. Brandt, I.C. Ramos, (http://www.refis.nl)
11. H. Okamura, T. Dohi, Software Reliability Engineering (ISSRE), 100, (2013)
12. T. Fujiwara, M. Kimur, JSEA, **7**, 396, (2014)
13. V.S. Kharchenko, O.M. Tarasyuk, V.V. Sklyar, V.Y. Dubnitsky, *International Computer Software and Applications (COMPSAC 2002)*, 541, (2002)

Metric-based method of software requirements correctness improvement

Svitlana Yaremchuk[1,*], *Nikolaos Bardis*[2] and *Kharchenko Vyacheslav*[3]

[1]Danube Institute, National University "Odessa Maritime Academy", 68600, Ukraine
[2] Department of Mathematics & Engineering Science Hellenic Military Academy, 16673, Greece
[3] Department of Computer Systems and Networks, National Space University named after N. E. Zhukovsky "KhAI", 61070, Ukraine

Abstract. The work highlights the most important principles of software reliability management (SRM). The SRM concept construes a basis for developing a method of requirements correctness improvement. The method assumes that complicated requirements contain more actual and potential design faults/defects. The method applies a newer metric to evaluate the requirements complexity and double sorting technique evaluating the priority and complexity of a particular requirement. The method enables to improve requirements correctness due to identification of a higher number of defects with restricted resources. Practical application of the proposed method in the course of demands review assured a sensible technical and economic effect.

1 Introduction and Background

Nowadays humanity utilizes a tremendous amount of software tools and software systems (SWS) for different domains and applications. Customers impose more complex and diverse requirements. Software produces more and more considerable influence on our life and our safety. In its turn, complexity of SWS increases [1]. Consequently, number of defects imposed by developers increases, the SWS reliability decreases, design costs rise. Huge losses and the decrease in the efficiency of human activity caused by the imperfection of SWS and design faults is one of the key challenges in modern IT engineering encouraging increased software reliability [2].

Motivation of research is, firstly, the need to analyze problems of reliability assessment and management. Secondly, in view of importance of software requirements correctness from the point of view of the SWS reliability we propose technique for assessment of complexity and improving of requirements' correctness.

2 Principles of Software Reliability Management. State of the Art

Researches worldwide developed a number of models, methods and software application to evaluate and manage the SWS reliability [3-11]. However, these methods don't take into due account variety and specific features of SWS development business processes and require great expenditures for adaptation and implementation. The above factors prevent to implement widely available model methods and reliability aids in software engineering routine practice, making such implementation impossible in a number of cases. Therefore, the first principle of the (SRM) provides adaptability, simplicity and low costs contributing to easy implementation of the assessment technique into business processes of software corporations.

The SWS lifecycle (LC) SWS consists of a sequence mutually dependent processes. These are stages of requirements' construction, design, codification, testing and the SWS maintenance. At these stages experts of various profile and qualifications create unique artifacts. They are requirements' specification, architecture and detailed SWS design, initial codes, testing procedures and versions, ready software product. The developed SWS reliability is pre-determined by correctness of each individual artifact, depends on its complexity and forms at each stage of its LC. However, existing reliability models don't take into sufficient account complexity and evaluate reliability at each LC stage fragmentarily. Therefore, the second SRM principle provides thorough complexity evaluation and artifacts reliability management throughout the entire SWS LC.

Different production data accumulates throughout the SWS LC stages. It includes user demands in textual, tabulated, graphical and verbal formats, architecture and design descriptions, algorithms, initial software code, testing scenarios, incoming data massive, video files of the SWS behavior under various conditions, technical

* Corresponding author: svetlana397@yandex.ru

 DOI: 10.1051/itmconf/20170903009

documentation, etc. This multi-type structured and non-structured data of various SWS versions and assemblies are stored and actively used by developers during the long term of operation. This data, also called as Big Data may be measured in terabytes. Methods and applications for Big Data processing and analysis to improve SWS reliability are not sufficiently developed. Therefore, the third SRM principle provides for development Big Data analysis methods and application for SWS LC.

Competitive market compels software corporations to reduce SWS development costs with sharp need to provide the demanded reliability level. The developers often face the problem how to spend limited resources in the most efficient manner? What artifacts subsets should be verified in the available resources don't allow to afford complete verification? What particular pages of e.g. 200 available should be checked to identify maximum of errors, controversies, inaccuracies or incompleteness? What of thousands modules should be tested to detect maximum number of defects? So, the fourth SRM principle consists in maximum possible reliability improvement under restricted resources condition.

Each SWS development starts from collection and analysis of requirements. This process includes clarifying and documenting what should be done by an application and what features it should possess. This process involves all the concerned parties – investors, purchasers, users, from the one side, and managers, analysts, designers, codifiers, testers, i.e. SWS developers, at the other side. In the course of extended joint work errors may occur with any party category at both sides. As a result, incorrect requirements contain defects of various types, such as erroneous data, unclearness, ambiguities, incompleteness, doubling, and contradictions. Imperfect requirements may cost very much. As source [12] marks, requirements defects are 20-50 times more expensive to correct once the defects got into development process. Inspection of one full-scale software project showed that problems with requirements led to US$ 600.000.000 development budget exceeding, 8 years time delay and reduced functionality. Another research [13] showed that it takes about 30 minutes to correct error detected at the stage of requirement processing, whereas it takes 5 – 17 hours to correct an error detected in the course of system testing. As identified in the [14] eliminating requirements' defects are 100 times expensive than those detected in the course of requirements development, if indicated by the customers. Evidently, methods directed to identify as much defects, as possible, in requirements saves plenty time and costs. Therefore, method of improvement of SWS requirements correction is suggested. Below standards and terms applicable at the stage of requirements construction are described.

3 Standards and Terms Applicable at the Stage of Requirements Construction

Standards provided in [15, 16] lost their actuality and are inapplicable. Actual standard is IEEE/ISO/IEC 29148-2011 "Systems and software engineering – Life cycle processes – Requirements engineering" [17]. Terms definitions are enlisted in compliance with the said standard.

SWS Project (software system project) – a full set of developed agreed artifacts including requirements specification, architecture and detailed design, initial code, test procedures and versions, technical documentation, etc.

Requirements are a document specifying users' needs and aims, condition, features and capabilities of the software product.

Concerned parties are individuals, groups and/or organizations taking an active part in the project.

Requirements Analyst is a person in requirements developers team in charge of the requirements extraction, analysis, definition, approval and their management.

Requirements Development is a process aimed to determine project scope, users class and representatives, requirements' extraction, analysis, specification and approval.

Requirements Analysis is Process of categorizing data concerning requirements, requirements evaluation to determine desirable quality, requirements representing in various formats, detailed requirements extraction from requirements of higher levels, requirements priority discussion, etc.

Software requirements specification, SRS – result of a system requirements documenting in structured, accessible and controllable format. According to source [18], requirements specification is a set of technical documentation containing detailed SWS under development and its functionality, requirements to external interfaces, non-functional requirements, glossary, processes and subject zone models, other SWS data, as may be required by parties in concerned.

Requirements Attributes – description data of requirements with unique requirements numbers, data sources, logical groundings, priorities, persons in charge, SWS versions numbers, etc.

Requirement's Elements – any elements composing the requirement – constant values, variables, arithmetic and logical operations, etc.

Requirement complexity – difficulties met in understanding and realization of a requirement related with great quantity of components and links with other requirements and business rules.

Business rule – Policy, guiding principles or provisions which determine or restrict certain aspects of business activities.

Requirement Priority – numerical and/or quality evaluation of requirement attribute, taking into account importance and term of its realization. According to source [18], priorities construe a way to solve conflict between the requirements competitive struggle for restricted resources/

Requirement metric – mathematical representation of numerical evaluation of a particular requirement's feature.

4 Method of SWS Requirements Correctness Improvement

Modern SWS include plenty multi-type requirements. They include functional requirements, requirements to user and software interfaces, hardware interfaces, communications interfaces, requirements to capacity, data safety storage, system safety, etc. Therefore, the SWS specification is a bulky document of hundreds pages. Theoretical approach supposes total review of all these requirements. In practice, "even in medium-size specification experts review thoroughly only the first part, some of the most stubborn may look through half, but it is unlikely that anyone reaches the end" [18]. Entire SWS requirement review is impossible due to lack of time, financial and human resources. In this view, the offered method or requirements correctness improvement consists in extracting the most priority and complicated requirements from their entire set to be thoroughly reviewed to detect the highest possible quantity of defects in prevailing restricted resources situation. Priorities are assigned to requirements after their collection and documenting jointly by the manager, analyst and customers' representatives. Requirements complexity is determined by means of metrics.

4.1. Overview of Requirement Metrics

Publications [12, 18, 20, 21] and standards [17, 19] have been studied for analysis of existing metrics. These standards contain numerical evaluations for about 50 metrics. They include metrics for completeness; precise detailed requirements; non-elementary requirements; requirements not coordinated with other requirements; undetectable and untestable requirements; missed or skipped requirements; non-obvious requirements; defective requirements detected for 1 hour of checking; detailed requirement value; metrics for deleting, adding or modification of the requirement. Their analysis shows that such metrics are designated for quality control, requirements review and analysis efficiency. However, there is no metric evaluating requirement complexity in these standards.

Authors of work [12] apply requirement defect metric as a number of defects per documentation page, provide its interpretation and permissible value as 0.2 defects per page of the requirements' testing description, i.e. one defect per 5 pages. Unclear definition since a number of requirements per page is unknown.

Authors of work [18] stress upon correctness and traceability as the most important feature of requirements without outlining evaluation formulae. The authors define correctness as a precise description of desirable functionality. Traceability is defined as a possibility of analysis both in backward direction to initial sources of requirements and forward to design elements, initial code, tests, scope of application and other SWS elements. The elements include requirements of different types, business rules, architecture and design components, initial code modules, test versions, reference files, traceability is of great importance for requirements management. It enables to determine all the workable artifacts which should be modified once the requirement changes. As per complexity, the authors mark complexity of requirements detection process, complexity of their statement in foreign language, complexity of understanding and technical implementation of requirements with complicated Boolean logics (combination of operators AND, OR, NOT) having, however, refrained from providing numerical evaluation aids for these features.

Authors of [20, 21] propose more advance set of metrics. In work [20] they are numerical metrics of unique, understandable, misinterpreted, modified requirements and requirements under testing. The authors also propose two newer metrics, closer in sense, but different in evaluation manner:

1) correctness $MC = \frac{R_c}{R_t} * 100\%$, with R_c - number of requirements received the same true interpretation by more than 1 referee; R_t - total quantity of requirements;

2) quality $MQ = \frac{R_t - (R_f + R_d + R_m)}{R_t} * 100\%$, with R_f - number of identified and corrected requirements with defects, R_d - number of deleted requirements with defects, R_m - number of modified requirements.

Authors of work [21] added metrics of ambiguity, logical linkage and capability to check the requirements starting from the initial source. Yet, the work does not contain complexity metrics.

Thus, the review of publications and standards showed the lack of metrics enabling to evaluate requirement complexity. The referred enlisted metrics describe construction process and various requirements features, saving the complexity. With such metric missing the most complicated requirements cannot be identified and selected for inspection and review to detect higher defects quantity with prevailing restricted resources situation. The [18] notes that the requirement complexity is evaluated by expert in a subjective manner. Analysis results and author's personal experience in SWS development enable to state that requirements complexity is not numerically evaluated in the software engineering. Evaluations are not applied to improve requirements correctness and the SWS reliability. Therefore, it is necessary to develop a metric for precise numerical evaluation of requirements' complexity.

4.2. Proposed Metric for Requirement Complexity Evaluation

The new metric should take into account internal and external complexity inherent with the requirement. Internal complexity is determined by quantity of components. External complexity is represented by links connecting a particular requirement with other requirements, business rules, etc. These links are illustrated at Figure 1 representing an entity relationships

DOI: 10.1051/itmconf/20170903009

(ER) diagram of logical contacts of the requirement with other elements.

The ER – diagram shows that the requirement may include one element, or more; may affect, or does not affect several dependent requirements; may depend, or not, on some affecting requirements; may depend (or not) on some affecting business rules; may depend, or not, on some affecting entities, strictly specified for each SWS being developed. The higher is quantity of elements in requirement and links of the requirement with other requirements, business rules and other entities, the higher is the requirement and its complexity rate by metric.

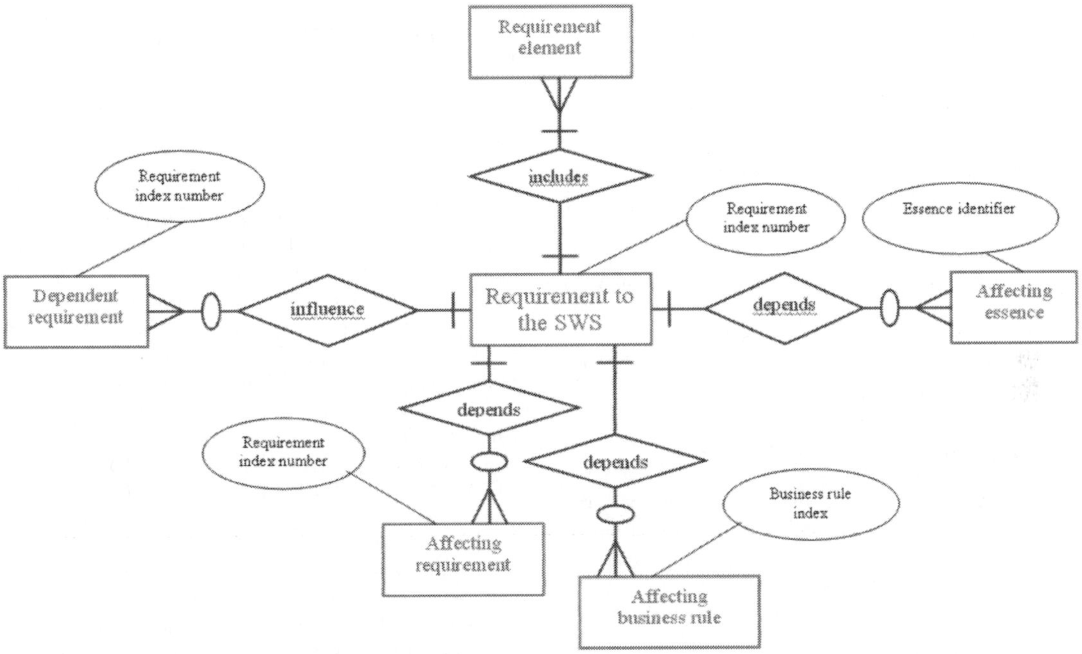

Fig. 1. ER – diagram of logical interconnections of the requirement with other components.

Table 1 contains proposed requirement complexity metric structure and examples of its evaluation. Data required to evaluate requirement complexity by metric is, as follows:

1. Individual requirement index within the system being under development;
2. Number of elements composing the requirement (e.g, to calculate tax rate (variable 1st element) multiplied (operation 2nd element) by 18% rate (constant value, 3rd element) from income value (variable, 4th element) deducting (variable, 5th element) maternity leave (variable, 6th element) and amount payable under sick leave (variable, 7th element);
3. Number of affecting business rules (e.g. global – currency exchange rate, regional – applicable tax rate, corporate – terms to earn quarter bonus in the corporation);
4. Number of affecting and dependent requirements;
5. Number of other affecting entities (e.g. hardware capacity, network traffic speed).

Table 1. Requirement Complexity Metric Structure.

Requirement Complexity Characteristics					Requirement Complexity Metric 6=2+3+4+5
Unique Requirement Index	Internal Requirement Complexity	External Requirement Complexity			
	Number of Components composing the Requirement	Number of Affecting Business Rules	Number of Affecting and Dependent Requirements	Number of Affecting and Dependent Entities	
-1-	-2-	-3-	-4-	-5-	-6-
F63	2	1	5	1	9

F64	3	0	2	2	7
F65	2	0	2	0	4
F66	5	0	1	2	8
F67	2	0	1	0	2
F68	4	3	0	1	8
Mean Requirements' Complexity					6,3

Mathematical representation of the requirement complexity evaluation by metric (requirement complexity metric, RCM) looks, as below:

$$RCM = EQ + IBRQ + IDRQ + IDEQ \quad (1)$$

EQ – the elements quantity (column 2), *IBRQ* - the influencing business rules quantity (column 3), *IDRQ* – quantity of the affecting and dependent requirements (column 4), *IDEQ* - quantity of the influencing and dependent entities (column 5).

Such is the proposed requirement complexity evaluation metric. Calculation of metric consists in summing up the numerical characteristics in internal and external complexity of a particular requirement. The metric enables to evaluate the requirement complexity in complexity units. Obtained values enable to sort the requirements by their complexity level.

4.3. RCM Metric Practical Application

The proposed *RCM* metric has been applied for functional requirements complexity evaluation within a particular corporate system of accounting and analysis. This medium-size system has been developed in Ukraine for the Danube Institute of National University "Odessa Maritime Academy" in 2016. The realized system consisted of 65 components and a database of 26 interlinked tables. Total size of the system amounted to 40000 lines of initial code. System specification encompassed 120 functional requirements. Requirements analysts calculated complexity rates applying the RCM metric and basing on Table 1. The rates in question have been obtained in the course of collection, review and documenting the requirements. The evaluation process was rather easily performed in MS Excel and did not demand additional time or additional software and depended only on organization aspect and personal qualities of experts. Requirements mean complexity RC_{avg} has been calculated being 6,3 complexity units. Total system complexity amounted to 120 · 6,3 = 756 complexity units. The obtained value has been applied to compare complexity of existing and previously developed systems enabling to update development terms and budget. It was also revealed that 15 % requirements had complexity characteristics exceeding 10 units (*RC>10)*. The analysis showed that those requirements had been hard enough to be realized. Detailed review of the requirements in question enabled the experts to form a complete idea of the complexity of the system being developed and to involve necessary resources. Certain complicated requirements have been modified or decomposed into easier. Some requirements, after discussion with the customers have been deleted from specification. After these activities total complexity of the system reduced to 83 units, i.e. by 11%.

4.4. Algorithm of SWS Requirement Correctness Improving

The offered method is based on logical admission that the more is requirement complexity, the more actual and potential defects it contains, such as contr5adictions, discrepancies, incompleteness, inaccuracies which at further stages of system design may likely become defects. Really, the more elements are assembled into a requirement, the higher is chance of errors in elements. The more entities are interconnected with the requirement, the harder it is to describe the requirement clearly, correctly, to codify it properly and to modify it in the course of the SWS maintenance.

The method may be applied once requirement construction process is completed and priorities of their realizations are assigned. Method utilized the proposed RCM metric to evaluate the requirements' complexity. *The method's incoming data* contains characteristics of each requirement.

1. Unique requirement index;
2. Detailed text description of the requirement;
3. Priority of realization of the requirement *Pr* (high *Pr=1*, medium *Pr=2*, low *Pr=3*).

This data is accumulated in the course of requirements construction and does not demand any additional activities and costs. Method procedure includes seven (7) steps:

Step 1. Full set of requirements should be sorted in the order from higher to lower.

Step 2. Subset requirements with $Pr = 1$ R_{pr1} should be derived from R_{full} set. Complexity values should be calculated using metric RCM according to expression (1). The requirements should be additionally sorted according to complexity decreasing criterion. Obtained *List 1* should contain rising rank to each requirement starting from 1 and further to n.

Step 3. Requirements R_{pr1} should be verified. Real and potential requirements' defects should be detected and eliminated (contradiction, mismatching, incompleteness, inaccuracies, in ambiguity).

Step 4. Resources permitting, derive requirements subset R_{pr2} with priority $Pr = 2$. Complexity values should be calculated using metric RCM according to expression (1). The requirements should be additionally sorted according to complexity decreasing criterion. Obtained *List 2* should contain rising rank to each requirement starting from $n+1$ and further to k. Further proceed to steps 3, 5, 6,7. With lack of resources proceed to Steps 6 and 7.

 DOI: 10.1051/itmconf/20170903009

Step 5. Resources permitting, derive requirements subset R_{pr2} with priority $Pr = 3$. Complexity values should be calculated using metric RCM according to expression (1). The requirements should be additionally sorted according to complexity decreasing criterion. Obtained *List 3* should contain rising rank to each requirement starting from $k+1$. Further proceed to steps 3, 6, 7. With lack of resources proceed to Steps 6 and 7.

Step 6. Requirements Correctness Evaluation. Having completed the verification quantity of requirements should be counted by categories, including total verified R_{total}; defective, fixed R_{fix}; defective deleted R_{del}, potentially defective, modified R_{mod}; skipped R_{gap}. The *requirements defectiveness* index **RDI** may be calculated using the formula (2) below

$$RDI = \frac{R_{fix} + R_{del} + R_{mod} + R_{gap}}{R_{total}} \quad (2)$$

The requirements correctness index, **RCI,** should be calculated using the formula (3) below:

$$RCI=1-RDI \quad (3)$$

Step 7. **Requirements correctness improvement.** Comparing index values of *RCI, RDI* with known under previous projects (if available) and/or permissible corporative values. Referring to the comparative results conclusion should be drawn about achieved requirements' correctness and finalizing or further proceeding with their construction stage. E.g. permissible corporative value for $RCI_{norm} = 0.92$ with $RDI_{norm} = 0.08$, i.e. not more than 8 defective of 100 total requirements are permitted. Should $RCI \geq 0.92$, the requirements are correct enough and fit for further development stages. Otherwise ($RCI < 0.92$), the requirements are not correct and demand re-proceeding and repeated performance of the procedure under offered methodology.

This step finalizes the method. Table 2 shows an example of ranking the requirements.

Table 2. Requirement Ranking Example.

Individual Requirement Index	Requirement Priority	Requirement Complexity Grade by Metric (1, ∞)	Requirement Rank (1, ∞)
F63	1	9	1
F64	1	7	2
F65	2	4	4
F66	2	8	3
F67	3	5	6
F68	3	8	5

The top rank for requirements with priority 1 is caused with the highest evaluated complexity rate (complexity rate = 9, rank = 1). As complexity rate decreases the rank decreases as well (complexity rate = 8, rank = 2). The highest rank for requirements with priority 2 should be equal to next after last assigned rank. The top rank for priority 2 requirements is equal to the rank with index following the last assigned rank. It is assigned to the requirement with the highest complexity index (complexity index = 7, rank = 3).

Figure 2 shows the UML-diagram of activities under the method. The UML diagram represents logical consistence of actions within the method.

After the method is completed the final data is obtained as a through *ranked list of requirements and defectiveness and correctness indexes*. The through ranked list enables to perform individual verification of the top priority and most complex requirements and to increase quantity of detectable potential and actual defects. Defectiveness and correctness indexes indicate either requirements' fitness to further stages of processing, or necessity of their reprocessing with further repeated verification. The method enables to reduce defectiveness and to improve correctness of requirements with prevailing restricted resources situation.

5 Case Study. Conclusions

The offered method has been applied in the course of inspection of functional requirements of corporate system, as referred above. Through ranked script of the requirements of top priority and the highest complexity has been applied. Methodology enabled to increase quantity of detectable actual and potential defects by 9% in average in comparison with total verification of the requirements disregarding their complexity, improve requirement correctness to the same extent and to avoid sensible costs which might be incurred to eliminate such defects at further stages of development.

The most important SRM principles have been outlined in this work:

1. *Compatibility* of models methods and technologies with SWS development business processes;
2. *Complexity evaluation* for SWS development process and artifacts reliability management throughout the SWS LC;
3. Development of methods and applications for analysis of *Big Data* SWS LC;
4. Maximum possible reliability improvement with *restricted resources* situation.

Requirement complexity evaluation metric (RCM) is offered within the SRM framework. The metric calculation consists of summing up quantitative characteristics of internal and external requirement complexity. The metric evaluates requirement complexity in complexity units. The obtained evaluation enables to determine total system complexity and adjust terms and costs of development. Modifications or decomposition of complex requirement enables to reduce total system complexity.

Method of improving requirements correctness has been offered within the SRM framework. The method applies an admission that more complex requirements contain more actual and potential defects. The method applies RCM metric to evaluate requirements complexity. The method applies a double sorting procedure basing on requirements' priority and complexity evaluation. The method is advantageous comparatively to the total verification due to possibility to select the most complicated requirements with top priority by means of generating their through ranked script. The method enables to improve requirements' correctness and defectiveness. The method enables to evaluate achieved indexes of correctness and defectiveness. The application of the method in the course of requirements verification provided substantial technological and economical effect.

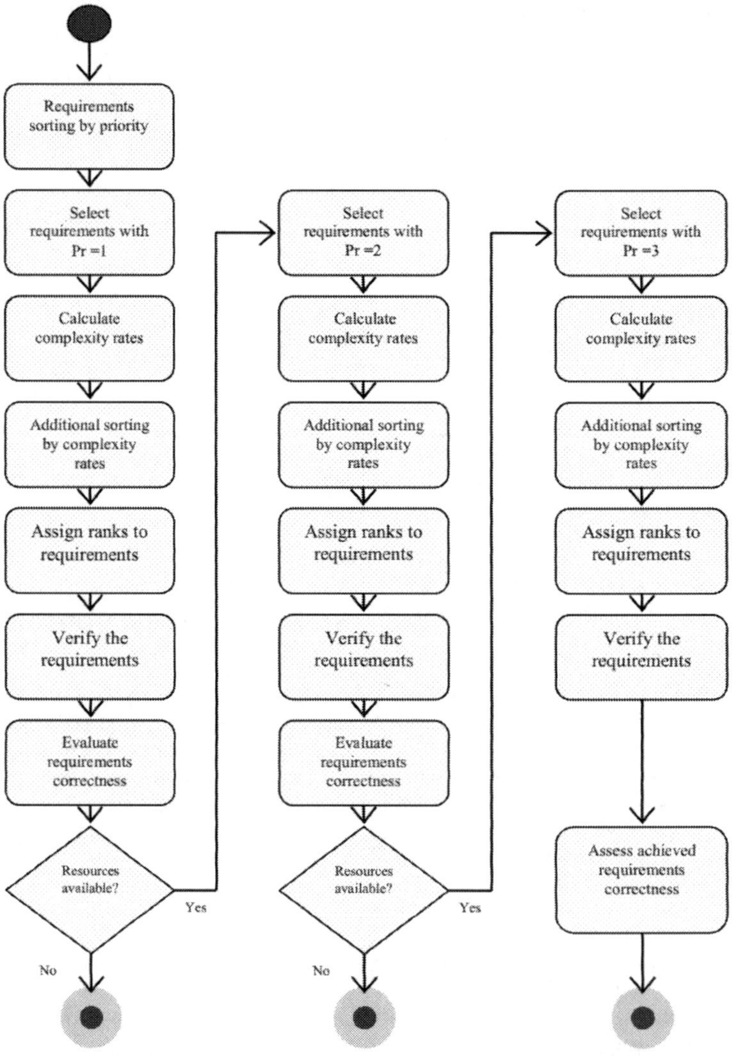

Fig. 2. UML - diagram of activity of SWS requirements correctness improving technique.

The method has been developed taking into account two principles: the principle of artifacts complexity evaluation and the principle of maximum possible reliability improvement in restricted resources situation. Further researches are planned in two directions. The first lies in achieving and providing organic building-in with different SWS development business processes by means of making method patterns with defining roles and activities of the developers. The second direction

DOI: 10.1051/itmconf/20170903009

lies in development of methods and applications for Big Data analysis occurring at all stages of the SWS LC.

References

1. E. Eric. Domain-Driven Design - Tackling Complexity in the Heart of Software. Addison Wesley, 560 p. (2013)
2. Frischknecht Ch. Top 10 Software Fails Of 2014. Access mode: http://www.tricentis.com/blog/2014/12/18/top-10-software-fails-of-2014 (2014)
3. G. Carrozza, R. Pietrantuono, S. Russo. Defect Analysis in Mission-critical Software Systems: a Detailed Investigation. J. Softw. Evol. and Proc. ; 2012/07/12, Vol.**2**, pp. 1-28. DOI: 10.1002/smr (2012)
4. Cotroneo, D., Pietrantuono, R., Russo, S. Testing techniques selection based on ODC fault types and software metrics. The Journal of Systems and Software, Vol.**86**, pp. 1613-1637 (2013)
5. Cotroneo, D., Pietrantuono, R., Russo, S. Combining Operational and Debug Testing for Improving Reliability. The Journal of Systems and Software, Vol. **62**, no. 2, pp. 408- 423 (2013)
6. V.S.Kharchenko, V.V.Sklar, O.M.Tarasyuk. Methods for modeling and evaluation of the quality and reliability of the software. Kharkov: Nat. Aerospace. Univ."KhAI", 159 p. (2004)
7. Yaremchuk, S., Kharchenko, V. Complexity-based Prediction of Faults Number for Software Modules Ranking Before Testing: Technique and Case Study. Proceedings of the 12th International Conference, ICTERI 2016. Kyiv, Ukraine, Vol. **1614**, pp. 461–474 (2016)
8. V. Yakovyna, D. Fedasyuk, O. Nytrebych, Iu. Parfenyuk, V. Matselyukh. Software reliability assessment using high-order Markov chains. International Journal of Engineering Science Invention. — Vol. **3**, Issue 7. – pp. 1–6 (2014)
9. Yakovyna V. S. Software failures prediction using RBF neural network. Researches of the Odessa national polytechnical university. — Vol. **2**(46). – pp. 111–118 (2015)
10. D. A. Maevsky, S. A. Yaremchuk, L. N. Shapa. A method of a priori software reliability evaluation. Reliability: Theory & Applications. — Vol. **9**. – № 1(31). – pp. 64 – 72 Access mode: http://www.gnedenko-forum.org/Journal/2014_1.html (2014)
11. S. A. Yaremchuk, D. A. Maevsky. The Software Reliability Increase Method. Studies in Sociology of Science. Vol. **5**, No. 2 pp. 89 – 95. Access mode: http://www.cscanada.net/index.php/sss/article/view/4845 (2014)
12. 12. Eric J. Braude. Software Engineering An Object-Oriented Perspective. John Wiley and Sons, 575 p. (2011)
13. Kelly, John C., Joseph S. Sherif, Jonathon Hops. An Analysis of Defect Densities Found During Software Inspections. Journal of Systems and Software Vol. **17**(2), pp. 111-117 (1992)
14. Hofmann, Hubert F., Franz Lehner. Requirements Engineering as a Success Factor in Software Projects. IEEE Software Vol. **18**(4):58-66 (2001)
15. IEEE 830-1998: IEEE Recommended Practice for Software Requirements Specifications. Los Alamitos, CA: IEEE Computer Society Press (1998)
16. IEEE 1233-1998: IEEE Guide for Developing System Requirements Specifications. Los Alamitos, CA: IEEE Computer Society Press (1998)
17. IEEE/ISO/IEC 29148-2011: Systems and software engineering – Life cycle processes – Requirements engineering (2011)
18. Wiegers K., Beatty J. Software Requirements, 3rd Edition. Microsoft Press, 673 p. (2013)
19. IEEE 1061-1992: A Software Quality Metrics Methodology. Los Alamitos, CA: IEEE Computer Society Press (1992)
20. Shahid Iqbal, M. Naeem Ahmed Khan. Yet another Set of Requirement Metrics for Software Projects. International Journal of Software Engineering and Its Applications Vol. **6**, No. pp.19-28 (2012)
21. Mohd. Haleem, Mohd.Rizwan Beg, Sheikh Fahad Ahmad. Overview of Impact of Requirement Metrics in Software Development Environment. International Journal of Advanced Research in Computer Engineering & Technology (IJARCET) Vol. **2**, No 5, pp. 1811-1815 (2013)

DOI: 10.1051/itmconf/20170903010

Comparison between innovative techniques of photogrammetry

Vincenzo Barrile[1,*], *Giuliana* Bilotta[2], and *Alice* Pozzoli[3]

[1]DICEAM Department, Faculty of Engineering Mediterranean University of Reggio Calabria, 89100 Reggio Calabria, Italy
[2]Department of Planning, IUAV University of Venice, 30135 Venice, Italy
[3]DICA Department, Politecnico di Milano, 20133 Milan, Italy

Abstract. Main objective of this work is to suggest some easy procedures to solve typical non-linear problems of analytical photogrammetry, which are also frequent in digital photogrammetry. In particular, we propose a two-steps procedure based on two phases: Relative Orientation and Absolute Orientation, both characterized by non-linear functions. It is firstly necessary to form the model, finding the solution of the Relative Orientation, and then to reconstruct the object considering the solution of the Absolute Orientation. Once validated the technique of orientation of two images, a procedure for the orientation of three images is proposed. In particular, the introduction of a third image allows avoiding human decision to find the final solution. In order to find an easy solution for three-image orientation, each model coming from two images of a triplet is analysed, computing the relative orientation bby using an exhaustive research of preliminary values of its parameters. This non-conventional approach supplies the orientation of two images, taking into account a priori information among four base solutions. Once each model is relatively registered, the absolute orientation is computed by using a linear parameterization. We did a comparison between the method of the three photogrammetric images and a method of Computer Vision.

1 Introduction

The main function of photogrammetry is the transformation of data from the image space to the object space. This transformation can be made in a direct way, with collinearity equations, or in two steps, with the formation of a model and, only in a second time, with the reconstruction of the original object [1].

It is well known that model formation and object reconstruction require the solution of the problems of Relative Orientation and Absolute Orientation respectively, which have both non-linear functional models. Therefore, they need respectively an exhaustive research of the preliminary values of parameters for the model formation, and a transformation of the parameters able to transform the problem of the object reconstruction in a simple linear problem. This way of working makes the orientation procedures much more flexible, and allows wide applications in close range photogrammetry, which is actually growing in importance.

2 From images to object via model

In the procedure of object reconstruction, it is necessary to take into account that at least two images are needed for reconstructing an object. In particular, a relation of roto-translation with scale variation constitutes the link between the coordinates of the point Q (x,y,z) in an image, and the coordinates of the corresponding point P (X,Y,Z) in the object space. Both reference systems are traditionally Cartesian reference systems, but it is possible to achieve the same results, with minor changes, using a different reference system, suitably linked to the previous ones. Let us show the above-mentioned relation:

$$\begin{vmatrix} x^{\circ} \\ y^{\circ} \\ -c \end{vmatrix}_{ij} = \hat{\lambda}_{ij}\hat{R}_j \left(\begin{vmatrix} \hat{X} \\ \hat{Y} \\ \hat{Z} \end{vmatrix}_i - \begin{vmatrix} \hat{X}_0 \\ \hat{Y}_0 \\ \hat{Z}_0 \end{vmatrix}_j \right) \quad (1)$$

Where, as shown in Fig. 1:

x^0, y^0, c = image coordinates and principal distance

$\hat{X}_0, \hat{Y}_0, \hat{Z}_0$ = coordinates of projection center

$\hat{X}, \hat{Y}, \hat{Z}$ = object coordinates

λ = scale factor, variable point by point

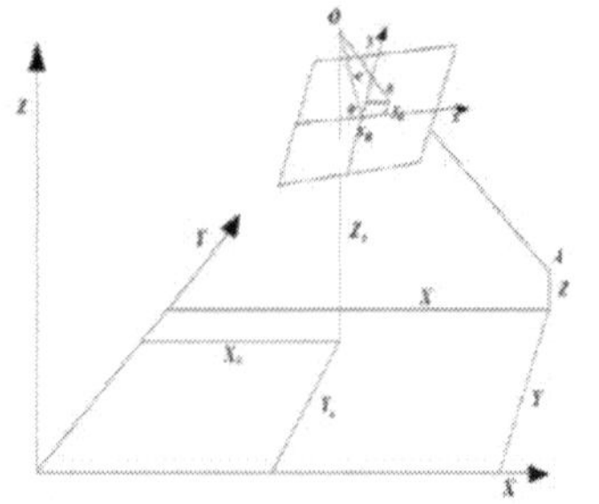

Fig. 1. Reference photogrammetric systems.

* Corresponding author: vincenzo.barrile@unirc.it

The photogrammetric technique is based on a transformation of a perspective (or a couple of perspectives) in a quoted orthogonal projection. In this transformation, we have non-linear equations and, before starting the plotting, we need information about the preliminary value of unknown parameters. Our main aim is to find expressions working with parameters easy to be obtained. In order to orient two images in the 3D space, we can choose two different procedures: a "one step" or a "two steps". The first procedure is based on collinearity equations and needs *12 parameters*: X_1, Y_1, Z_1, X_2, Y_2, Z_2 (coordinates of the two projection centers), and ω_1, φ_1, κ_1, ω_2, φ_2, κ_2 (attituded angles of the two sensor). The second one separates the model formation (Relative Orientation) from the object reconstruction (Absolute Orientation). In this procedure, we define the problem of Relative Orientation by means of 5 parameters: φ_1, κ_1, ω_2, φ_2, κ_2 (Symmetric Relative Orientation), or b_y, b_z, ω_2, φ_2, κ_2 (Asymmetric Relative Orientation). On the contrary, to define the problem of Absolute Orientation we need 7 parameters: t_x, t_y, t_z (shift vector), λ (scale factor), Ω, Φ, K (Cardanic angles).

3 Model construction

3.1 Relative Orientation Parameters

Regarding the Relative Orientation, we make an exhaustive research of the preliminary values, solving a linearized problem in all its possible cases. Notice that an exact solution has been found, but it leads to an equation of order four, which supplies four plausible solutions, as we can easily achieve by repeating a linearized problem via an exhaustive research.

The Relative Orientation is based on the *Coplanarity Condition*. It shows that the point P_1 in the first image, and its homologue point P_2 in the second image, have a unique corresponding point Q on the object.

In case of Asymmetric Relative Orientation, we have to define b_y, b_z, ω_2, φ_2, κ_2, which are the parameters of position and attitude of the second image, compared to those of the first image. Notice that b_x is already defined in the Absolute Orientation, as the scale factor λ. In case of Symmetric Relative Orientation, we have to define φ_1, κ_1, ω_2, φ_2, κ_2, parameters which represent position and attitude of the two in the Absolute Orientation, as the global attitude angle Ω.

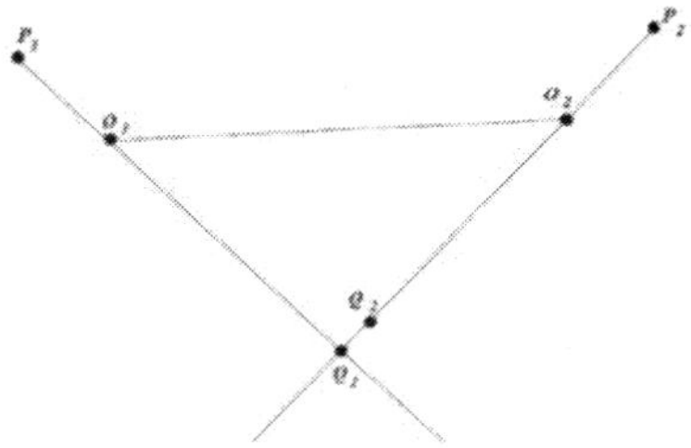

Fig. 2. Coplanarity condition.

3.2 Exhaustive Research

For the Relative Orientation, we should have previous information about the preliminary values of the parameters. It is not always possible to know them, before the plotting. Let us point out that non-conventional photogrammetry implies often camera acquisition without classical surveying measurement. If we consider the classical Symmetric procedure of Relative Orientation, we can make an exhaustive research of all possible preliminary parameters, because we work in a closed group (in the topological sense) of values compared to the rotations in the space.

Notice that in the Asymmetric procedure of Relative Orientation, we have two shift parameters to be searched, but the group of shifting is not a closed one, so we had to use a different way to find the preliminary values. However with the following relations is possible to transform the Symmetric Relative Orientation parameters in the Asymmetric ones, and vice versa:

$$b_x = cos\varphi_1 cos k_1 \qquad \varphi_1 = arcsin b_z$$
$$b_y = cos\varphi_1 sin k_1 \qquad k_1 = arcatn\frac{b_y}{b_x}$$
$$b_z = sin\varphi_1 \qquad (2,3,4,5,6)$$

$$R_2^T(\varpi_2\varphi_2k_2|b_xb_z) = R_2^T(\varpi_2\varphi_2k_2)R_1(\varpi_1\varphi_1k_1) \quad (7)$$

$$R_2^T(\varpi_2\varphi_2k_2) = R_2^T(\varpi_2\varphi_2k_2|b_xb_z)R_1^T(\varpi_1\varphi_1k_1) \quad (8)$$

The convergence of linearization of trigonometric functions is acceptable as far as values lower or near Π/4. Therefore, we decided to explore all the admissible values for rotation angles with a step of Π/4, as shown in Table 1.

As known, if the φ angle is around ±Π/2, we cannot individuate the k rotation, which is fixed equal to zero. Indeed in the polar zones (we assumed their range in a circle of one degree), the two angles are identical or quasi identical, and this fact produced singularity or ill-conditioning.

Table 1. Exhaustive Research for Symmetric Relative Orientation parameters.

	φ_1	k_1	ω_2	φ_2	k_2
Π/2	○			○	
Π/4	•			•	
0	•	•	•	•	•
Π/4	•	•	•	•	•
Π/2	○	•	•	○	•
3Π/4		•	•		•
Π		•	•		•
5Π/4		•	•		•
3Π/2		•	•		•
7Π/4		•	•		•

Where ° $k_1 \equiv 0$ if $\varphi_1 \equiv \pm\Pi/2$ and/or $k_2 \equiv 0$ if $\varphi_2 \equiv \pm\Pi/2$

DOI: 10.1051/itmconf/20170903010

The exhaustive research explored $5 \times 8 \times 8 \times 5 \times 8 = 1\ 2\ 8$ possible configurations. For each case, a linear system was solved, using the values of this configuration (case), as preliminary values of the parameters of the Symmetric Relative Orientation.

Examples were carried out in all the middle points of the possible configuration. Considering the 5 parameters of the Symmetric Relative Orientation, the angles κ_1, ω_2, κ_2 are defined in a complete rotation (8 configurations), whilst φ_1, φ_2 are defined in a half rotation (5 configurations), which led to the above mentioned 12800 cases.

Each linear system solution gave us the estimate parameters for the Symmetric Relative Orientation. The convergence to admissible values is when σ_0 is small enough. Considering only the distinct solutions, we found four analytical acceptable configurations.

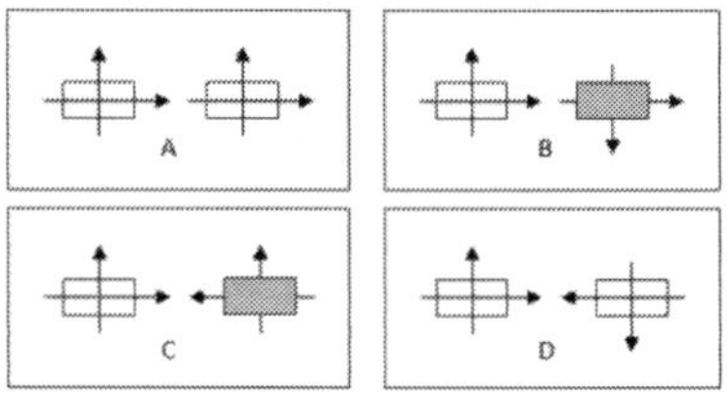

Fig. 3. The four final possible configurations.

4 Object reconstruction

4.1 Absolute Orientation

Starting from a rototraslation in the space, a rational alternative to classical Rotation Matrix is the *Rodriguez Matrix*.

$$\mathbf{R} = (\mathbf{I} - \mathbf{S})^{-1}(\mathbf{I} + \mathbf{S}) \tag{9}$$

Where **I** is the *identity matrix* of 3x3 dimensions, and **S** is an emisymmetric matrix defined as follows:

$$\mathbf{S} = \begin{vmatrix} 0 & c & -b \\ -c & 0 & a \\ b & -a & 0 \end{vmatrix} \tag{10}$$

This *emisymmetric matrix* **S** permits to find the exact solution of the absolute orientation problem, thanks to the solution of a linear system, after a suitable substitution of variables. We start from the conventional rototraslation in the space, with a global scale variation λ, and we compute the expected value in a way to eliminate the shift vector *t*:

$$\begin{aligned} y_i &= \lambda R_i x + t \\ \bar{y} &= \lambda R \bar{x} + t \\ \hline y_i - \bar{y} &= \lambda R(x_i - \bar{x}) \end{aligned} \tag{11}$$

Indeed, it is possible to calculate t lately with the following expression:

$$t = \bar{y} - \lambda R \bar{x} \tag{12}$$

Making the square of the second expression in the formula number (11), we also find an easy expression to calculate the scale factor:

$$y^T y = \lambda^2 x^T R^T R \ = \lambda x^2 x^T x \Rightarrow \lambda = \sqrt{\frac{y^T y}{x^T x}} \tag{13}$$

Now we have to make a substitution of variables in a way to transform the non-linear problem in a linear one:

$$y_i = \lambda R x_i \tag{14}$$

With a further substitution, we obtain our linear system of equations:

$$y_i = R\sqrt{\frac{y^T y}{x^T x}} x_i = R\ _i x \tag{15}$$

Thus, after these simple substitutions, we obtain a linear solution, showing the direct proportion between the model coordinates $x = x(u°, v°, w°)$ and the object ones $y = y(X, Y, Z)$:

$$y_i = R\ _i = (Ix - S)^{-1}(I + S)x_i \ \Rightarrow \ (I - S)y_i = (I + S)x_i \tag{16}$$

$$\begin{vmatrix} 1 & -c_j & b_j \\ c_j & 1 & -a_j \\ -b_j & a_j & 1 \end{vmatrix} \begin{vmatrix} \hat{X} \\ \hat{Y} \\ \hat{Z} \end{vmatrix}_i = \begin{vmatrix} 1 & c_j & -b_j \\ -c_j & 1 & a_j \\ b_j & -a_j & 1 \end{vmatrix} \begin{vmatrix} u° \\ v° \\ w° \end{vmatrix}_i \tag{17}$$

Reorganizing matrices and vectors, in a way which collects in a unique vector the three unknown parameters, coming from the above mentioned emisymmetric matrix, we obtain the following final equation:

$$\begin{vmatrix} 0 & (\hat{Z}_i - w°_i)_j & -(\hat{Y}_i - v°_i)_j \\ -(\hat{Z}_i - w°_i)_j & 0 & -(\hat{X}_i - u°_i)_j \\ (\hat{Y}_i - v°_i)_j & -(\hat{X}_i - u°_i)_j & 0 \end{vmatrix} \begin{vmatrix} \hat{a}_j \\ \hat{b}_j \\ \hat{c}_j \end{vmatrix} + \begin{vmatrix} \hat{X}_i - u°_i \\ \hat{Y}_i - v°_i \\ \hat{Z}_i - w°_i \end{vmatrix}_j = 0 \tag{18}$$

4.2 Exact Solution of the Absolute Orientation

In our procedure for the Absolute Orientation, the object reconstruction does not need preliminary parameters, because we can reach the exact solution, by solving the linear system, mentioned in the above paragraph. We tested this procedure, considering 208 possible configurations. These cases come from an object rotation following the global attitude angles (Ω, Φ, K), with a step of Π/4. Exam was performed analyzing the rotation in the space of a cube with 27 control points, regularly distributed.

5 Three-images procedure

5.1 Three-image orientation through exhaustive research

In the following, an automatic procedure of external orientation of three images will be proposed. This procedure doesn't require previous knowledge of the unknown orientation parameters. Indeed, exploring the 3D space with a step of Π/4, it is possible to find all the preliminary values of the unknown orientation parameters. This idea want to avoid the linearization of the orientation functions supplying the lack of

 DOI: 10.1051/itmconf/20170903010

information about the position and the attitude of an image. The procedure is based on the classical two-step approach of photogrammetry, i.e. relative and absolute orientation. References [2] and [3] show a solution of relative orientation problem based on exhaustive research of the preliminary values of parameters proposed. First, given a tern of images, all possible combinations of pairs of images are considered; in such a way, their relative orientation is computed, by using an exhaustive research of the approximate values of the unknown parameters. By using this unconventional approach, each pair of images gives four analytically acceptable solutions, among which, only the decision of the user would allow to select the proper one. Actually, to start the Absolute Orientation, we have to select manually one among the founded four configurations. The idea of introducing a third image allows to avoid human intervention to find the final solution. In fact, after obtaining the four possible solutions related to each pair of images, it is necessary to perform a linkage between the models. By following this procedure, sixteen possible combinations are obtained; however, only two of these combinations, the specular ones, are consistent with the observed object. Finally, the searched solution is obtained by orienting the two possible solutions in an absolute reference system.

5.2 Bridging the models

The step of the Model Construction gives four admissible solutions, as above said, and produces four distinct models (called A, B, C, D). In case of three partially overlapped images, this step can be repeated two times. Indeed the model 1-2 can be formed by the images 1 and 2, and the model 1-3 can be formed by the images 1 and 3. A 3D S-transformation allows to make the bridging of these models, taking into account all the four models obtained according to the admissible solutions founded in the Relative Orientation. The procedure leads to sixteen different small blocks.

The majority of these blocks are completely unlikely; indeed the sigma naught of the 3D S-transformation adjustment is enormous. This fact is reasonable because if and only if both models (1-2 and 1-3) are congruent between themselves, the bridging can be carried out successfully.

6 Numeric experiments

To verify precision, accuracy and reliability of these techniques, a program in Fortran 95 language (compiled and assembled with Lahey-Fujitsu Fortran 95 version 5.6) was written, implemented and tested. In all the examples, we introduced random errors, with standard deviation of 20 μm, as usual in photogrammetry. In the following flowchart (Fig. 4), we will summarize the global procedure for the orientation of two images, then we will present a brief explanation of these programs.

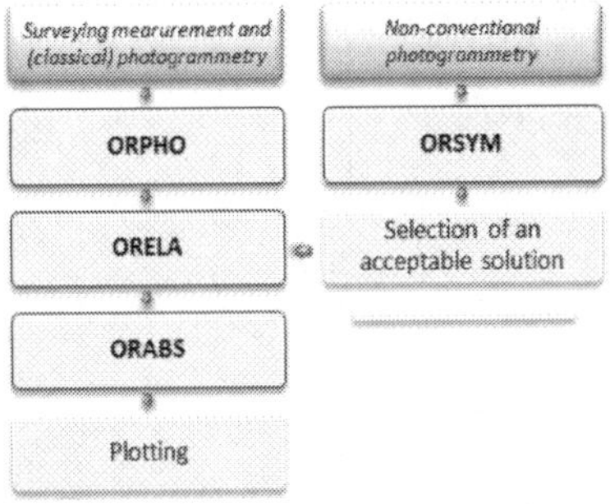

Fig. 4. Global procedure for the orientation of two images.

ORPHO_ it converts Cardanic angles in Eulerian angles and viceversa. This is a very large transformation used in close range photogrammetry, because it is essential for the image orientation, when the rotation angles are acquired by surveying measurements.

ORSYM_ it calculates the preliminary values for the Symmetric Relative Orientation. It solves 12800 linear problems, exploring all possible configurations in the space, with a step of Π/4. The same program, choosing one of the four distinct solutions, allows to calculate the preliminary parameters for the Asymmetric Relative Orientation.

ORELA_ it calculates the adjusted parameters of the Asymmetric Relative Orientation, starting from its preliminary ones. If these preliminary values are unknown, at the data acquisition, it is possible to get them from the results of the previous program. On the contrary, if they are already known, it is possible to transform the Eulerian angles into the Cardanic ones, by means of ORPHO.

ORABS_ it calculates the adjusted Absolute Orientation parameters. They are calculated with a simple substitution of variables, able to transform the non-linear problem of the Absolute Orientation in a linear one.

Before to conclude we wish to presents some results of these experiments. We considered robust statistical index (mode, median, 1st and 3rd quantiles), able to analyze distribution free problems.

The following tables and figures show the difference among the nominal values and the preliminary ones.

	Ω	Ψ	K
mode	17	1	0
percentile 0,25	10	5	10
median	21	13	22
percentile 0,75	38	25	39
max	93	56	86

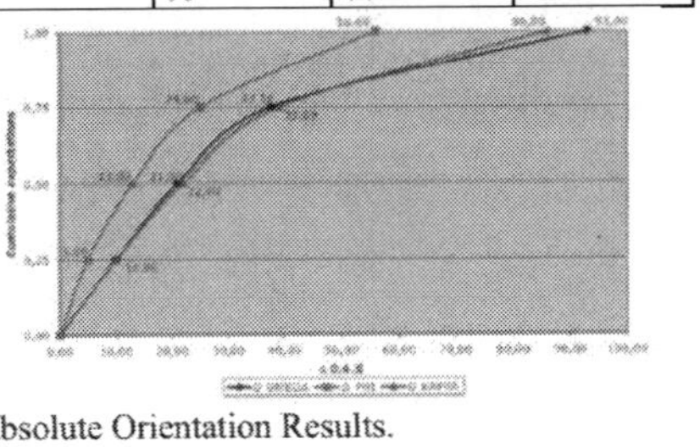

Fig. 5. Absolute Orientation Results.

 DOI: 10.1051/itmconf/20170903010

	$\Delta\varphi_1$	$\Delta\kappa_1$	$\Delta\omega_2$	$\Delta\varphi_2$	$\Delta\kappa_2$
mode	8	5	8	7	7
percentile 0,25	18	11	33	16	22
median	46	41	91	36	53
percentile 0,75	84	99	176	68	139
max	287	569	937	286	827

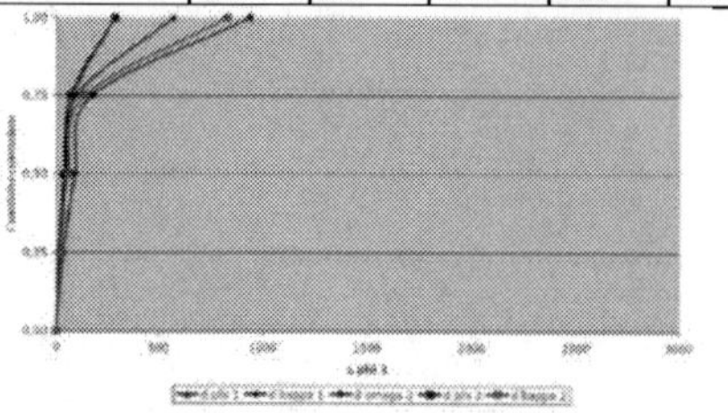

Fig. 6. Symmetric Relative Orientation results.

	$\Delta\varphi_1$	$\Delta\kappa_1$	$\Delta\omega_2$	$\Delta\varphi_2$	$\Delta\kappa_2$
mode	96	491	208	14	275
percentile 0,25	36	224	362	19	250
median	68	432	760	47	514
percentile 0,75	102	682	1301	98	862
max	165	2205	2937	251	1895

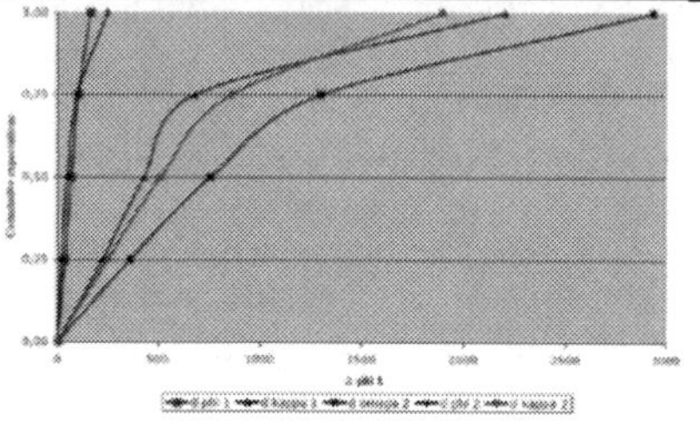

Fig. 7. Symmetric Relative Orientation results (Polar regions).

In the light of the proposed experiments, new programs, which would permit an easy solution for the Relative and Absolute Orientations, gave very satisfied expected results. This procedure has great potentiality for non-conventional data acquisition (e.g. amatorial images, images coming from unknown and old sources, equipped vehicles, robots, and many other applications in close range photogrammetry). The advantage of a linear Absolute Orientation should also taken into account. Moreover, even if the solution achieved in the Relative Orientation requires an exhaustive research, it is again quick and easy, and seems to solve positively the problem how to acquire the preliminary values of the Relative Orientation parameters.

7 A global procedure for 3 images orientation

As above shown, the global procedure for the orientation of two images can be summarized in the flowchart of Fig. 4. With the new procedure, we eliminate any human intervention after the starting inputs. For that reason we unify all the Orientation programs in one called ORTRE. This program can run automatically and is able to find the adjusted parameters of the Absolute Orientation.

In the following flowchart (Fig. 8) we want to show how all the global procedure run after the starting inputs of three images which is the same as for the orientation of two images.

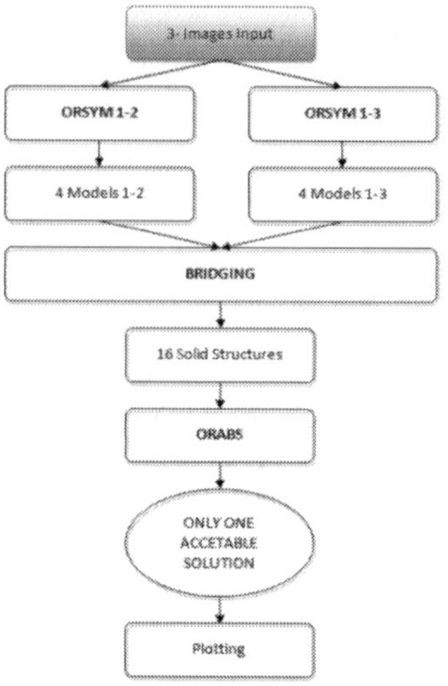

Fig. 8. Image Orientation – Global Proced. (ORTRE Program).

8 Computer Vision

8.1 Method of Computer Vision

Regarding the second method, we present the results of testing and comparison between the two methods applied and implemented.

Objective of the work is to estimate, more precisely as possible, a 3d model of marine surfaces. Specifically, we presented the same innovative proposal – rigorous photogrammetry type that makes use of three frames acquired simultaneously from different locations – and a methodology of Computer Vision.

The advantages of the Computer Vision technique, as is known, are low costs, automaticity and immediacy: we can get an accurate and usable result for further processing and analysis processes. Following the acquisition of the images, the reconstruction of the 3D model from digital images was performed with commercial software, based on special algorithms of the Structure from Motion technique (SfM). The SfM is successfully used in a wide range of applications, for generation of photogrammetry, survey and topographical mappings, to control activities, for the evaluation of damage due to disasters as well as for the preservation of evidence for legal disputes resolutions.

8.2 Application and study area

While waiting to comparing a hydraulic model confluence already experienced in Milan, we present the results of experiments made comparing, on a sea area of Reggio Calabria [4], the model obtained through the methodology of Computer Vision with the model obtained from the rigorous photogrammetric technique with three cameras.

The software used was PhotoScan in the Professional version 1.2.4, developed by Agisoft House LLC of St.

Petersburg (Russia). For imaging we used a HP Z800 workstation, with two CPU Xeon Hexa Core X5650, 64 GB Ram, mass storage 128 GB SSD and additional Hard Disk 2TB, two video cards NVIDIA Quadro FX 4800, each with 1, 5 GB RAM, and OS Windows 7 Professional 64-bit. The computerized processing procedure to a 3D model frames took place in a specific order in different phases. The first phase was the alignment of the images and the construction of the sparse point cloud (cloud points basis). This step identifies the common points in both frames using the image matching algorithms. The points selected in the various photos must necessarily have some features in common to be able to be properly overlapped. The result is an overall view of all shooting points positioned along the three Cartesian axes (local reference system) and the points constituting the cloud of anchorage points. The construction of the denser point cloud (dense cloud points) represents the second stage of processing, in which starting from the cloud of sparse points we get a more detailed and dense cloud of points.

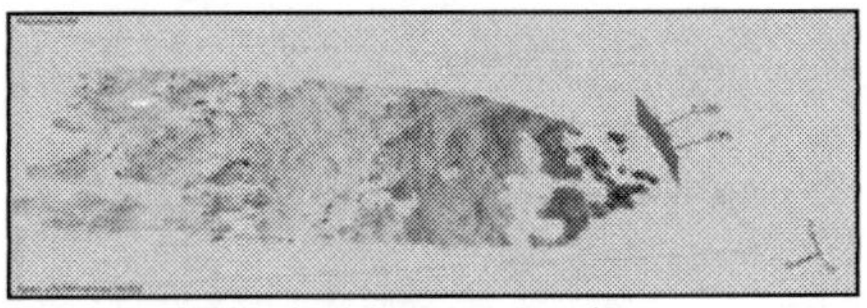

Fig. 9. Texture obtained with point shooting.

The 3D model with texture can be scaled and geo-referenced using the data collected during the campaign of shooting and establishing, on the frames and on the model of those markers points, for scaling the model. For validating results obtained in experiments carried out by the use of the pair of digital cameras and commercial software Agisoft PhotoScan (a "low-cost model"), we realized a metric three-dimensional model of the sea surface, in the same area and at the same moment, experiencing the use of three digital cheap camcorders resistant to weathering NILOX Mini Action Cam. We placed devices triangularly at equal distances of 1 m, performing the processing of the acquired data using the rigorous and experimental photogrammetric methodology with three cameras already exposed in the preceding paragraphs of the article.

In terms of accuracy, estimation of the leading photogrammetric orientation parameters, accuracy of results, we found a substantial similarity of results of the two models.

Fig. 10. 3D model (rigorous photogrammetric technique).

Table 2. Comparison of the values obtained from the two models

	Δ [m]	σ [m]
Technique 1	0.05	0.07
Technique 2	0.03	0.06

Fig. 11 shows comparisons between the two techniques in terms of accuracy obtained by setting the target positioned on the homologous elements, such as the edges of the reef and the "test" fixed points arranged appropriately on the sea surface.

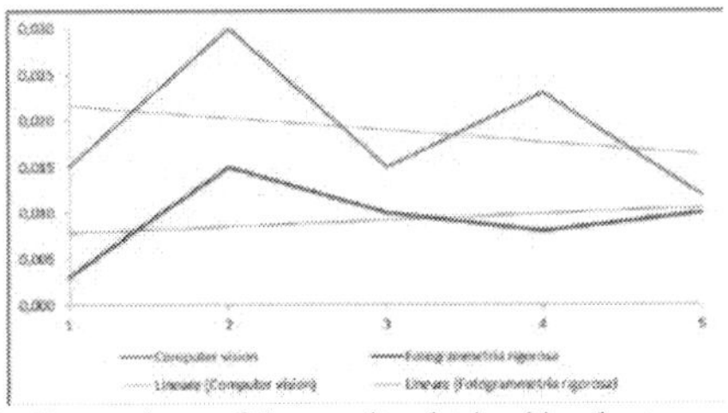

Fig. 11. Comparison of the results obtained by the two techniques.

9 Conclusions, further developments

We proposed a solution to solve for the problem of object reconstruction from three images not requiring any initial approximations for orientation parameters. This first procedure is based on the classical two steps approach of photogrammetry, i.e. relative and absolute orientation. All possible combinations of pairs of images in the triplet are considered and relatively registered by an algorithm running on exhaustive research [2], [3].

Advantages of this method are twofold. First, the three-image approach is based on a more reliable configuration with respect to simple relative orientation of a pair of images. Secondly, the method is very suitable to provide initial values of the unknowns to solve for machine vision or complex photogrammetric problems. This way of working makes the orientation procedures much more flexible, and allows wide applications in close range photogrammetry, which is actually growing in importance. Indeed not only classical close range photogrammetry, (e.g. for architectural and archaeological surveying as well as for industrial applications), but also data acquisition by equipped vehicles, and robots are nowadays typical data, largely diffused. Especially in the last cases, the need of real time (or quasi-real time) data processing and validation is very high. For these reasons, model formation and object reconstruction require the solutions of the problems of Relative Orientation and Absolute Orientation respectively, avoiding to waste time in an "a posteriori" search of preliminary values. Having regards to the promising results obtained, we may experience the technique of Computer Vision also on the same study area described in the previous sections of this paper.

References

1. K. Kraus, *Photogrammetry*. Bonn: Dümmler, **1**, **2** (1993)
2. L. Mussio, A. Pozzoli, "Non-Linear Problems of Analytical Photogrammetry", *IAPRS*, **34**, Part 6/W11, pp. 210-215 (2003)
3. L. Mussio, A. Pozzoli, "Quick Solutions

DOI: 10.1051/itmconf/20170903010

particularly in Close Range Photogrammetry", *IAPRS*, **34**, Part 6/W12, pp. 273-278 (2003)
4. V. Barrile, D. Lamari, V. Gelsomino, "Tecniche imaging per la modellazione 3D della superficie marina", *Atti del 61° Convegno nazionale SIFET*, Lecce (2016)
5. Y.L. Abel-Aziz, H.M. Karara, "Direct Linear Transformation into Object Space Coordinates in Close-Range Photogrammetry", *Proc. of Symp. on Close-Range Photogramm.*, pp. 1-18 (1971)
6. S. Finsterwalder, "Die Geometrischen Grundlagen der Photogrammetrie", *Jahresbericht der Deutschen Mathematiker-Vereinigung*, **6**, Teubner, Leipzig (1899)
7. H.G. Fourcade, "A new method of aerial surveying", *Trans. Roy. Soc. of South Africa*, **14**, 1, pp. 93-112 (1926)
8. S. Hattori, Y. Myint, "Automatic Estimation of Initial Approximations of Parameters for Bundle Adjustment", *PE&RS*, **61**, 7, pp. 909-915 (1995)
9. C. Heipke, "Automation of Interior, Relative and Absolute Orientation", *ISPRS J Photogramm Remote Sens.*, **52**, pp. 1-19 (1997)
10. H.-P. Pan, "A Direct Closed-Form Solution to General Relative Orientation", Technical Report on *Photogrammetron.*, pp. 1-20 (1996)
11. E. Kruppa., "Zur Ermittlung eines Objektes aus zwei Perspektiven mit inner Orientierung", *Sitz.-Ber. Akad. Wiss., Wien, Math. Naturw. Kl.* (Abt. IIa 122), pp. 1939-1948 (1913)
12. H. C. Longuet–Higgins, "A computer algorithm for reconstructing a scene from two projections", *Nature*, **293**, 10, pp. 133-135 (1981)
13. H.-P. Pan, "A direct closed-form solution to general relative orientation of two stereo views", *Digit. Sign. Process.*, **9**, 3, Academic Press, pp. 195-211 (1999)
14. T. Sarjakoski, "Concept of a completely digital stereoplotter", *The Photogramm. Journ. of Finland*, **2**, pp. 95-100 (1981)
15. H. Schmid, "An analytical treatment of the orientation of a photogrammetric camera", *Photogr. Eng.*, **20**, pp. 765-781 (1954)
16. G. Schut, "Analytical aerial triangulation and comparison between it and instrumental aerial triangulation", *Photogrammetria*, **12**, pp. 311-318 (1955-56)
17. P. Stefanovic, "Relative Orientation – a new approach", *ITC Journal*, pp. 417-448 (1973)
18. E.H. Thompson, "A rational algebraic formulation of the problem of the relative orientation", *Photogramm. Rec.*, **3**, 14, pp. 152-159 (1959)
19. P.H.S. Torr, D.W. Murray, "The Development and Comparison of Robust Methods for Estimating the Fundamental Matrix", *Int J Comput Vision*, **24**, 3, pp. 271-300 (1997)
20. R.I. Hartley, and A. Zisserman, "Multiple View Geometry", *Comp. Vis.*, Cambridge University (2000)
21. K. Rinner, R. Burkhardt, *Photogrammetrie. In Handbuch der Vemessungskunde*, (Hsgb. Jordan, Eggert, Kneissel), **3**, a/3, Metzlersche Verlagbuchhandlung, p. 2286 (1972)
22. P.J. Rousseeuw, A.M. Leroy, *Robust Regression and Outliers Detection*, Wiley, New York (1987)
23. G.H. Schut, "On Exact Linear Equation for the Computation of Rotational Elements of Absolute Orientation", *Photogrammetria*. **17**, 1, pp. 34-37 (1961)
24. J. K. Semple, G.T. Kneebone, *Algebraic Projective Geometry*, Oxford Science (1952)
25. P. Stefanovic, "Relative Orientation – a new approach", *ITC Journal*, pp. 417-448 (1973)
26. J. Weng, T.S. Huang, N. Ahuja, *Motion and Structures from Image Sequences*, Springer (1993)
27. O.D. Faugeras, Q.T. Luong, S.J. Maybank, "Camera self-calibration: Theory and experiments", *Proceedings of the 2nd ECCV*, pp. 321-334. (1992)
28. M.A. Fischler, R.C. Bolles, "Random Sample Consensus: A Paradigm for Model Fitting with Applications to Image Analysis and Automated Cartography", *Comm. of the Ass. for Computing Machinery*, **24**, pp.381-395 (1981)
29. R.I. Hartley, "Estimation of relative camera position for uncalibrated cameras", *Proceedings of the 2nd ECCV*, pp. 579-587 (1992)
30. R.I. Hartley, "Lines and Points in Three Views – a Unified Approach", *Proc. of "Image Understanding Workshop"*, Monterey, CA, pp. 1009-1016 (1994)
31. I. Niini, "Comparison of the projective block adjustment method versus the bundle method", *IAPRS*, **33**, Part B3, pp. 643-650 (2000)
32. C. Ressl, "An introduction to the relative orientation using the trifocal tensor", *IAPRS*, **33**, Part B3, pp.769-776 (2000)
33. M.E. Spetsakis, J. Aloimonos, "A Unified Theory of Structure from Motion", *Proc. of Image Understanding Workshop*, Pittsburg, Pennsylvania, pp. 271-283 (1990)
34. P.H.S. Torr, A. Zisserman, S. Maybank, "Robust Detection of Degeneracy", *Proceedings of the 5th ECCV*, Boston, pp. 1037-1044 (1995)

ITM Web of Conferences **9**, 03011 (2017) DOI: 10.1051/itmconf/20170903011
AMCSE 2016

Design of a Microstrip Bowtie Antenna for Indoor Radio-Communications

Hector Fraga-Rosales, *Mario* Reyes-Ayala*, *Genaro* Hernandez-Valdez, *Edgar Alejandro* Andrade-Gonzalez, *Jose Raul* Miranda-Tello, *Felipe Alejandro* Cruz-Perez and *Sandra Lirio* Castellanos-Lopez

Metropolitan Autonomous University, Electronics Department, 02200 Azcapotzalco, Mexico City, Mexico

Abstract. In this paper, a microstrip bowtie patch antenna (MBPA) for wireless indoor communications is carried out. Here, a microstrip transmission-line feed network was designed in order to match the MBPA. The proposed antenna uses a ground plane with the aim of narrowing down the back lobes in comparison with bowtie sheet antennas, which radiation pattern is omni-directional. The far-field pattern of the antenna was simulated using a finite-element numerical algorithm and obtained by interpolation employing near-field equipment. The experimental results are described in detail intending to agree well with the simulated predictions. The antenna was designed, measured and built and its far field performance was evaluated with a 2.11 GHz resonant frequency. The azimuth and elevation antenna patterns, antenna gain and, the matching frequency were the main parameters obtained to analyze the antenna behaviour. The antenna has a gain approximately equal to 8.77 dBi and its beam-widths are higher than 100° in E plane.

1 Introduction

Microstrip antennas have some advantages in order to be used in a lot of mobile communications systems [1-3]. The main advantages of this kind of antennas can be summarized as follows: they are compatible with mobile terminal equipment where a thin profile and a light weight might be necessary; they are relatively low-cost than other manufacturing techniques; their bandwidths are narrow; they have normally moderate quality factors; and, they have enough antenna gain, that is needed in the indoor radio-communications links [4-6].

Microstrip antennas can be made to emulate wire antennas; monopoles, dipoles, Yagi-Uda and logarithmic array antennas are widely configuration examples. Besides, rectangular and circular patches give a very narrow-band response approximately equal to 2%. Then, bowtie antenna offers a moderate bandwidth (5%) in comparison with rectangular and circular choices. Other approaches like fractal, slotted and defected structures have been developed intending to increase the bandwidth, but there are some tradeoffs to take into account, because this sort of antennas require a very high manufacturing accuracy, their electrical features are not stable and, its price is frequently more expensive [1-3, 7-9].

Bowtie antenna can be divided into two categories, sheet and microstrip patch antennas. Bowtie sheet antennas have omnidirectional radiation patterns and as a consequence of this, they have a lower antenna gain [1]. Thu use of ground planes in microstrip bowtie antennas changes the omnidirectional antenna pattern into a directional one.

In this paper, a Microstrip Bowtie Patch Antenna (MBPA) is presented and it is proposed for indoor communications systems. According to the considerations mentioned before, the MBPA is chosen in order to achieve a low-cost microstrip antenna with a 5% bandwidth and a moderate antenna gain in a wide beamwidth. The antenna will be used in educational and low-cost mobile handsets or other terminal equipment for transportation vehicles; but, it will be almost used in low power radio applications.

The article is divided into four sections: the section one is an introduction to microstrip bowtie antennas; the section two is dedicated to describe the design methodology of the MBPA; in the third section of the paper, the main simulated and measured results are presented and discussed; and, the conclusions of this work are described in the section four.

2 Design of the antenna

The MBPA design is based on the triangular patch antenna theory. Uni-planar bowtie antennas like MBPA usually combine the use of two triangular antennas, but it is possible to use circular magnetic walls. This antenna can be built in a single side or in both sides of the Printed Circuit Board (PCB). In this work, a MBPA is designed and probed, this antenna uses a ground plane with the aim of achieving a directional radiation pattern with a very low back lobes, see the Figure 1.

Each triangle has a non-perfect magnetic wall, where the effect of this wall can be estimated using empirical methods. The design procedure of the MBPA presented

* Mario Reyes-Ayala: mra@ correo.azc.uam.mx

here is based on the assumption that the triangles have perfect magnetic walls.

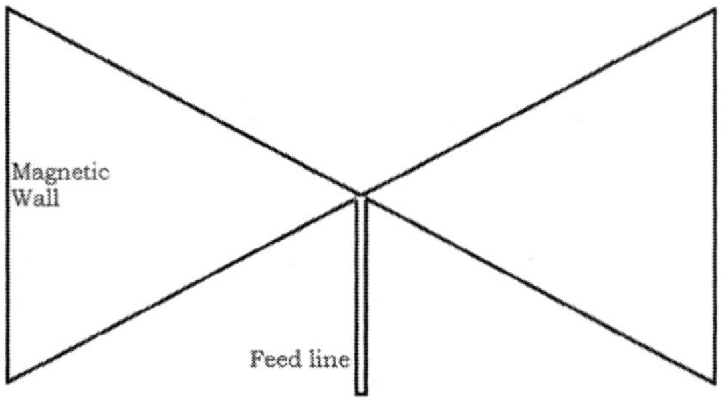

Fig. 1. Biconical antenna.

The resonant frequency corresponding to the many different modes in the MBPA can be calculated by equation (1) [8-9]..

$$f_{m,n} = \frac{2c}{3a\sqrt{\varepsilon_r}}\sqrt{m^2 + mn + n^2} \quad (1)$$

Where m, n determine the $TM_{m,n}$ dominant mode; c is the light speed, m/s; ε_r is the dielectric constant of the substrate; and, a is the length of the antenna arm, m.

It is possible to increase the accuracy of the resonant frequency if the dielectric constant and the length of the arm are replaced by their effective values. The effect of the dielectric constant of the substrate is the major cause of this consideration.

$$\varepsilon_{\text{eff}} = \frac{\varepsilon_r + 1}{2} + \frac{\varepsilon_r - 1}{2}\frac{1}{\sqrt{1 + \frac{12d}{a}}} \quad (2)$$

$$a_{\text{eff}} = a + \frac{d}{\sqrt{\varepsilon_r}} \quad (3)$$

Where ε_{eff} is the effective dielectric constant of the substrate; a_{eff} is the effective length of the arm, m; and, d is the thickness of the substrate, m.

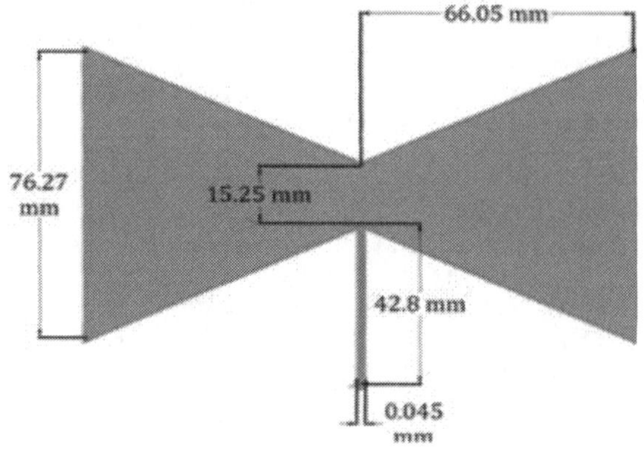

Fig. 2. Layout of the antenna.

The geometry of the MBPA is illustrated in the Figure 2, where the antenna has been matched by a microstrip transformer to a 50Ω coaxial line and the triangles are folded in the patch.

This design and simulations were carried out for a resonant frequency nearly equal to 2.5 GHz considering the use of a FR-4 substrate. In the results section, the antenna was performed with a different substrate, which dielectric constant is approximately equal to 6.1. The final resonant frequency was 2.11 GHz approximately, as has been mentioned above.

The dimensions of the proposed antenna are shown in the Table 1, where the resonant frequency is included in both: with the ideal and the effective dielectric constants and the effective length of the antenna arms.

Table 1. Parameters obtained in the antenna design.

Parameter	Value
a	76.27 mm
u	66.05 mm
c	15.25 mm
f_r	2.5 GHz
ε_{eff}	4.22
a_{eff}	77 mm
f_r	2.52 GHz

The microstrip transmission line is necessary to match the antenna [1-3], where the thickness of the substrate is approximately 1.544 mm and the length of this strip line is equal to a quarter of the wavelength, see Figure 3. The coaxial connector is located at the end of the microstrip line on the ground plane.

Fig. 3. Profile of the microstrip antenna.

3 Results

In this section, the most important electrical and radiation features of the antenna are presented. In the first stage, the antenna was modeled and simulated using High Frequency Simulation Structure (HFSS). The Figure 4 illustrates the field pattern on the antenna surface in the dominant TM_{01} mode.

DOI: 10.1051/itmconf/20170903011

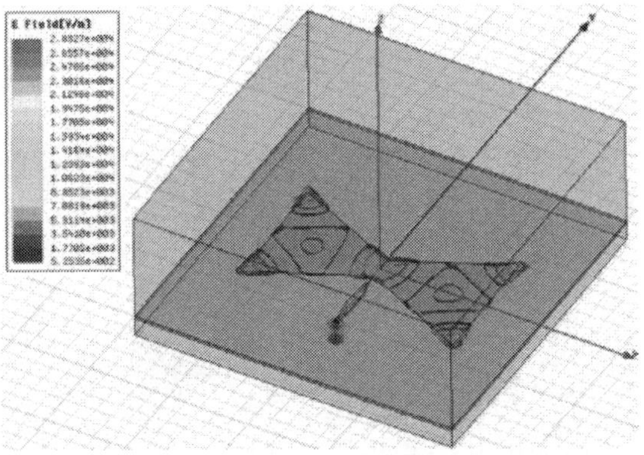

Fig. 4. Simulated model of the antenna.

The simulation process includes the optimization of the structure. In this stage, the antenna patterns and S_{11} parameter are the most important results. The Figure 5 shows the matching interval of the antenna (from 2.05 GHz to 2.1 GHz).

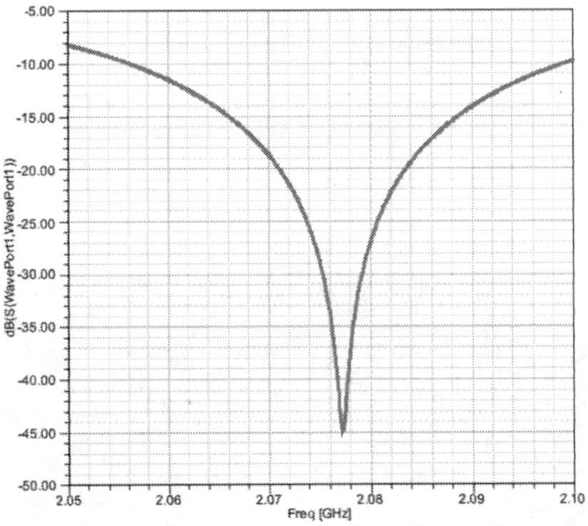

Fig. 5. S_{11} Parameter in the matching frequency interval of the antenna.

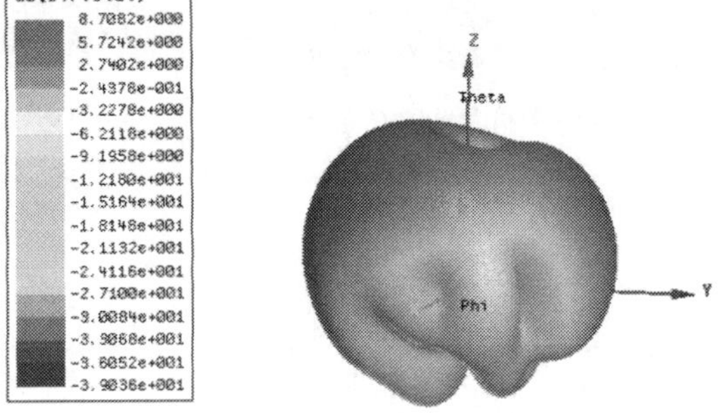

Fig. 6. Simulated 3-D antenna pattern at the far-field zone.

The simulated 3-D antenna pattern at the far field zone is shown in the Figure 6, where the corresponding antenna gain is closed to 8.7 dB.

The 3-D antenna pattern can be divided in two radiation patterns. These patterns could be seen from Figure 7, for $\phi = 0°$. The simulated radiation pattern in the E-Plane for $\phi = 90°$ is plotted in the Figure 8. In both antenna patterns, it is clear that the beamwidth is greater than 100°, this characteristic is desirable in almost all the mobile radio-communication systems.

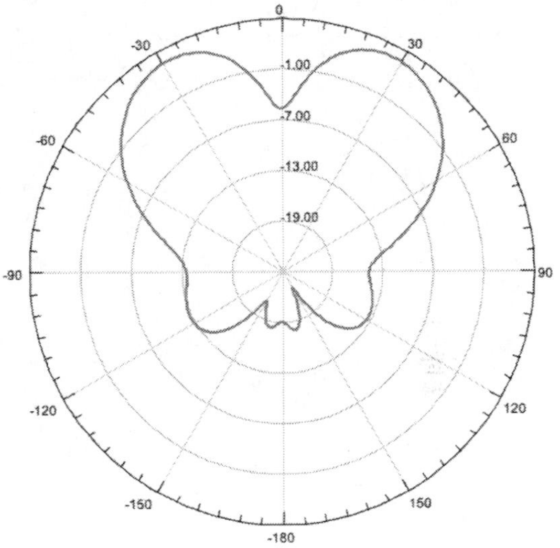

Fig. 7. Simulated radiation pattern in the E-Plane, for $\phi = 0°$.

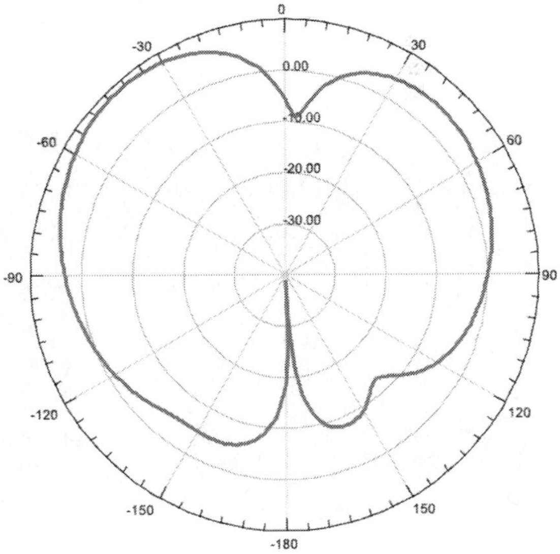

Fig. 8. Simulated antenna pattern in the E-Plane for $\phi = 90°$.

Based on the simulation results, the MBPA was built using a PCB with a dielectric constant $\varepsilon_r \approx 6$, as it is shown in Figure 9. The computational tool (HFSS) employs a finite-element numerical method, which offers a better agreement between simulated and experimental results. Other software choices in simulation of electromagnetic structures like Numerical Electromagnetic Code (NEC) or ENSAMBLE use the method of moments, where the antenna is modeled with surfaces instead of volumetric objects.

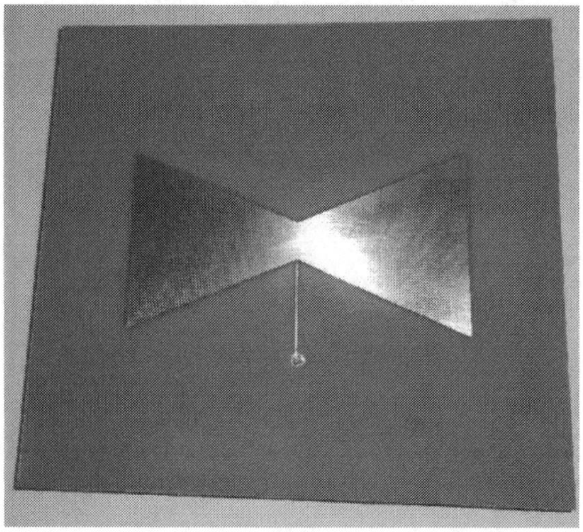

Fig. 9. MBPA printed in a PCB with a dielectric constant nearly equal to 6.

The return loss of the proposed antenna is shown in Figure 10. The measured frequency interval where the antenna matching is less than -10 dB is almost the same in comparison with the corresponding in the simulation results.

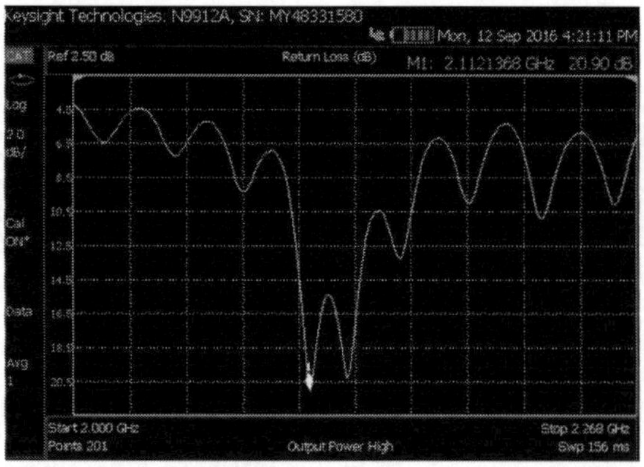

Fig. 10. Measured return loss for MBPA.

The measured far-field 3-D radiation pattern of the MBPA is shown in Figure 11. This result was obtained by interpolation using near-field evaluation equipment. It is obvious that simulation and measured antenna patterns look similar. The size of the antenna shown in the Figure 9 is compatible with the near-field equipment, which is a good choice in comparison with anechoic chambers.

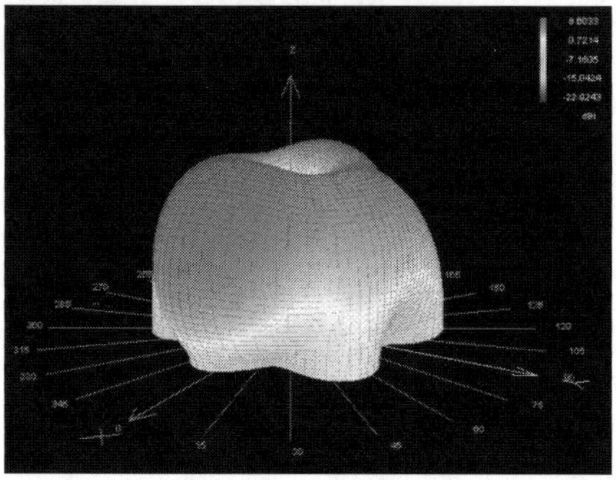

Fig. 11. Measured 3-D antenna pattern for MBPA.

The antenna has also a second resonant frequency at one half of the main resonant frequency (1.09 GHz instead of 2.11 GHz). This feature can be used in applications for a lower carrier frequency with the aim of reducing the electrical size of the structure, see Figure 12.

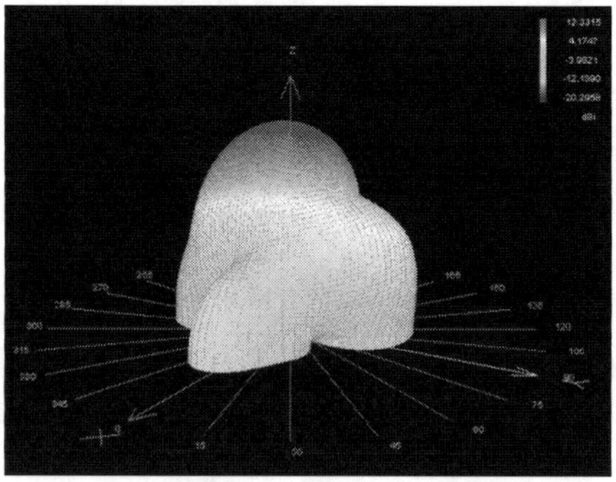

Fig. 12. Measured 3-D antenna pattern for MBPA (1.09 GHz).

4 Conclusions

A microstrip bowtie patch antenna was designed and implemented for indoor radio-communications. The proposed antenna has been simulated and measured in order to compare the results, where both numerical results are closed. The antenna gain (≈ 8.6dBi) is enough to make feasible the link budget in indoor radio-communications systems. In addition to this, the

 DOI: 10.1051/itmconf/20170903011

beamwidth is also compatible with this kind of applications.

The measured S_{11} parameter shows a good agreement with the simulation results and it is clear that the MBPA has a higher bandwidth in comparison with rectangular and circular microstrip patch antennas.

Finally, it is important to emphasize that the MBPA could be used at 1.09 GHz, as a consequence of the total length of the structure. In this case, the beamwidth is narrowed down to 60° approximately.

References

1. C. A. Balanis, *Antenna Theory*, 811-876 (2005).
2. D. M. Pozar, *Microwave Engineering*, 137-146 (1995).
3. J. Volakis, *Antenna Engineering Handbook*, **7**, 1-29 (2007).
4. K. R. Mahmoud, PIER, **100**, 105-117, (2010).
5. C. N. Ng, S. Uysal and M. S. Leong, *RAWCON'98*, 205-207 (1998).
6. B. Harchandra, R. Singh, IJETT, **18**, 171-175 (2014).
7. K. W. Loi, S. Uysal, M. S. Leong, IEE Proc. Microw. Antennas Propag., **145**, 137-140 (1998).
8. M. Roslee, K. S. Subari, I. S. Shahdan, *RFM'2011*, 430-433 (2011).
9. C. Guo, R. C. Liu, *GPR'2010*, 1-5 (2010).

Relationship among resonant frequencies of Sierpinski multiband fractal antennas

Ivan R. Gonzalez-Rangel[1], *Genaro* Hernandez-Valdez[1,*], *Edgar A.* Andrade-Gonzalez[1], *Mario* Reyes-Ayala[1], *J.R.* Miranda-Tello[1], *Jose* Serrano-Chavez [1]

[1] Electronics Department, UAM-A, Mexico City

Abstract. In this paper, the relationships between the different resonance frequencies of Sierpinski fractal antennas of four-iterations are studied. In particular, Sierpinski fractal antennas with operating frequencies of the initial triangle of 250 MHz, 350 MHz and 530 MHz were designed and built. The antennas are made of copper tablets with bakelite substrate. The performance of the designed antennas is measured in terms of return losses. The return losses are obtained experimentally with a "RFX" system that measures antenna parameters in conjunction with a network analyzer. These results are compared with numerical simulations of commercial finite-element program that analyzes high frequency electromagnetic structures "HFSS". Experimental and simulation results show that there is approximately a factor of 2 between the resonance frequencies of the first and second iterations and the second and third iterations.

Keywords: Fractal, multiband, Sierpinski

1 Introduction

For the design of antennas, fractal antennas are important to solve challenges related with lower-cost, multi-band applications, broadband, and miniaturization [1], [12]. The operation parameters of a fractal antenna strongly depend on the type of fractal geometry and materials used in its construction [1], [10-13].

Today, with the miniaturization of wireless devices there is an increased demand for smart antennas. The designs of these antennas don't have an easy way to get operating requirements. This is a challenge because the operating characteristics and performance of the antennas are affected significantly by their small sizes. Furthermore, operating in more than one frequency band is also a challenge in the design of the antennas [5-10].

Fractal antennas use fractal geometry in their design; these forms allow that a single antenna has the ability to simultaneously resonate at more than one frequency band. Fractal geometry maximizes the perimeter of the antenna in a given area. This reduces the size of the area of an antenna maintaining the same perimeter. The constructions of some fractal shapes have properties that allow them to have a multiband behaviour. Some ideals fractal shapes are self-similar. A classic example of self-similar fractal geometry is the Sierpinski triangle [3-6].

In this research, the methodology proposed in [5] is used to properly build monopole-type multiband fractal antennas using Sierpinski fractal geometry with four iterations. For building these antennas, copper is used for the active antenna element, bakelite as substrate, and SMA type feeder. The contribution of this work is the study of the relationship between the resonance frequencies of the various iterations of the Sierpinski fractal antenna. The performance of the antennas studied is measured in terms of return loss. The return loss is measured experimentally using a measuring system parameter antenna in conjunction with a network analyzer. These results are compared with those obtained by finite element simulation; specifically, the High Frequency Electromagnetic Structures simulation package is used to this end (HFSS).

2 Sierpinski Geometry

The construction of the Sierpinski triangle is based on an initial equilateral triangle that represents the zero-th iteration. In the first iteration, it is drawn another equilateral triangle that is rotated 180 geometrical degrees, whose sides have a length equal to 0.5 times the length of the sides of the initial triangle. This triangle is suppressed from the initial triangle; the result of this procedure is shown in the second geometric shape of Fig. 1. So, in the first iteration there are three equilateral triangles inside the triangle of the zero-th iteration. If in each of these triangles the above process is repeated, the second Sierpinski iteration is obtained, this case is shown in the third geometric shape of Fig. 1. This process is repeated N times according to the total number of desired iterations. The last geometric shape of Fig. 1 represents the Sierpinski triangle with $N = 4$ iterations.

* Corresponding author: ghv@correo.azc.uam.mx

 DOI: 10.1051/itmconf/20170903012

2.1 Sierpinski triangle area

After each transformation of the Sierpinski triangle, a fragment of the initial triangle is removed. The area of the triangle decreases three quarters of the initial area in each iteration, that is, the resulting area decreases by a factor ¾. Equation (1) calculates the area resulting from the *n*-th iteration.

$$A_n = A_0\,(3/4)^n \tag{1}$$

where A_0 is the area of the initial triangle and n is the iterations number. It is noted that increasing the number of iterations, the triangle area decreases, therefore, with a large number of iterations the area tends to zero [8].

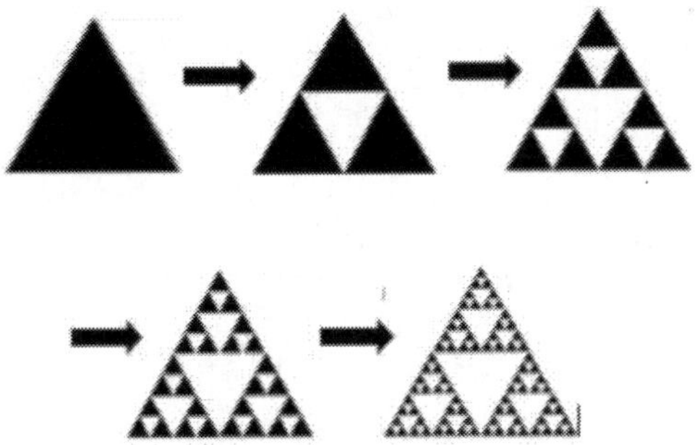

Fig. 1. Sierpinski triangle transformations for 4 iterations.

2.2 Sierpinski triangle perimeter

As discussed above, the initial triangle is transformed into three self-similar shapes after each iteration. Fig. 1 shows that each of the sides of the triangles of the current iteration has a length equal to 0.5 times the length of the sides of the triangles of the previous iteration. After deleting the triangles that are rotated 180 geometrical degrees, the perimeter of the resulting fractal figure can be calculated with equation (2).

$$P_n = P_0\,(3/2)^n \tag{2}$$

where P_0 is the initial perimeter of the triangle and n is the iteration number. It is noted that as the number of iterations increases, the perimeter of the antenna tends to infinity [8].

3 Sierpinski Monopole

The Sierpinski monopole is a fractal antenna based on the Sierpinski triangle. The number of iterations is directly related to the number of frequency bands in which the antenna can operate. However, having a very large number of iterations results in both a significant decrease in the area of the antenna and considerable increase in its perimeter, such that the effectiveness of the antenna will be reduced [4-7]. This is why the number iterations should not be too large. The height of each one of the triangles that there are in the antenna corresponds to a frequency band. These heights are shown in Fig. 2.

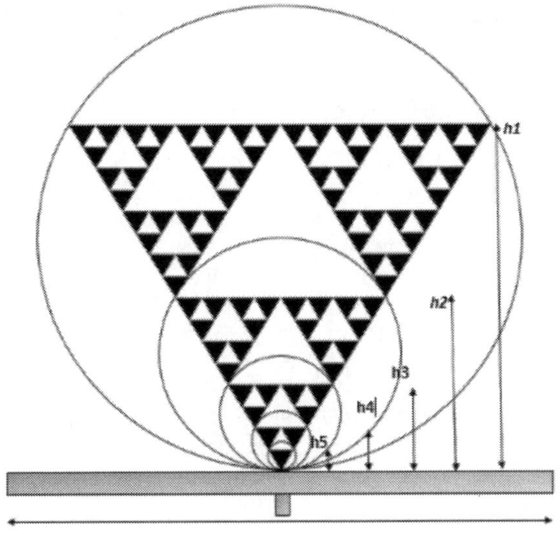

Fig. 2. Monopole-type Sierpinski multiband fractal antenna.

The resonance frequency of the zero-th iteration of the antenna is calculated using (3).

$$f \approx k\frac{c}{h} \tag{3}$$

where f is the resonance frequency of the first iteration of the antenna, c is the value of the speed of light in vacuum, h is the height value of the larger triangle of the antenna, and k is a constant with value 0.142. The factor k is smaller than 1, so this determines how small is a fractal antenna compared against a conventional antenna. This paper considers three Sierpinski antennas with resonant frequencies for the zero-th iteration equal to 250 MHz, 350 MHz, and 530 MHz.

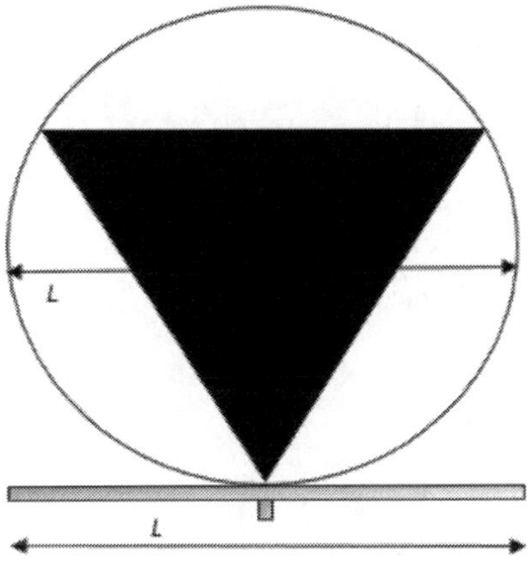

Fig. 3. Ground plane and the diameter of the zero-th iteration of the Sierpinski fractal antenna.

The Copper antennae are made over a bakelite plate with a relative permittivity $\varepsilon_r = 4.8$ and thickness of 1.565 mm. In the resulting triangle, after the first

DOI: 10.1051/itmconf/20170903012

iteration, it is considered a circle with diameter L which touches its vertices, as it is illustrated in Fig. 3. The ground plane is on the opposite side of the bakelite plate and has an area equal to $L \cdot L$, according to the frequency of each of the three developed antennas.

4 Simulation Results

We used HFSS (High Frequency Structural Simulator) software of ANSYS, which is a simulator for structures and devices that respond at high frequencies. This program maintains high standards of precision; it performs an analytical solution that is based on finite element integral equations and hybrid methods for solving a wide range of microwave, radio frequency and digital high speed applications.

The HFSS simulator offers a wide range of materials, so it is important to choose the correct type of material used in each structure in order that the simulation results are closer to the experimental results.

The antenna design within the simulator comprises a part of copper, which is the antenna structure with a thickness of 0.035mm; a substrate that simulates the printed circuit board (a substrate of bakelite was chosen with a thickness of 1.565 mm). The simulator uses the default parameters of this material, one of these parameters is the relative permittivity (ε_r=4.8). Copper was chosen for the ground plane, and an SMA connector was chosen as a feeder.

The frequency sweep was adjusted to the range from 10 MHz to 8 GHz. The simulator has three types of frequency sweeps, the first one uses the discrete Fourier transform, the second method uses the fast Fourier transform, and the last method uses interpolation. For this work, the method of the fast Fourier transform was used. An option was also included to normalize ports with an impedance of 50 ohms.

The simulations were performed with a resolution of 10 MHz to obtain the resonant frequencies of the antennas with a high level of accuracy. In Fig 4, the simulated structure in HFSS is presented.

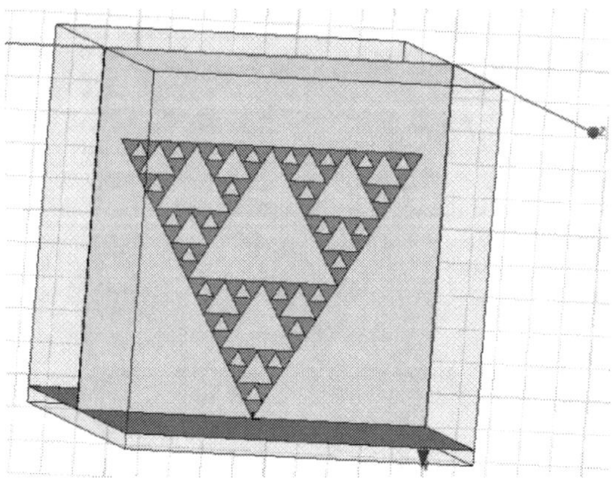

Fig. 4. Simulated structure in HFSS.

Simulation results for antennas Sierpinski with maximum height of 8 cm, 12 cm, and 16.5 cm are shown in Figs. 5, 6, and 7, respectively. These heights correspond to the frequencies of the zero-th iteration equal to 530 MHz, 350 MHz, and 250 MHz, respectively.

Authors in [5] state that, if both the contribution of the center triangle shaped holes to the antenna performance in each iteration are neglected (white triangle shaped regions in Fig. 2) and it is assumed that the current flowing from the feeder is concentrated over a region that is comparable in size to the wavelength, a behavior similar to five-scaled bowtie antenna (each one operating at its resonant frequency) is expected. Due to the fact that the scale factor among the five heights shown in Fig. 2 is 2, authors in [5] concluded that a similar behavior between the resonant frequencies of the Sierpinski multiband antenna is expected. Figs. 5-7 shows that the bands in which the Sierpinski antenna resonates in the n-th iteration are adjusted by a certain factor to the resonant frequency of the $(n+1)$-th iteration [4-7]. For instance, Fig. 6 shows that between the resonant frequencies of the following iterations: zero-th and first, first and second, second and third, and third and fourth, there exist an approximate factor of 3.1, 1.8, 1.68, and 1.47, respectively. These results differ from the expected value of 2 that is justified in [5]. In the next section, our simulated results are experimentally verified.

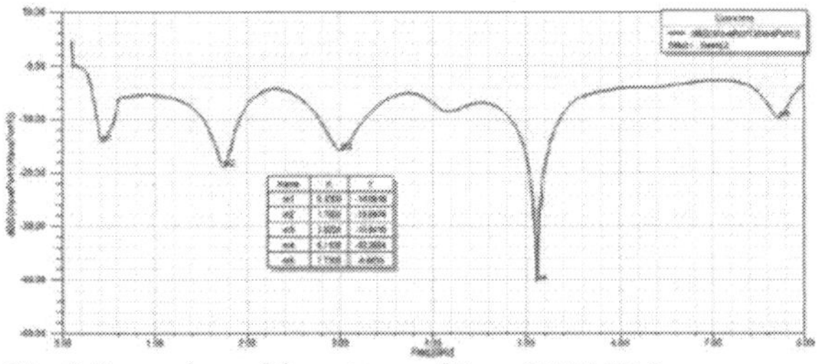

Fig. 5. Return loss of the antenna of 8 cm (530 MHz).

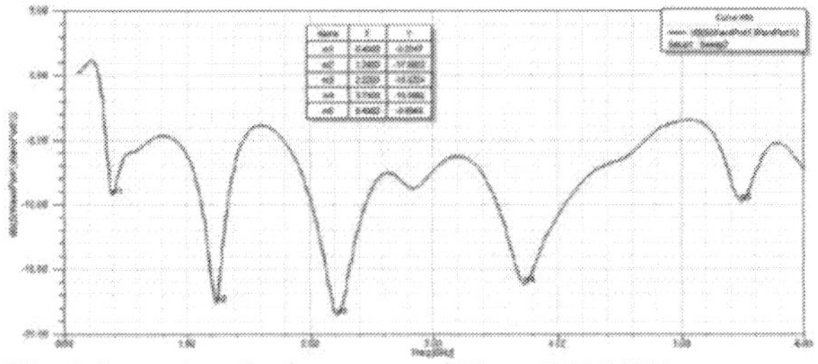

Fig. 6. Return loss for the antenna of 12 cm (350 MHz).

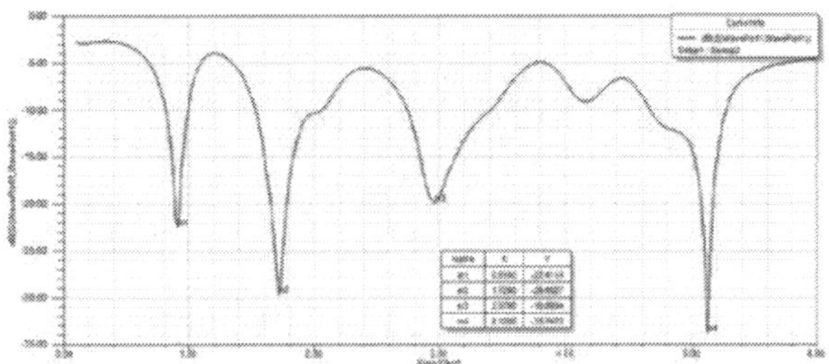

Fig. 7. Return loss for the antenna of 16.5 cm (250 MHz).

DOI: 10.1051/itmconf/20170903012

5 Experimental Results.

This section presents the experimental results of the return losses of Sierpinski antennas designed and simulated in previous sections. These antennas are experimentally analyzed using a measure device of antenna pattern, which measures the electromagnetic field in two orthogonal directions within the near field of the antenna under study. These measurements are used to estimate the far field distribution using flat opening for transforming the angular spectrum or spectral transformation plane wave (PSW). Then a second custom algorithm adjusts the projection of the far field to remove predictable coupling effects in measurements. This algorithm also takes into account the compensation response. This measurement system is capable of operating from 300 MHz to 6 GHz with a resolution of 10 MHz. The results obtained are shown in Figs. 8-10.

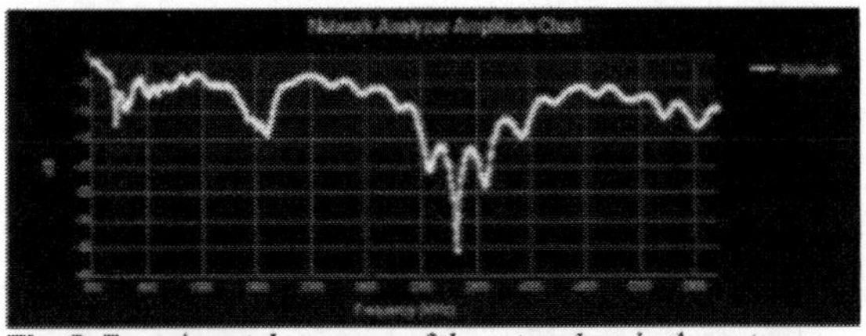

Fig. 8. Experimental measure of the return loss in the antenna of 8 cm (530 MHz).

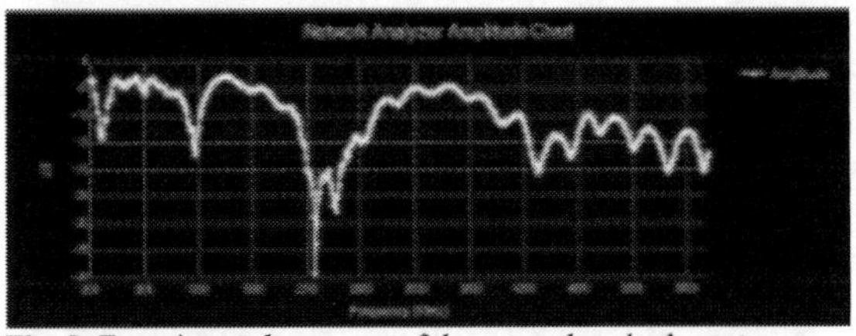

Fig. 9. Experimental measure of the return loss in the antenna of 12 cm (350 MHz).

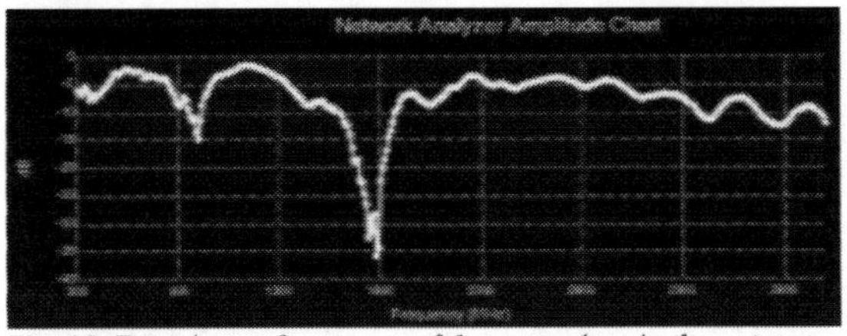

Fig. 10. Experimental measure of the return loss in the antenna of 16 cm (250 MHz).

In the fifth column of Tables I, II, and III is shown the factor between the resonance frequencies of two adjacent iterations. In particular, from the values reported in this column, it is observed that, regardless of the designed resonant frequency of iteration zero, the factor between the resonance frequency of the first and second iteration is close to two. Note that, in general, this factor depends on the iteration number, resonant frequency of the iteration zero, longitude and permittivity constant of the substrate.

Figs. 8, 9, and 10 show the return loss obtained experimentally for antennas with a maximum height of 8 cm, 12 cm, and 16 cm, respectively. The experimental and simulated results presented in Figs. 5-10 are summarized in Tables I, II, and III. Tables I, II y III, correspond to the Sierpinski fractal antenna with a maximum height of 8 cm, 12 cm and 16.5 cm, respectively. In the third column of Tables I, II y III, the simulated resonant frequencies of the different iterations of the fractal Sierpinski antennas are shown. On the other hand, in the fourth column of Tables I, II and III the experimental results of the resonance frequencies of the different iterations of Sierpinski antennae under study are shown. In these tables a high correlation between experimental and simulation results is observed; this is especially true for the second and third iteration.

Table 1. Sierpinski fractal antenna of 8 cm (530 MHz).

Iteration	Height (cm)	Simulated frequency (GHz)	Experimental frequency (GHz)	Factor f_{n+1}/f_n
0	8	0.43	0.51	3.72
1	4	1.75	1.90	1.91
2	2	3.02	3.62	1.52
3	1	5.11	5.52	1.05
4	0.5	7.7	5.82	---

Table 2. Sierpinski fractal antenna of 12 cm (350 MHz).

Iteration	Height (cm)	Simulated frequency (GHz)	Experimental frequency (GHz)	Factor f_{n+1}/f_n
0	12	0.40	0.41	3.10
1	6	1.24	1.27	1.87
2	3	2.23	2.37	1.87
3	1.5	3.74	4.44	1.27
4	0.75	5.50	5.64	---

Table 3. Sierpinski fractal antenna of 16.5 cm (250 MHz)

Iteración	Altura (cm)	Frecuencia simulada (GHz)	Frecuencia experimental (GHz)	Factor f_{n+1}/f_n
0	16.5	---	---	---
1	8.25	0.91	0.90	2
2	4.12	1.72	1.78	1.92
3	2.06	2.97	3.42	1.87
4	1.03	5.12	3.78	1.11

6 Conclusions

The ratios between the resonance frequencies of the various iterations of the fractal Sierpinski multiband antenna were investigated. To this end, three Sierpinski antennas of four iterations with operating frequencies in the iteration zero of 250, 350 and 530 MHz were designed, simulated, and built. The results show that only between the first to second, and second to third iterations there is a factor around two. For any other pair of adjacent iterations, the factor between the resonance frequencies depends on the designed value of the frequency of the iteration zero and other design parameters of the antenna (thickness and dielectric constant of the substrate, etc.)

References

1. Balanais Constantine A. *Modern antenna handbook* (Wiley. Canada, 2008).
2. Andreas F. Molisch. *Wireless Communications* (IEEE PRESS, 2006).
3. Cardama Aznar Ángel. *Antenas* (Alfaomega, 2004)
4. Muhammad Waqas. "Multiband Sierpinski Fractal Antennas". *IEEE 13th International Multitopic Conference, 2009. INMIC 2009.* 1-6. (Dec 2009).
5. C. Puente, J. Romeu, R. Pous and A. Cardama, "On the Behavior of the Sierpinski Multiband Fractal Antenna". *IEEE Transactions on Antennas and Propagation*, Vol. 46, No. 4, (Abril 1998).
6. MReza Khorramian, Bijan Zakeri, Ata Ebrahimzadeh, "Design and Analysis of a Multiband Sierpinski Type Monopole Fractal Antenna". *IEEE Conference Publications. 21st Telecomunications Forum*, 675-677, (2013).
7. C. Puente, J. Romeu; R. Pous, X. Garcia, F. Benitez. "Fractal Multiband Antenna Based on the Sierpinski Gasket". *IET Journals & Magazines Electronics Letters*, vol. 32, 1-2. (1996).
8. González R. Ivan R. "Antena fractal de Sierpinski para operar en las bandas de 900 MHz y 1.8 GHz," *Pistas Educativas No. 120*,143-159, (October 2016).
9. Naresh Kumar, N.; Saran Krishna, K.; Choukiker, Y.K.; Behera, S.K., "Compact modified Sierpinski fractal monopole antenna for multiband wireless applications," *IEEE Conference Publications*, 1264-1266. (2014).
10. Yuan-hai Yu; Chang-peng Ji., "Research of fractal technology in the design of multi-frequency antenna," *IEEE Conference Publications*, 1-4. (2011).
11. Nayak, P.B.; Verma, S.; Kumar, P., "Multiband fractal antenna design for Cognitive radio applications," *IEEE Conference Publications*, 115-120. (2013).
12. Werner, D.H.; Ganguly, S., "An overview of fractal antenna engineering research," *IEEE Journals & Magazines*, 38-57. (2003).
13. Dwivedi, R.P.; Upadhyay, D., "High gain dual band antenna using fractal geometry for mobile communication," *IEEE Conference Publications*, 50-55. (2015).

DOI: 10.1051/itmconf/20170903013

Research Optimization for the Utilization-Exploitation of Hellenic Army Vacant Warehouses

Theodoros Zikos[1*], *Dimitrios* Zaires[2], and *Nikolaos V.* Karadimas[3]

[1]Colonel, Hellenic Army General Staff, Thessalonica, Greece
[2]S.Lieutenant, Hellenic Army General Staff, Orestiada, Greece
[3]Lecturer of Informatics, Hellenic Army Academy, Mathematics and Engineering Sciences, Vari, Greece

Abstract. The aim of this paper is the research optimization of all factors concerning the utilization and exploitation of empty warehouses belonging to the Hellenic Army, pointing out that in periods of economic crisis the Hellenic Army has the capability of self-sustainability at every level of command. This approach analyzes the acknowledge of the objectives and the strategies employed by the Hellenic Army Logistics Department and also the analysis of the environment (internal and external), approaching factors that lead to the necessity of the exploitation, such as the adequacy of the warehouses, the reduction of human resources, the difficulty of maintaining the facilities by allocated credits and the capability of their evacuation. Formal methods like SWOT (Strengths, Weaknesses, Opportunities, Threats) analysis, Political/Law, Economic, Social, Technological, Demographic, Global Environment analysis, structural-industry analysis of the competitive environment and value chain analysis are used for the accurate determination of all optimal options which will lead to the best possible solution for the development and utilization of the aforementioned facilities. This optimization which invests in the extroversion of the Hellenic Army enhances our vision for the full utilization – operation of the Hellenic Army Empty Warehouses.

1 Introduction

This paper is to analyse all factors concerning the utilization and exploitation of empty warehouses belonging to the Hellenic Army pointing out that in periods of economic crisis the Hellenic Army has the capability of self-sustainability at every command level and propose all optimal solutions for the their fully operation. In this context for research optimization and the assessment of all factors followed the strategic management process [1], so that these solutions are functional at all levels of command of the Hellenic Army (Departmentorate – business – functional levels).

2 PEST-DG Analysis

The Political/Law, Economic, Social, Technological, Demographic, Global Environment (PEST-DG) analysis [2] contributed with four ways to "strategic analysis" for the utilization – operation of Hellenic Army empty warehouses:

1. The factors recorded, are not only a source of information, but could also serve as model variables that will help identify the most important influences of the external environment.
2. It leads the Logistics Department in selecting a minimum number of key factors affecting it and its competitors, hence its strategic location.
3. Used to identify "long-term drivers of new changes".
4. Leads to examine the factors that historically and prospectively affect a business and the industry in which it operates (impact analysis).

In particular, the economic crisis and the existing financial situation of the country does not permit for the next few years the appropriation of the necessary credits for maintaining the existing facilities, the construction of new ones, the obtainment of new arms systems and the creation of stock while the existing one keeps decreasing.

Another very significant factor which affects the maintenance of the facilities prevailing throughout the year is weather conditions. Strong winds which most of the times destroy (lift) the roofs, heavy rainfall and snowfalls during the winter destroy the clear plastic and the gutters creating static issues for the buildings. Other factors that the PEST-DG analysis emerge positive and negatives are:

- finding of financial resources,
- the exploitation of inactive facilities,
- incentives for providing more logistics services,
- exploitation of the National Strategic Reference Framework (NSRF) programs,
- lack of experience of relevant collaboration of the Unit with individuals,
- the ensuring deterrence of storing illegal products (smuggled etc.) due to the frequent controls by

* Corresponding author: hellas.ted@gmail.com

the competent services within the military facilities if they are rented to individuals.

3 Structural-Industry Analysis of the Competitive Environment

According to the model of the five forces of Porter [3] or model of structural analysis of an industry the five forces that determine competition are:

1) *The intensity of competition between existing firms in the industry*. This analysis should lead to the recovery of those factors that should focus the interest of the Hellenic Military Logistics Department.
2) *The new companies entering the industry:* The new entry threat in existing markets increases as decreasing the "entry barriers» such us, the amount and cost of capital required, access to distribution channels, cost advantages regardless of size, expected retaliation, National laws or government intervention.
3) *Bargaining Power Suppliers:* The power of suppliers increases when there is centralization of suppliers, the "replacement cost" of a supplier to another is very high, is highly likely that "vertical integration forward" on the part of the supplier, customers of a supplier are not important, and the supplier is not looking forward to a long-term cooperation with them.
4) *Bargaining Power of Buyers:* When there is a centralization of all buyers, or where the quantities purchased are great, when there are alternative sources of supply (undifferentiated product) and there are many companies that produce many products.
5) *The Threat of Substitute Products*. This risk has three forms:
   a) a real risk of similar replacement product.
   b) a new production process can put aside the use of a product.
   c) certain products may compete completely different.

As each company, so the Hellenic Military Logistics Department has to responds to the following questions: “Replacement products (Empty Warehouses) render inoperable our products or just give the impression to the consumer of greater benefits or higher value?”, “How easy is it for the consumer replacement decision?”, “What is that cost?”, “How the replacement risk can be reduced?” In this context, the Hellenic Army has the capability as an alternative optimal choice to offer the vacant warehouses for rent to individual companies.

So, the Hellenic Army after the identification of the main factors in the external environment and the assessment of the competitors in relation to whether the main factors influencing their own place (in the past, present and future) reached the following conclusions: Collaboration with logistics companies (3PL and 4PL), which in many cases may see the move of the Unit as competitive.

4 Value Chain Analysis

In this phase, an indicative value chain is analysed [4], started with the Resource-Capabilities Audit and emerge the logistics work context of the Hellenic Army Logistics Department such us the existing experience in material management, the storage evacuation capability etc.

4.1 The Main Functions

According to international Logistics standards [5], [6], which the Army follows, a supply unit can perform the following tasks:

- Receive, store and preserve/maintain the materials from the supply internal and external sources,
- Supply the operational units with materiel,
- Receive and handle the returnable materiel by the operational units by separating the inspected one from the useless one and forward it for recycle-repair etc.,
- Complies materiel stocks with statutory orders of superior squads and relevant plans,
- Make regular and occasional stocktaking so that there will be at any moment absolute accordance between the audit and the physical inventory,
- Cover the operational needs - the Logistics.

Additionally, the common ways to secure a military base [7] where the units and their facilities are located can be achieved with garrison and security departments, military dogs and their companions, system of circumferential visual surveillance, alarm systems in the warehouses, perimeter wall and barbed wire. In this context, main factors involved in storage [5] are the flow of spare parts - materials from the supply sources, the existing management of the materials, the existing facilities (warehouses) and the warehouse equipment. In the case of warehouse equipment, in packaging materials are crates palettes, etc., in storage materials are shelves, tubs, chests of drawers, etc., in materials for warehouses protection are insulations, water leaks, etc. and finally are the storage conditions, as well.

4.2 Operations - Management of Materials

The placement of the materials in the warehouses of a single management takes place according to the class of the material. The class of every material concerns the general category to which the material belongs. For example, a class of material is generally pneumatic tires - air chambers or engines. A similar procedure was followed for grouping all the materials in order to be placed in the warehouses of every management. The level of the stock in the departments is substantially low. Furthermore, it is estimated that in the near future this situation will not change, on the contrary, it is estimated that due to the financial situation of the country the level of the stock will be diminished.

DOI: 10.1051/itmconf/20170903013

4.3 Existing Facilities

The existing facilities, where other military units or even public or private companies such as individual logistics companies could storage their materials, are located in the main distribution centers next to major transport hubs, airports and ports. They are composed by:

- Administrative areas, which houses transaction and administrative support offices.
- Storage rooms, where management of materials is located and include the warehouses.
- Storage sheds, where bulky materials are stored in layers.
- Back-up facilities, such as export center, squads for receiving foreign materials – trade etc.

4.4 Storage Optimization Capability

Through the optimization of the warehouses the following is succeeded:

1. Better monitoring of the materials since the multiple positions for every material are limited.
2. Reduction of movement of personnel and means.
3. More effective maintenance of the materials.
4. Better maintenance of the warehouses, which contain materials only, whereas, those which do not contain materials are led to obsolescence due to the almost minimal appropriated credits.

Under these conditions, a gradual transfer of the materials of every management has taken place by concurrent arrangement of the materials at the new warehouse and at the same time actions were initiated for the transfer of the next materials in order to avoid any problems. The transfer of the materials of the management takes place after a full inventory of the materials which can be moved and placed in secure places appropriate for storing the corresponding materials.

4.5 Secondary Functions

Human Resources: The existing staff of the units is expected for the military personnel to be maintained or reduced according to the transfers-assignments of the executives. For the civilian personnel is expected to be reduced due to retirements and the suspension of recruitments mainly of technical staff and workers.

The Engineering Department of the Army has the knowledge and the expertise to cover the requirements for the maintenance of the facilities but due to financial situation of the country and the big number of the existing bases makes this not possible today. Simultaneously, they are not permitted the appropriation of the necessary credits for the procedure of the maintenance.

These factors combined with the adequacy of the warehouses and the capability of their evacuation reinforcing the need of finding alternative resources for their maintenance and their fully exploitation and employment.

5. Recourse Validation - Utilization of Vacant Warehouses

Therefore, it is understandable that the creation of competitive advantage depends on how each company manages the "value system" [4]. Labelling of operations contributing to creating and sustaining competitive advantage, which can be either reduced cost with stable prices thus increasing profit margins than the competition or product differentiation by achieving higher prices than the competition and therefore higher margin.

1. *Analysing the Cost Efficiency:* Cost efficiency is important for companies competing in the basis of continuous decline of the cost of their products.
2. *Analysing the Cost Effectiveness:* Effectiveness is the most important instrument of business competing by differentiating their products. The assessment of effectiveness is related to how well a company adapts its products to the existing needs of consumers. This assessment should be done only in those activities of the "competitive strategy" (competitive business level strategy), and aims at finding the "cost drivers or value" of each action.

This means that Hellenic Army's vacant warehouses can be exploited by individual logistics companies, by other military units or by the personnel of the Armed Forces by paying a fee. This leads us to the utilization of the vacant warehouses. For this purpose, a research was conducted at the local logistics companies in order to collect information about the modern methods they apply to the maintenance, as well as to receive suggestions regarding their use.

The investigation showed that the companies maintain their facilities from the income they make via the storage of materials-products. It is indicative that a warehouse with dimensions of 500 m2 similar to the one in the Hellenic Army Unit's is rented for the amount of 2.500 euros on a monthly basis. Additionally, the logistics companies face a serious problem with their safety and thus they spend a lot of money on security. However, the most important is that they showed an interest at the prospect of renting warehouses from the Army to individuals.

6 SWOT Analysis

It is considered imperative before proposing the necessary optimal actions to be taken for the exploitation of the warehouses, to define within the unit the strong and weak points as well as to distinguish the opportunities given at the exterior space as well as the threats that may emerge. This traditional study is known as Strengths, Weaknesses, Opportunities, Threats (SWOT) analysis and is the environment, which has already preceded by assessing each element separately.

Table 1. The SWOT analysis table.

	Positive Factors	Negative Factors
Internal Factors	**Strengths** • Adequacy of the storage rooms. • Safety of the military facilities. • The personnel are trained • Geographical position of the Unit • Fully organized and operational logistics system.	**Weaknesses** • Staff reduction of the Unit. • Problems of maintaining the existing facilities. • Bureaucratic approval procedures to use military facilities by individuals.
External Factors	**Opportunities** • Finding financial resources. • Exploitation of inactive facilities. • Reconstruction of the spirit 'management; to the personnel of the Unit. • Extroversion. • Incentives for providing more logistics services. • Collaboration with logistics companies (3PL and 4PL). • Exploitation of the NSRF programmes.	**Threats** • Individuals' entry to military facilities with what this might mean for security issues for the camp. • Lack of experience of relevant collaboration of the Unit with individuals. • Ensuring deterrence of storing illegal products (smuggled etc.) due to the frequent controls by the competent services within the military facilities. • Logistics companies (3PL and 4PL) to see the move of the Unit as competitive.

A major *advantage* of the unit and the main source of its power is the big number of storage rooms. The existence of warehouses of different dimensions and construction gives the opportunity for storing different kinds of materials as well as better grouping of the materials. The management of the materials is conducted through logistical equipment which helps all the operations of the Unit. Finally, it can provide multifaceted safety not only in safeguarding the material which is in the storage rooms but also protecting from external threats.

A great *weakness* is considered to be the lack of cash flow for the maintenance of the facilities and the bureaucratic approval procedures to be used military facilities by individuals.

The economic crisis constitutes a very big opportunity for concluding trade agreement regarding the exploitation of the empty warehouses since individuals, companies and agencies (of the public and private sector) are looking for inexpensive storage rooms. Under these circumstances a unit should invest in extroversion.

The *threats* emerging from the exploitation of the empty warehouses are the entry of unknown personnel to the unit if they are given for rental as well as the monitor of the materials which will be stored in order to ensure deterrence of illegal materials (smuggled, forbidden, etc.). Finally, the logistics companies (3PL and 4PL) will react negatively when they see the Unit as a competitor which claims a share in the market.

7 Optimization Proposal

The strategic management process for the utilization – operation of the Hellenic Army Empty Warehouses leads us to format strategies, implement them and evaluate the results. The evaluation of "basic skills" and the "distinctive competencies" of the enterprise is required in order to understand the "strategic skills". As a result of this procedure, is attempted the approach of all optimal solutions for the fully utilization and exploitation of the Hellenic Army Empty Warehouses. These are:

1. *Logistics Companies:* They showed an interest at the prospect of renting warehouses from the Army
2. *Conferral of the warehouses to other Units:* No necessity and face problems of concerning their structure, monitoring and use, maintenance due to lack of funding.
3. *Used by the personnel of the Armed Forces by paying a fee:* There was an interest in using these warehouses by the personnel of the Armed Forces by paying a fee (Not fully exploited).

The exploitation of the warehouses by the Hellenic Army units suggests also the best possible provision of information for their operational use. Our competitive advantage is the extroversion [8] of Hellenic Army, the lower cost we provide, our fully organized and operational logistics system, the social acceptance, our ethical Departmentorate responsibility and devotion to duty.

8 CONCLUSION

This paper consists a research optimization for the exploitation of the Hellenic Army Empty Warehouses, aiming to achieve optimal solutions by analysing the strategic level of the logistics Department. In addition, identifies all optimal alternative options for the decision analysis procedure [9] created with a view to take the full advantage – exploitation of the existing vacant warehouses.

Research in Utilization - Exploitation of the Empty Warehouses of the Hellenic Army until now limited and can be furthered to combine aspects such as multiple-criteria decision analysis with the use of the Analytic Hierarchy Process (AHP), which is a structured technique for organizing and analysing complex decisions in the functional level.

Moreover, the literature is lacking when it comes to objectives other than optimization and decision making of the vacant military facilities. In this context, the area that will garner attention for the use of the Military Warehouses from individual logistics companies is the Project Management. Network Analysis (Activity On Node, Activity on Arrow etc.) and Work Breakdown Structure (WBS) analysis can be used to determine the precise actions and the event times of the project with the complete satisfaction of stakeholders (Hellenic Army – Logistics Companies).

Warehouses are one of the most important components in the supply chain management. The main roles that they play in the operational planning of the military logistics management includes the material flow along the supply chain to accommodate variability caused by factors such as batching transportation, and the value-added-processing with activities such as kitting, packaging, labeling, and product customization (clarified as a material, spare part, autonomous material etc.).

In this paper, the optimization proposal for the strategic management process of the utilization – operation of the Hellenic Army Empty Warehouses is based on the given advantageous geographical position of the Units. In this context and taking into consideration the amount of the old existing warehouses it is imperative to extend this study in two axes: The first one to remain only in the exploitation of some storage facilities and the second to seek their expansion in permanent installations or movable storage rooms and warehouses by deciding which is the best effectiveness solution. This future work concerns facilities of the same military camp and the two axes must be determined both simultaneously, with the development of one or more feasibility studies.

The development of these feasibility studies should be focused in the field of the strategic location of the facilities by minimizing the total expected opportunity loss of the change in the operational use of the warehouses and in the field of financial management in order to exploit the existing financial resources and try to approach the new forms of investment presented in this optimization proposal (in the functional level). For the implementation of the project it must be considered multiple criteria such as the cost and the risk in the uncertain business environment that will lead us achieve the Fully Employment of the warehouses of the Hellenic Military.

References

1. G. Johnson, K. Scholes, R. Whittington, *Fundamentals of Strategy*, Pearson Education, 2009
2. M. Hitt, R. Ireland, R. Hoskisson, *Strategic Management: Competitiveness & Globalization: Concepts and Cases*, 11th ed, Cengage Learning, 2015
3. M. Porter. The Five Competitive Forces that Shape Strategy, Harvard Bus. Rev. 2008, pp.86-104
4. M. Porter, *Competitive Advantage*, The Free Press, NY, 1985
5. B. Donath, J. Mazen, C. Dubin, P. Patterson, *The IOMA handbook of logistics and inventory management*, Inst. Mngmt and Ad, John Wiley & Sons, Inc., NY, 2002.
6. D. Simchi-Levi, B.J. Chen Xin, *The Logic of Logistics: Theory, Algorithms and Applications for Logistics Management*, Springer, 2014
7. B.A. Wayland, *Security for Business Professionals: How to plan, implement, and manage your company's security program*, Butterworth-Heinemann, pp 52-60, 100-106, 2014
8. T.A. Shimp, Advertising, *Promotion, and Other Aspects of Integrated Marketing Communication* London: South Western-Cengage Business, 2010
9. Th. Zikos, D. Zaires, N.V.Karadimas, Automated Procedure for Utilization - Operation of Hellenic Army Empty Warehouses, (to be published).

DOI: 10.1051/itmconf/20170903014

ERP Systems Created to Support Academic Management in Contexts of Geographic Dispersion: A Case Study in Regional Higher Education

Moreno Gabriel[1], *Salinas* Martha[2], *Anzola* Diego[3] And *López* Juan[4]

[1] Politécnico Gran Colombiano, Engineering Faculty, Bogota, Colombia
[2] Corporación Universitaria Minuto de Dios, Cundinamarca, Colombia
[3] Lumon SAS, Software Development, Bogota, Colombia
[4] Lumon SAS, Software Development, Bogota, Colombia

Abstract. In response to extension of coverage goals in higher education, academic programs offering has been growing in regions geographically dispersed which management requires solid information systems able to adapt to the context needs and particularities. This paper presents a descriptive case study conducted in one of the head offices of Corporación Universitaria Minuto de Dios operating in 34 municipalities of cundinamarca department in colombia, highlighting challenges evidenced in the process of design and implementation of a custom ERP system. The case of study provides wide contextual details about selection and design of components, customization, integration and evaluation of the ERP system. Through this analysis are documented (a) the challenges in integration, escalability and interoperability of ERP systems in higher education that are evidenced in a context of regionalization with geographic dispersion variables; and (b) the differential features provided by a custom ERP system respect to the currently used systems in higher education.

1 Introducción

The literature records significant progress regarding research on creation and implementation of ERP systems (entreprise resourcing planning), understood as systems that facilitate decision making thanks to the integration of the information related to all the areas of a company [1]. The studies concerning focus on issues related to implementation procedures, the critical factors of success or failure, the selection of suppliers, among others.

However, research on the implementation of ERP in higher education is still incipient and its studies are related to the issues referred previously. The investigation conducted obeys to the creation and implementation of a custom ERP system for the institution in the frame of a particular context of regionalization of higher education.

This document registers the process of investigation, the analyzed variables, the exercises of participative and collective construction of the modules of the ERP, the challenges in the integration, scalability and interoperability of the modules and the evaluation developed.

It is important to highlight that the developed system constitutes nowadays the mechanism of academic programming in a headquarters of the institution, which attends about 17000 student in 34 places of operation, in a territory of 24.210 km.

2 The research method

In order to describe the process of design and implementation of a custom ERP system, a descriptive case study was developed characterized for being a way of investigating an empirical topic that answers to the questions "how and why", when the researcher has little control over events and when the researcher focuses on a contemporary phenomenon developed within some context of real life.

Data collection for the case study was conducted between august 2014 and december 2015 developed through focus groups with program coordinators and academic managers. Additionally, the reports of Academic Council headquarters and institutional reports generated on the design and implementation of were analyzed. A non-probabilistic and intentional sample

DOI: 10.1051/itmconf/20170903014

with a targeted selection of information was built, that is to say, the informative content representing the educational leaders who were implementing the designed system.

For the period of the research was necessary to develop a total of 10 focus groups during the process and five reports of Academic Council and three institutional reports were analyzed , which were compared with the literature studied on the design, implementation and adoption of systems in higher education. The information was systematized through association with each of the components or modules of the system matrices, and variables customization, integration and evaluation system.

3 The case study

La Corporación Minuto de Dios is a multi-campus system consisting of seven rectories and vicerrectories scattered in various regions of colombia, whose missionary purpose has been, for twenty-two years, bringing quality education to regions with preferential option for those with access difficulties. It currently has a range of 72 academic programs with face-to-face methodology and 30 traditional distance programs methodology.

Researchers have selected cundinamarca offices because it contributes with the 19 % coverage of the institution, attending around 17,000 students, in 34 municipalities in the department of cundinamarca, with an administrative team of approximately 320 employees and 840 teachers, to the date of the study, implying academic management under complex variables of volume and geographic dispersion. Planning, monitoring and control is centralized at the offices of academic affairs which are located in Bogota, and through an academic- administrative organization divided into four regional centers: Zipaquirá, Soacha, Girardot y Madrid. Each of them attending the operation of municipalities in their respective geographical area of influence.

Since 2001, the institution has used two systems all over the country for its management, BANNER and SAP. The first one, for the registration and academic control of students and teachers and also including prospecting processes, academic record, enrollment, academic history and reports; the second one, used for the administrative and financial management support management functions , accounting, finance, banking, sales, purchasing, inventory control, facilities management, customer relations and reports.

The experience of cundinamarca during its six years of constitution revealed the absence of mechanisms to support academic planning, whose result should be recorded periodically in the academic system of the institution called BANNER.

The academic planning process covers the subprocess of registration and selection of teachers from different municipalities of operation; curricular planning (elective and mandatory courses for each academic term); definition of quotas for each course offered with their respective allocation of teachers needed for decision making regarding hiring, schedule planning and decision making for the optimization in the use of facilities. Academic planning was made through excel with the impossibility of a validation mechanism, consistent and efficient control, and with a centralized administrative and academic organization.

4. Literature review

The literature presents the advantages and challenges registered in the implementation of ERPs and possible areas of research around the implementation of systems in higher education. The ability of to articulate and interconnect different areas and processes within an organization around an integrated solid and shared data system, the importance of proper design, implementation and use of systems in organizations to strengthen the proper decision-making, among other issues have motivated theoretical and empirical research on the subject.

Some of the most determinant results of the investigation developed up to the date go around the context needs, the aspects influencing the decision to adopt this type of systems, and the aspects related to the adoption of systems. [1], [2]–[5]

4.1 ERP systems in higher education

ERP systems have been used in higher education to support the administrative, organizational and accounting functions of the institutions, specially for the administration of students and teachers, the facilities, courses and units admnistration, the management of schedules, the students monitoring and the financial control of the institution [3], [4], [6].

Some of the main benefits of ERP implementation in higher education are: they register the improvement to the access of information for institution planning and management, better services for students, teachers and leaders, costs reduction. ERP systems enable the aptitude to control all the functions and operations of the institution thanks to the use of unified information architectures. The quality and efficiency increasing of the information needed by diverse stakeholders, communication and coordination, administrative activities centralization are also some of their benefits.

Due to the challenges that the adoption and ERP implementation represents in the universities, it becomes necessary to highlight the complexity of these institutions. There is a high percentage of faults registered associated with the inability to satisfy the expectations and needs of the users of these system. Theres is a challenge due to the fact that the solutions do not adapt completely to support specific activities. The "better practices" extrapolated of other scenes not necessarily are adapted for the educational field by his high degree of complexity and heterogeneity [7], [8].

4.2 Variables associated with management in contexts of geographical dispersion

The study has addressed the phenomenon of geographical dispersion from the contributions of O'Leary and Cummings. The theoretical dimensions empirically justified for the characterization of the dispersion in work teams and their relationships constitute one of the main differentials in the creation of a modelled system orientated to the academic management, in a context determined by the need of management of at least 34 dispersed work teams in a territory of 24.210 km ^{2}.

The proposed dimensions are registered in the following table from O'Leary & Cummings[9]:

Table 2. Key Outcomes and Theoretical Mechanisms Distinctly Associated with Each Dimension of Dispersion

Characteristic	Description	Example Outcome	Example Mechanism
Spatial	Geographic distance among team members	Reduced spontaneous **communication** (Burke et al. 1999; Dennis et al. 1988; Saunders et al. 2004)	Decreasing the likelihood of face-to-face interaction
Temporal	Time difference among team members	Reduced real-time **problem solving** (Grinter et al. 1999; Herbsleb et al. 2000; Malone and Crowston 1994)	Decreasing the likelihood of synchronous interaction
Site (Configurational)	Locations where team members work	Increased **coordination complexity** (Sarbaugh-Thompson and Feldman 1998; Yoo and Alavi 2001; Zigurs et al. 1988)	Increasing the number of dependencies which must be managed
Isolation (Configurational)	Locations where team members work alone	Decreased **awareness** (Amstrong and Cole 2002; Dennis 1996; Tan et al. 1998)	Increasing the remoteness of isolated team members
Imbalance (Configurational)	Locations with uneven distribution of team members	Increased **intragroup conflict** (Allmendinger and Hackman 1995; Kabanoff 1991; Mannix 1993)	Increasing majority influence and the potential for negative subgroup dynamics

Fig. 1. Multidimensional Conception of Geographic Dispersion.

The current academic information system of the institution does not contemplate this analysis of variables, and it was found that it is not parametrized the totality of the tutoring centers. Since the record does not correspond to the reality it has to be included an additional process of unification of information. Now then, only an designed to the needs might addressed these considerations.

4.3 ERP System customization

Studies regarding ERP use in higher education are limited, it does not rely on a comprehensive model that allows to evaluate the impact of the information systems, this fact is reinforced due to the absence of conceptual frames of reference that help to understand better the complexity of the educational sector[10]–[13].

Nevertheless, opposite to the customization of the systems, there is highlighted necessity to recognize that during the implementation of an information system, the management of the human risk and organizational is crucial for its success, even more that the technical risk.

The resistance indicators, the suitable integration of all the sources and the suitable analysis of needs and values of the institutional subcultures must be recognized to resist the barriers that could be generated before the change.

From the technical point of view, the customization involves the development of three steps in the implementation of an : (i) the evaluation of the needs and requirements; (ii) the review of the different available solutions and decision on the system that better adjusts to the needs; (iii) the implementation and customization of the selected system to adapt it to the needs of the institution or the design of the life cycle of the applications for its construction and implementation[3], [4], [7].

It is outlined in the literature as functional and technical factors of success, the strong communication and coherence between the different departments of the university, the adoption of strategies of knowledge management from the conception of a life cycle of the , the modular nature of these systems, problematic for the universities since they must adapt to the system more than the system to the organization [3], [8]

4.4 Financial and operational issues

ERP implementation implies a series of direct and indirect costs such as the software applications and tools, the investment in hardware, the personnel, the trainings and tutorings and the costs associated with the maintenance and future updates of the system.

In case of the modular systems, these costs increase while a new module or package is incorporated. Another factor that can increase the costs consists on the development of the systems or sources of information that the institutions have already had at the moment of the implementation [2][3][10], [11].

For an institution, the adoption or creation of ssystems must allow the reduction of costs.

5 Study results

In this section, the relevant results of the study are registered and they are related to the following areas: (a) decisions according to the adoption and evaluation of the current academic in the institution; (b) procedures of customization for the design of a complementary system; (c) aspects of integration of the modules of the ERP ; (d) benefits and challenges of the implementation of the taylor-made ERP.

5.1 Adoption and evaluation of the current academic ERP in the institution

Since 2001 the institution has implemented information systems to support decision making corresponding to the processes of self assessment and strategic planning. Additionally, it possesses several information systems, two of them corresponding to systems named SAP and BANNER. The last one is a system of academic management of students and teachers (management of planning, admissions, student's record, list of courses, teacher staff, facilities, academic programming, inscription, academic history, grades).

There is not any registration of formal assessments to the actual information system; nevertheless, the self assessment of the management of technological services (gst) have given as a result plans of improvement mainly oriented to strengthen the areas of management for the development and acquisition of new technological applications for the academic management and the investigations, control and audit of the academic information system, follow-up to the processes managed across the academic information system orientated to the quality of the information and functional training of the academic information system.

Banner is an internationally recognized, functional and efficient information system, which adoption was complex with some kind of emptinesses in the information, as well as some absences in the customization. It responds to the nature and structure of the headquarter but it has not achieved the conditions of geographical dispersion in Cundinamarca headquarter cause it does not identify all the operations centers and information must be reprocessed through excel.

Such as manual work consumes administrative time, it has a high risk of human error, it subtracts management capacity and does not allow adequate and timely decisions related to management resulting in high financial costs and low quality.

The exposed considerations have motivated the institution to look for new and innovative applications adequate for the proper context of higher education and the conditions related to the nature of this university.

5.2 Procedure for the customization of a complementary system for the institution.

It was concluded a necessity for the offices of Cundinamarca for a system for academic planning. The custom system needed to integrate the components for the decision making around coverage, facilities disposition, teachers recruiting and selection by municipality with attention to the variables of geographical dispersion, information processing for schedule, allocations management and resources and spaces optimization. In general, all the necessary information processing made previously to the register in banner.

The study gave as a result a proposal of a procedure consisting of three steps to customize an system for an institution with geographical dispersion:

1st. Step: to evaluate needs and requirements of the institution respect to the existing information systems. It's necessary to identify the components, issues and elements which are fullfiling the necessities adecuately, the ones which are not covered, and the ones which need adjustments. Product of these evaluation one way is to begin the customization of a new system appropriate to the characteristics of the institution and coherent with conditions like geographical dispersion.

2nd. Step: to develop a custom system covering the software construction process: analysis, design, development, testing, implementation and star up. To achieve this objective the following variables are considered:

a) Dimensions of dispersión: geographical distance among team members, time diference among team members, locations where team members work, locations where team members work alone, locations with uneven distribution of team member. These spatial, temporal, site, isolation, and imbalance dimensions define specific charactersitics of the modules, roles and users of the system.

b) Software architecture specification attending the purpose and scope of the product, perspective, components and functions, kinds of users and characteristics, operating environments, desing and implementation restrictions, user interfaces, hardware, software and communication, users documents, assumptions and dependencies.

3rd step: modules testing and permanent software settings. Each component needs periodic testing based on direct fieldwork with each team or users group having in mind the following modules integration aspects:

Academic planning needs at least four integrated modules of *curriculum management* (study programmes with courses, hours and credits); *teachers management*

DOI: 10.1051/itmconf/20170903014

(teachers recruitement, personal and profesional information management, work plans) *pre-planning* (assisted programming of groups and schedules) and *facilities management* (having in mind dispersion dimensions).

6 Conclusions and future research

The research allowed to evaluate the existing and implementable entreprise resourcing planning systems for a higher education institution. It was necessary to create a new complementary and customized ERP because of the conditions of dispersion not satisfied by the evaluated systems. It was necesary to find the mechanisms for a custum software development through institutional research and a software development consulting.

The most important results are related to (i) needs identification and met based on dimensions of dispersion; (ii) practices on saving resources thanks to facilities and human resources management obtained with the system; (iii) significant reduction in planning process time from three months to two weeks with the consecuent satisfaction of students thanks to clarity and oportunity in schedules and programming; (iv) capacity of making better decisions related to contracting based on real and controlled information.

The most relevant challenges are around information quality delivered by dispersed teams. It's evident the resistance to change and control. That's why is important to face and enhance capacity of correction *in situ,* technical and functional leadership, teams training and support.

Higher education institutions must make an appropriate data handling and management to enhance quality. Global changing and technologies advance are available for any kind of entreprises, all of them demanded by society to offer services and products more efficient, and education is not an exception.

The results presented show that a custom ERP system can find higher efficiency since it's a system with ability to adapt to the institution instead of wasting efforts in adapting to an existing system. Although it represents similar adaptative challenges, it was demonstrated that minimize frustration risks as the creation is collective, participatory and based on technical rigurosity and necessary support.

Future research could attend a better understanding of intelligence business in education. Interdisciplinar frameworks created by educational management and software engineering experts. It's important to approach high complexity of this kind of institutions and take advantage of systems in the enhancements of education quality.

References

[1] A. Rabaa y W. Gable, G. & Bandara, «ERP Systems in the Higher Education Sector : A Descriptive Case Study», *Proc. 20th Australas. Conf. Inf. Syst.*, pp. 456-470, 2009.

[2] D. Swartz y K. Orgill, «Higher Education ERP : Lessons Learned», *Educase*, pp. 1-12, 2000.

[3] B. M. B. Kalema, O. O.Olugbara, y R. M.Kekwaletswe, «Identifying Critical Success Factors: the case of ERP Systems in Higher Education», *African J. Inf. Syst.*, vol. 6, n.° 3, pp. 68-70, 2014.

[4] L. Zornada y T. B. Velkavrh, «Implementing ERP systems in higher education institutions», *27th Int. Conf. Inf. Technol. Interfaces, 2005.*, pp. 307-313, 2005.

[5] A. Abugabah y L. Sanzogni, «Enterprise resource planning (ERP) system in higher education: A literature review and implications», *World Acad. Sci. Eng. Technol.*, vol. 71, n.° 11, pp. 49-53, 2010.

[6] D. F. Rico, «ERP in higher education», *Retrieved April*, vol. 16, p. 2007, 2004.

[7] A. Aljohani, A. Peng, y M. Nunes, «Critical factors leading to ERP replacement in Higher Education Institutions in Saudi Arabia : A case study», 2013.

[8] A. a Rabaa, «Identifying Critical Success Factors of ERP Systems at the Higher Education Sector», *ISIICT 2009 Third Int. Symp. Innov. Inf. Commun. Technol.*, pp. 133-147, 2009.

[9] M. Oleary, «The Spatial , Temporal , and Configurational Characteristics of Geographic Dispersion in Teams T HE S PATIAL , T EMPORAL , AND C ONFIGURATIONAL», n.° October, 2016.

[10] I. Guitart y J. Conesa, «Analytic Information Systems in the Context of Higher Education: Expectations, Reality and Trends», *2015 Int. Conf. Intell. Netw. Collab. Syst.*, pp. 294-300, 2015.

[11] S. Takahashi, A. Sugiyama, y Y. Kohda, «Knowledge, Information and Creativity Support Systems», vol. 416, 2016.

[12] O. Marjanovic, «Addressing the ICT-related challenges of Business Intelligence education», *Proc. ITI 2011 33rd Int. Conf. Inf. Technol. Interfaces*, pp. 151-156, 2011.

[13] C. A. de Oliveira, «Notes about COOL: analysis and highlights of complex view in education.», *Campus -- Wide Inf. Syst.*, vol. 29, n.° 4, pp. 304-311, 2012.

AUTHOR INDEX